Emerging Materials for Additive Manufacturing

Emerging Materials for Additive Manufacturing

Editors

Swee Leong Sing
Wai Yee Yeong

MDPI • Basel • Beijing • Wuhan • Barcelona • Belgrade • Manchester • Tokyo • Cluj • Tianjin

Editors
Swee Leong Sing
National University of Singapore
Singapore

Wai Yee Yeong
Nanyang Technological University
Singapore

Editorial Office
MDPI
St. Alban-Anlage 66
4052 Basel, Switzerland

This is a reprint of articles from the Special Issue published online in the open access journal *Materials* (ISSN 1996-1944) (available at: https://www.mdpi.com/journal/materials/special_issues/AM_3D_printing_materials).

For citation purposes, cite each article independently as indicated on the article page online and as indicated below:

LastName, A.A.; LastName, B.B.; LastName, C.C. Article Title. *Journal Name* **Year**, *Volume Number*, Page Range.

ISBN 978-3-0365-6642-9 (Hbk)
ISBN 978-3-0365-6643-6 (PDF)

Contents

About the Editors

Swee Leong Sing

Swee Leong Sing is an Assistant Professor in the Department of Mechanical Engineering, National University of Singapore (NUS). Prior to joining NUS in August 2021, he was a Presidential Postdoctoral Fellow at the Singapore Centre for 3D Printing and School of Mechanical and Aerospace Engineering, Nanyang Technological University, Singapore after being awarded the prestigious fellowship in 2020. Swee Leong has been active in the 3D printing field for more than a decade. He obtained his BEng (Hons) in Aerospace Engineering and PhD in Mechanical Engineering with a topic in additive manufacturing (AM) in 2012 and 2016, respectively. Swee Leong's research focuses on using advanced manufacturing techniques as enablers for materials development and to create strategic values for industries. He is also active in inter-disciplinary research and translational work. His research has been awarded the Best PhD Thesis Award by MAE, NTU, Singapore as well as the Springer Theses Award from Springer Nature, Germany in 2017. In 2022, he was also awarded the Young Professional Award by ASTM International and named a Highly Cited Researcher by Clarivate. Swee Leong has worked on numerous 3D printing projects with government agencies, universities, research institutes and industrial collaborations. Swee Leong has filed multiple patents pertaining to 3D printing and materials. He has published 1 book, 5 book chapters and more than 60 peer reviewed journal and conferences articles. Swee Leong currently has an h-index of 33 with more than 5000 citations (Web of Science, 15 January 2023).

Wai Yee Yeong

Wai Yee Yeong is Professor of Mechanical Engineering in the School of Mechanical and Aerospace Engineering at Nanyang Technological University. She is the winner of the TCT Women in 3D Printing Innovator Award 2019 and was named one of the Singapore 100 Women in Tech List 2021. Prof. Yeong is named as one of the Highly Cited Researchers by Clarivate in 2022. Her work is well-recognized, with an h-index of 60 and more than 14,000 citations on Google Scholar. She has filed multiple patents and knowhows, with a key interest in Bioprinting and 3D printing of new materials. On the academic fronts, Prof. Yeong is the Associate Editor for 2 international journals and has authored 4 textbooks on 3D printing. Prof. Yeong serves as Program Director in the Singapore Centre for 3D Printing and HP-NTU Digital Manufacturing Corp Lab. In 2022, she was awarded the prestigious NRF Investigatorship (Class of 2022) in her pursuit for ground-breaking and high-risk research.

Editorial

Emerging Materials for Additive Manufacturing

Swee Leong Sing [1,*] and Wai Yee Yeong [2]

1 Department of Mechanical Engineering, National University of Singapore, 9 Engineering Drive 1, Singapore 117575, Singapore
2 Singapore Centre for 3D Printing, School of Mechanical & Aerospace Engineering, Nanyang Technological University, 50 Nanyang Ave, Singapore 639798, Singapore
* Correspondence: sweeleong.sing@nus.edu.sg

Additive manufacturing (AM) has grown and evolved rapidly in recent years. There are many exciting research and translational works in many areas of application, such as biomedical [1,2], aerospace [3,4] and electronics [5–7]. These advancements are typically coupled with materials development, which has resulted in more functionalities added to 3D printed parts, such as multi-material fabrications [8–10] and integration with machine learning or digital twins [11–13]. Such enhancements in functionalities have enabled the evolution of AM from a rapid prototyping tool to an actual manufacturing solution.

In this Special Issue, state-of-the-art research and review articles on emerging material systems for AM are collected, with a focus on the process–structure–properties relationships. In total, two reviews and thirteen original research articles are included. In their review article, Minasyan and Hussainova discussed the recent developments of ceramic particulate-reinforced aluminium alloys produced by laser powder bed fusion [14], while Hou et al. elaborated the use of monitoring systems for powder bed fusion processes with a focus on metals in their comprehensive review [15]. For original research, Gatões et al. studied the fabrication of different stainless steels using selective laser melting, a type of laser powder bed fusion technique [16]. In their study, Mally et al. benchmarked the mechanical properties of ferritic steels produced by selective laser melting with relevant forged parts [17]. Using selective laser melting as well, Koh et al. studied the fabrication of silica-reinforced steel matrix nanocomposites [18]. Lim et al. studied the bone conduction capacity of highly porous titanium scaffolds with different designs produced by selective laser melting [19]. Chen et al. studied the effect of laser scanning speed on the microstructure and mechanical properties of K418 nickel-based alloy produced by laser powder bed fusion [20]. Böhm et al. evaluated the feasibility of using a mixture of two aluminium alloys to eliminate solidification cracks formed during laser powder bed fusion [21]. Chen et al. studied the fabrication of bimetallic structures using TiNi-based shape memory alloy by laser-directed energy deposition [22]. Also using laser-directed energy deposition, Menon et al. attempted to quantify the process using multi-fidelity surrogate-based process mapping [23]. Hein et al. studied the effect of heat treatment on metastable β titanium alloy produced by laser powder bed fusion [24]. Romani et al. studied the metallization of recycled glass fibre-reinforced polymers that are processed by UV-assisted 3D printing [25]. Hailu et al. studied the effect of structure design on the performance of functionally graded materials produced by the MultiJet Fusion technique [26]. Marczyk et al. analysed the use of concrete–geopolymer hybrids reinforced with aramid roving for 3D concrete printing [27]. Yao et al. evaluated the feasibility of colour 3D printing by studying the pigment penetration in powder-based additive manufacturing [28].

Citation: Sing, S.L.; Yeong, W.Y. Emerging Materials for Additive Manufacturing. *Materials* **2023**, *16*, 127. https://doi.org/10.3390/ma16010127

Received: 1 December 2022
Accepted: 13 December 2022
Published: 23 December 2022

Author Contributions: S.L.S. and W.Y.Y., writing—original draft preparation, writing—review and editing. All authors have read and agreed to the published version of the manuscript.

Funding: This research received no external funding.

Conflicts of Interest: The authors declare no conflict of interest.

References

1. Luis, E.; Pan, H.M.; Bastola, A.K.; Bajpai, R.; Sing, S.L.; Song, J.; Yeong, W.Y. 3D Printed Silicone Meniscus Implants: Influence of the 3D Printing Process on Properties of Silicone Implants. *Polymers* **2020**, *12*, 2136. [CrossRef] [PubMed]
2. Luis, E.; Pan, H.M.; Sing, S.L.; Bajpai, R.; Song, J.; Yeong, W.Y. 3D Direct Printing of Silicone Meniscus Implant Using a Novel Heat-Cured Extrusion-Based Printer. *Polymers* **2020**, *12*, 1031. [CrossRef] [PubMed]
3. Gradl, P.; Tinker, D.C.; Park, A.; Mireles, O.R.; Garcia, M.; Wilkerson, R.; Mckinney, C. Robust Metal Additive Manufacturing Process Selection and Development for Aerospace Components. *J. Mater. Eng. Perform.* **2022**, *31*, 6013–6044. [CrossRef]
4. Omiyale, B.O.; Olugbade, T.O.; Abioye, T.E.; Farayibi, P.K. Wire arc additive manufacturing of aluminium alloys for aerospace and automotive applications: A review. *Mater. Sci. Technol.* **2022**, *38*, 391–408. [CrossRef]
5. Huo, Z.; He, J.; Pu, H.; Luo, J.; Li, D. Design and printing of embedded conductive patterns in liquid crystal elastomer for programmable electrothermal actuation. *Virtual Phys. Prototyp.* **2022**, *17*, 881–893. [CrossRef]
6. Zhou, J.; Yan, H.; Wang, C.; Gong, H.; Nie, Q.; Long, Y. 3D printing highly stretchable conductors for flexible electronics with low signal hysteresis. *Virtual Phys. Prototyp.* **2022**, *17*, 19–32. [CrossRef]
7. Monisha, M.; Agarwala, S. Biodegradable materials: Foundation of transient and sustainable electronics. *Mater. Sci. Addit. Manuf.* **2022**, *1*, 15. [CrossRef]
8. Lee, J.M.; Sing, S.L.; Yeong, W.Y. Bioprinting of Multimaterials with Computer-aided Design/Computer-aided Manufacturing *Int. J. Bioprint.* **2020**, *6*, 245. [CrossRef]
9. Rao, J.; Sing, S.L.; Lim, J.C.W.; Yeong, W.Y.; Yang, J.; Fan, Z.; Hazell, P. Detection and characterisation of defects in directed energy deposited multi-material components using full waveform inversion and reverse time migration. *Virtual Phys. Prototyp.* **2022**, *17*, 1047–1057. [CrossRef]
10. Bandyopadhyay, A.; Zhang, Y.; Onuike, B. Additive manufacturing of bimetallic structures. *Virtual Phys. Prototyp.* **2022**, *17*, 256–294. [CrossRef]
11. Nguyen, P.D.; Nguyen, T.Q.; Tao, Q.B.; Vogel, F.; Nguyen-Xuan, H. A data-driven machine learning approach for the 3D printing process optimisation. *Virtual Phys. Prototyp.* **2022**, *17*, 768–786. [CrossRef]
12. Lu, C.; Jia, X.; Lee, J.; Shi, J. Knowledge transfer using Bayesian learning for predicting the process-property relationship of Inconel alloys obtained by laser powder bed fusion. *Virtual Phys. Prototyp.* **2022**, *17*, 787–805. [CrossRef]
13. Gong, X.; Zeng, D.; Groeneveld-Meijer, W.; Manogharan, G. Additive manufacturing: A machine learning model of process-structure-property linkages for machining behavior of Ti-6Al-4V. *Mater. Sci. Addit. Manuf.* **2022**, *1*, 6. [CrossRef]
14. Minasyan, T.; Hussainova, I. Laser Powder-Bed Fusion of Ceramic Particulate Reinforced Aluminum Alloys: A Review. *Materials* **2022**, *15*, 2467. [CrossRef]
15. Huo, Z.-J.; Wang, Q.; Zhao, C.-G.; Zheng, J.; Tian, J.-M.; Ge, X.-H.; Liu, Y.-G. Online Monitoring Technology of Metal Powder Bed Fusion Processes: A Review. *Materials* **2022**, *15*, 7598.
16. Gatões, D.; Alves, R.; Alves, B.; Vieira, M.T. Selective Laser Melting and Mechanical Properties of Stainless Steels. *Materials* **2022**, *15*, 7575. [CrossRef]
17. Mally, L.; Werz, M.; Weihe, S. Feasibility Study on Additive Manufacturing of Ferritic Steels to Meet Mechanical Properties of Safety Relevant Forged Parts. *Materials* **2022**, *15*, 383. [CrossRef]
18. Koh, H.K.; Moo, J.G.S.; Sing, S.L.; Yeong, W.Y. Use of Fumed Silica Nanostructured Additives in Selective Laser Melting and Fabrication of Steel Matrix Nanocomposites. *Materials* **2022**, *15*, 1869. [CrossRef]
19. Lim, H.-K.; Ryu, M.; Woo, S.-H.; Song, I.-S.; Choi, Y.-J.; Lee, U.-L. Bone Conduction Capacity of Highly Porous 3D-Printed Titanium Scaffolds Based on Different Pore Designs. *Materials* **2021**, *14*, 3892. [CrossRef]
20. Chen, Z.; Lu, Y.; Luo, F.; Zhang, S.; Wei, P.; Yao, S.; Wang, Y. Effect of Laser Scanning Speed on the Microstructure and Mechanical Properties of Laser-Powder-Bed-Fused K418 Nickel-Based Alloy. *Materials* **2022**, *15*, 3045. [CrossRef]
21. Böhm, C.; Werz, M.; Weihe, S. Practical Approach to Eliminate Solidification Cracks by Supplementing AlMg4.5Mn0.7 with AlSi10Mg Powder in Laser Powder Bed Fusion. *Materials* **2022**, *15*, 572. [CrossRef] [PubMed]
22. Chen, Y.; Rios, C.O.; McLain, B.; Newkirk, J.W.; Liou, F. TiNi-Based Bi-Metallic Shape-Memory Alloy by Laser-Directed Energy Deposition. *Materials* **2022**, *15*, 3945. [CrossRef] [PubMed]
23. Menon, N.; Mondal, S.; Basak, A. Multi-Fidelity Surrogate-Based Process Mapping with Uncertainty Quantification in Laser Directed Energy Deposition. *Materials* **2022**, *15*, 2902. [CrossRef] [PubMed]
24. Hein, M.; Dias, N.F.L.; Pramanik, S.; Stangier, D.; Hoyer, K.-P.; Tillmann, W.; Schaper, M. Heat Treatments of Metastable β Titanium Alloy Ti-24Nb-4Zr-8Sn Processed by Laser Powder Bed Fusion. *Materials* **2022**, *15*, 3774. [CrossRef]
25. Romani, A.; Tralli, P.; Levi, M.; Turri, S.; Suriano, R. Metallization of Recycled Glass Fiber-Reinforced Polymers Processed by UV-Assisted 3D Printing. *Materials* **2022**, *15*, 6242. [CrossRef]
26. Hailu, Y.M.; Nazir, A.; Lin, S.-C.; Jeng, J.-Y. The Effect of Functional Gradient Material Distribution and Patterning on Torsional Properties of Lattice Structures Manufactured Using MultiJet Fusion Technology. *Materials* **2021**, *14*, 6521. [CrossRef]

27. Marczyk, J.; Ziejewska, C.; Korniejenko, K.; Łach, M.; Marzec, W.; Góra, M.; Dziura, P.; Sprince, A.; Szechyńska-Hebda, M.; Hebda, M. Properties of 3D Printed Concrete–Geopolymer Hybrids Reinforced with Aramid Roving. *Materials* **2022**, *15*, 6132. [CrossRef]
28. Yao, D.; Yuan, J.; Tian, J.; Wang, L.; Chen, G. Pigment Penetration Characterization of Colored Boundaries in Powder-Based Color 3D Printing. *Materials* **2022**, *15*, 3245. [CrossRef]

MDPI

Review

Laser Powder-Bed Fusion of Ceramic Particulate Reinforced Aluminum Alloys: A Review

Tatevik Minasyan * and Irina Hussainova *

Department of Mechanical and Industrial Engineering, Tallinn University of Technology, Ehitajate 5, 19086 Tallinn, Estonia
* Correspondence: tatevik.minasyan@taltech.ee (T.M.); irina.hussainova@taltech.ee (I.H.)

Abstract: Aluminum (Al) and its alloys are the second most used materials spanning industrial applications in automotive, aircraft and aerospace industries. To comply with the industrial demand for high-performance aluminum alloys with superb mechanical properties, one promising approach is reinforcement with ceramic particulates. Laser powder-bed fusion (LPBF) of Al alloy powders provides vast freedom in design and allows fabrication of aluminum matrix composites with significant grain refinement and textureless microstructure. This review paper evaluates the trends in in situ and ex situ reinforcement of aluminum alloys by ceramic particulates, while analyzing their effect on the material properties and process parameters. The current research efforts are mainly directed toward additives for grain refinement to improve the mechanical performance of the printed parts. Reinforcing additives has been demonstrated as a promising perspective for the industrialization of Al-based composites produced via laser powder-bed fusion technique. In this review, attention is mainly paid to borides (TiB_2, LaB_6, CaB_6), carbides (TiC, SiC), nitrides (TiN, Si_3N_4, BN, AlN), hybrid additives and their effect on the densification, grain refinement and mechanical behavior of the LPBF-produced composites.

Keywords: laser powder-bed fusion; additive manufacturing; aluminum alloys; reinforcement; ceramic particulates; grain refinement; crystallographic texture; mechanical properties

Citation: Minasyan, T.; Hussainova, I. Laser Powder-Bed Fusion of Ceramic Particulate Reinforced Aluminum Alloys: A Review. *Materials* **2022**, *15*, 2467. https://doi.org/10.3390/ma15072467

Academic Editors: Swee Leong Sing and Wai Yee Yeong

Received: 18 February 2022
Accepted: 21 March 2022
Published: 27 March 2022

Publisher's Note: MDPI stays neutral with regard to jurisdictional claims in published maps and institutional affiliations.

1. Introduction

In many engineering solutions, product performance is determined by weight, which can be scaled down by material-efficient construction and the use of low-density alloys [1,2]. Due to exceptional strength/stiffness-to-weight ratio, low density, good damage tolerance, ability to be heat treated and the low cost, aluminum (Al) alloys are extensively used in many exclusive fields, such as: automotive, aerospace, marine navigation, rail transit, architectural construction, microelectronics and consumer applications [3–7].

In the meantime, owing to the moderate strength and relatively poor wear resistance of aluminum alloys, they are not applicable as structural materials for critical parts of aircrafts or satellites [8,9]; therefore, there is a need to improve the mechanical properties of aluminum alloys to be used for special applications. Along the modern industrial developments, the demand for complex-shaped products in diverse sectors is widespread. Problems related to traditional casting of aluminum alloys include coarse microstructures, a long process chain with limited flexibility [10], use of PM/casting molds [11] and a high rate of tool degradation [12].

Additive manufacturing (AM) provides an integrated way of item production [13]. Additive manufacturing, also known as 3D printing, refers to the layer-wise fabrication process of functional objects adopting nearly unlimited geometrical complexity, processing freedom, high level of accuracy and customization with elimination of traditional economy-of-scale constraints [14]. Furthermore, the material efficiency and design flexibility of AM technology meet the requirements for resource optimization, mass customization and accelerates the time to enter the market. In terms of dissimilar material joining and hybrid structures, AM is considered a versatile tool for complete spatial control of local material composition, microstructure and properties [15].

Among the most advanced AM technologies available, laser powder-bed fusion has gained increased attention in both the industrial and academic sectors. The essence of the process lies beneath the selective melting/solidification of the desired sections of consecutive powder layers by a precise (computer-controlled) high-energy laser beam directed by 3D CAD (computer-aided design) file [16–18]. Within the scanning process, the laser energy is supplied into the powder layer, and the powder particles–laser beam interaction takes place over a very short duration resulting in high heating/cooling rates [19–21]. The heat is absorbed by the powder particles following both bulk coupling and powder coupling mechanisms [11]. The laser-aided processing not only produces layers of fused powder, but also creates metallurgical bond with its preceding layer, which leads to a proper densification and competent mechanical behavior of the fabricated parts. Generally, the LPBF process can be ascribed with the following steps: scattering and absorption of laser waves by the powder particles, heat transfer, melting and coalescence of particles, generation of the melt pool and its solidification [22,23]. Due to a high cooling rate (up to 10^6 K/s), the microstructure of the fabricated samples can dramatically differ from the conventionally prepared counterparts [3,24]. During solidification, the melted material tends to undergo a significant non-equilibrium metallurgical process, demonstrating different modes of heat and mass transfer, causing the formation of unique microstructures [25].

During the laser treatment, each powder layer possesses its innate thermal history, generating a complex thermal cycle, which results in high residual stresses, periodic cracks, undesirable microstructural features and a lack of morphological uniformity [26]. Intricate physics governing the laser beam–feedstock interaction (energy absorption, heat and mass transfer), in situ chemical reactions, phase transformations and lack of insights of uncontrollable non-equilibrium metallurgical processes restrict the printability of many alloys by LPBF [13,27]. To date, most commercial aluminum alloys for important applications remain challenging for processing by LPBF due to feedstock particles' poor flowability, high affinity to oxygen, high laser reflectivity (hence low absorptivity), high material thermal conductivity, large solidification range and solidification cracking [4,10,14]. The 2xxx, 6xxx and 7xxx series of high-strength age-hardenable aluminum alloys contain elements that widen the solidification temperature range, leading to the segregation of phases with low melting point during epitaxial grain growth [28]. Moreover, the high thermal conductivity and high laser reflectivity of materials require excess heat to reach melting. This can cause vaporization of volatile alloying elements (Zn, Mg, etc.) and lead to heterogeneity within the completed part [10]. Hence, alloys with a large solidification range have a poor applicability to AM due to the formation of hot cracks at various process stages [23].

There are several near-eutectic Al–Si alloy grades suitable for LPBF and available on the market. These materials display an excellent fluidity, high thermal conductivity, low coefficient of thermal expansion (CTE) and outstanding castability [29]. Hypoeutectic Al–Si (7–12 wt.%)-Mg (>1 wt.%) alloys [10,30] possess the largest share among Al alloys applicable for LPBF process. The incorporation of silicon is a critical issue for Al–Si alloys, since Si reduces the melting point and narrows the solidification temperature range through the formation of a eutectic, thus inhibiting crack formation and propagation. Nevertheless, LPBF-fabricated Al–Si alloys generally face issues of low strength, low ductility, moderate fatigue and wear resistance, which limit their use as structural components [4,8], and, hence, there is an admitted necessity to develop novel aluminum alloys for LPBF. Owing to extremely quick solidification process inherent to LPBF, the majority of high-strength alloys, traditionally esteemed to be "non-weldable materials", suffer from hot cracking and porosity along the columnar grain boundary. However, even so determined "printable" alloys through LPBF possess a non-uniform microstructure and demonstrate poor mechanical performance [31].

For a wide acceptance of the alloys for industrial use, the materials must ensure a number of required properties. The ideal alloy must be highly matched for the extreme thermal conditions by means of decreasing fabrication defects. Meanwhile, it is crucial for it to possess a suitable microstructure along with specific mechanical properties, which are comparable to the existing peak-aged wrought alloys, and to maintain a major part of its strength at elevated or high temperatures [30]. To further improve the mechanical performance of LPBF-prepared aluminum alloys, a substantial amount of research has been devoted to the following:

(i) Studying the modification of existing compositions by minor alloying constituents to generate strengthening phases upon the fabrication process or during post-processing (heat treatment) [32]. (The effects of common modifying elements are given in Figure 1).
(ii) The addition of grain refiners (stable, non-soluble solid ceramic particulates) to reduce hot-tear susceptibility, grain growth and dislocation motion by developing aluminum matrix composites (AMC) [8,33]. The latter conveys a combination of properties of two or more physically distinct phases with the aim to produce parts with far superior properties to the individual components [34].
(iii) Heat treatment [35–37].

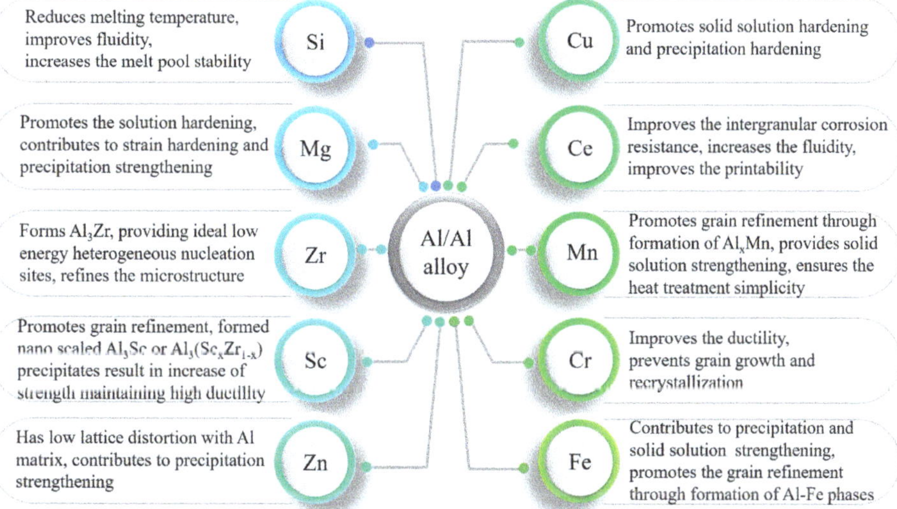

Figure 1. The influence of the main and modifying components on LPBF fabricated Al alloys [14,27,30,33,38–54].

AM processes are categorized as master forming technologies, where customized designed objects' properties are generated by the fabrication process itself. Therefore, the composition and aluminum alloy chemistry undertake a central role during the LPBF process [1]. Combining the advantages offered by AM with the favorable mechanical properties of aluminum alloys will create viable mass-market manufacturing strategies that will increase the adoption and implementation of both across the world [7].

In this review paper, the focus is placed on the laser powder-bed fusion of the ceramic particulate (boride, carbide, nitride and hybrid additive) reinforced aluminum alloys, concentrating on the effect of additives on the microstructure and grain refinement of the produced materials. Thereafter, the mechanical properties and the mechanisms responsible for their change are confronted to lead to a deeper understanding of the possible performance of ceramic particulate reinforced aluminum matrix composites (AMCs). The list of used reinforcements and their unique features during the LPBF process, as well as diagrams showing the strengthening, hardening and grain-refining effect of the added particulates, are specified. The properties and efficiency of AMCs prepared by the traditional or other additive manufacturing techniques are beyond the scope of this paper.

Reinforcement with Ceramic Particulates

The influence of rapid cooling during LPBF on the Al alloy microstructure is described by three factors: (i) constitutional changes due to a great level of undercooling; (ii) individual phase refinement, when the scale of microstructural refinement is strongly related to the velocity of the solidification interface; (iii) generation of phases in metastable state [10].

In contrast to coarse-grained cast Al alloys, LPBF-fabricated Al alloys exhibit a refined microstructure, reduced dendritic branching, decreased segregation patterns, extensions of solid solubility of alloying components, formation of metastable crystalline, quasi-crystalline, amorphous phases [10] and microstructural anisotropy [55].

Generally, the anisotropy in LPBF-fabricated parts is a major processing bottleneck triggered by the generation of coarse columnar grains with a preferential crystallographic texturing along the build direction [56]. The main microstructural characteristics in LPBF-fabricated hypoeutectic Al–Si alloys are columnar primary-Al grains and the eutectic Si phase. The formation of such columnar grains is induced by the high thermal gradients, which hinders nucleation ahead of the solidification front stimulating epitaxial grain growth during LPBF [57]. Epitaxially grown columnar grains are formed during partial (or complete) melting of the preceding solidified layers upon laser scanning of new layers and further develop through successive irradiated layers. Moreover, the formation of columnar grains can lead to intergranular hot tearing [58]. An effective solution is to provoke the equiaxed grain formation during cooling process, which is reached upon modulating the thermal gradient, cooling rate and alteration of cooling conditions [59,60].

One of the approaches for microstructure and properties optimization during LPBF processing is either ex situ or in situ inoculation. In situ reactions in the particle-reinforced composite systems prohibit the formation of interfacial compounds, support the nucleation and growth from the parent matrix phase to generate chemically more stable reinforcing compounds. The distribution of the in situ reinforcements is more homogeneous and provides a strong interfacial bonding with the matrix [61]. The chemical reaction between the reactants might also originate an extra thermal energy for the fusion, which can strengthen the matrix-reinforcement binding. Such assets lead to supreme material performances, allowing MMCs (metal matrix composites) to reach mechanical properties far superior to the ex situ reinforced or non-reinforced metals/alloys. However, due to a wide variety of technological challenges, these MMCs are seldom implemented for commercial applications. Successful design requires a large number of factors to be considered, such as powder compositions, presence of native oxide films on powder particles, powder flow, exothermicity of the in situ reaction and process parameters. The "in situ" formed elements, such as O, C and N, might dissolve in a metal matrix, causing significant embrittlement. Furthermore, additional heat released during the process might cause melt pool instability, leading to an intensive powder splash and evaporation [62,63].

Commonly, for grain refinement, the addition of stable grain refiners (inoculants) with the smallest possible lattice mismatch to aluminum is widely used in conventional casting processes. Refiners suppress the columnar solidification and promote the formation of a fine, uniform, equiaxed grain structure by stimulating heterogeneous nucleation and achieving the columnar-to-equiaxed transition [64]. The latter magnifies the total area of grain boundaries per unit volume, decreasing the residual liquid film thickness along the solidification process, and thus prohibits the formation and propagation of cracks [28]. The heterogeneous nucleation of α-Al during solidification takes place preferably on the inoculants, which provide the low-energy interfaces between a refiner and a matrix [65].

To determine the comparative values of interfacial energy, atomic matching throughout the interface is generally employed as an indicator. To reduce interfacial energy, the main requirements are coherent or semi-coherent interfaces and reproducible orientation relationships (ORs) between two crystals, as different lattice parameters cause distortion of the lattice, resulting in an excess strain energy, which is determined by a lattice mismatch (also called as lattice disregistry, δ) [58]. The selection of potent grain refiners with the smallest disregistry with the matrix crystal throughout a specific interface is favored [58]. If disregistry value is below 10%, both in situ formed and added inoculants have the ability to induce heterogeneous nucleation of Al grains [66].

Nucleant particles serve a dual role in the AMCs as refiners and reinforcements, and they can be classified in three categories: non-oxide ceramics, oxide ceramics and carbon-based compounds. Generally, the ceramic particulates of a high hardness, good thermal stability, relatively high laser absorptivity and compatibility with metals/alloys are suitable constituents for the preparation of high-performance AMCs [67]. To meet the demand to satisfy the "light weight and high strength" concept, novel AMCs are continuously under development [5,11,68].

For the conventional AMCs, relatively coarse ceramic particles with a size ranging from several tens to hundreds of micrometers are broadly utilized as reinforcements. However, reasoned by limited interfacial wettability between reinforcement and matrix, the large particles are susceptible to cracking during mechanical loading, causing reduced ductility and inducing premature failure of AMCs [69]. Consequently, both tensile strength and ductility of AMCs increase if the fine-sized reinforcements are used. On that account, the introduction of the nano-scaled ceramic particles can remarkably enhance the mechanical performance of AMCs [70,71].

However, the agglomeration of nanoparticles may cause unfavorable microstructural changes and affect the mechanical behavior of the composites, as well as affecting thermal and rheological behavior of the melt pool (increasing viscosity, especially in case of high volume of nanoparticles) and shifting the LPBF parameter window. The LPBF method enables effective fabrication of composites reinforced with ceramic reinforcements, taking into account the unique metallurgical nature of the process, high temperatures and thermal convection in a micron-sized molten pool [23,72,73].

2. Non-Oxide Additives

Non-oxide additives (borides, carbides, nitrides, etc.) are one of the most used reinforcements for Al alloys due to their high melting temperatures and chemical stability [74]. AMCs merge the ductility and toughness of aluminum with the high strength and modulus of the ceramic reinforcement [75], hence achieving an improvement of the overall characteristics and durability [12]. The low laser absorptivity of aluminum in the infrared range challenges the controlled melting, while the increase in the laser absorption of ceramic particulate decorated/mixed aluminum alloy at a laser wavelength of 1064 nm promotes the LPBF process. The introduction of ceramic particles to the pure alloy increases laser absorptivity of the overall powder mixture, as (i) non-oxide ceramic particles display high laser absorptivity and (ii) the added ceramic particles increase surface roughness of decorated powder, promoting multiple reflections of the laser in the powder bed [28]. As shown in Figure 2a–c, the ray absorption of the SiC/AlSi10Mg and TiB_2/AlSi10Mg powder mixtures is higher compared to pure AlSi10Mg alloy. There is a lower intensity of interactions between laser rays and particles of pure AlSi10Mg compared to SiC and TiC added composite powder. (Figure 2d–g) [76].

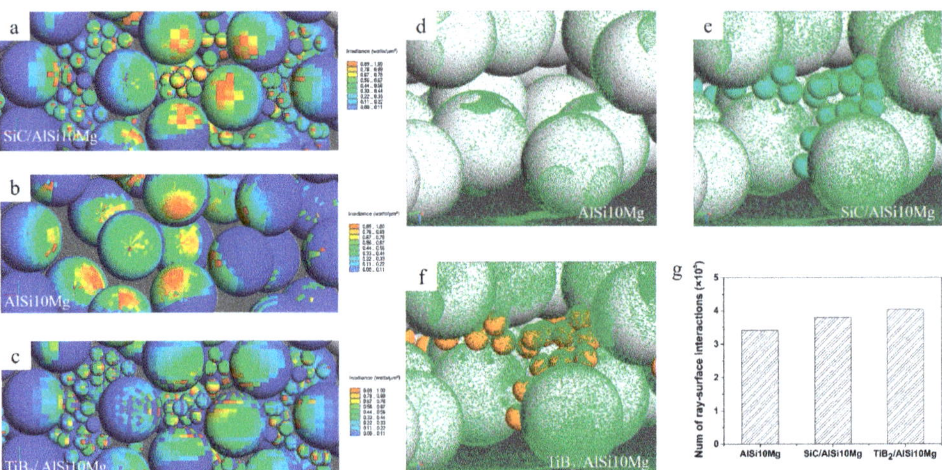

Figure 2. Irradiance distribution for SiC/AlSi10Mg (**a**), AlSi10Mg (**b**), TiB_2/AlSi10Mg (**c**) powder mixtures (top view). Illustration of track spot of each laser ray on the particle surface of AlSi10Mg (**d**), SiC/AlSi10Mg (**e**), TiB_2/AlSi10Mg (**f**) (side view) and numerical representation of laser–particle interactions (**g**) (reproduced with permission from [76]).

2.1. Borides: Grain Refining and Strengthening Effect of TiB_2, LaB_6, CaB_6

As one of the proven highly effective grain refiners for Al alloy, TiB_2 particles exhibits good thermal stability, good wettability and interfacial compatibility, in addition to the acknowledged crystallographic orientation relationship with Al matrix, contributing to a comprehensive mechanical performance of AMCs [59,73]. The addition of TiB_2 to AlSi10Mg increases the laser absorptivity of the powder bed by almost 1.5 times [76]. To provide even distribution, small particle size and adequate interfacial bonding of the TiB_2 particles, in situ fabrication approaches have been implemented, offering the advantages of a clean interface between ceramic particles and matrix alloy and fine morphology of in situ formed particles [5]. Both in situ and ex situ fabrication of TiB_2 reinforced Al alloys are discussed below.

In Ref. [77], 0.5–8 wt.% nano-sized TiB_2 particles were introduced into AlSi10Mg, which resulted in the elimination of columnar grains and refined elongated dendritic structures from 4.6 to 2 µm, as shown in Figure 3a–d and Table 1. Similar results were obtained in Refs. [59,73], as the introduction of 1–5 wt.% and 5.3 wt.% (3.4 vol%) TiB_2 to AlSi10Mg, respectively, led to remarkable grain refinement down to 1.55 µm (Figure 3e–g,i,j). However, the incorporation of only 1 wt.% TiB_2 into AlSi10Mg [78] did not demonstrate a dramatic difference between reinforced and pure alloy parts; however, the grain size distribution became distinctly narrow (Figure 3h).

A microstructure with average grain size of 1.38 µm for the vertical sector was observed [79] when 6.5 wt.% TiB_2 was added (Figure 3k). However, the increase in TiB_2 content to ~11.6 wt.% (almost two times) [80] did not result in further grain refinement (Figure 3l).

Table 1. Characteristics of boride (particulate) reinforced AMCs fabricated by laser powder-bed fusion.

System	Used Device, Process Parameters	Relative Density (%)	Average Grain Size (μm)	σ_y/σ_u (MPa)	$\varepsilon/\varepsilon_c$ (%)	Hardness (HV)	N
AlSi10Mg/ 1 wt.% TiB_2	SLM 150 HL P = 350–450 W v = 1800 mm/s d = 50 μm h = 50 μm E_v = 77.7–100.0 J/mm^3	99.95	~6.3	-	-	~126 HV0.2	[78]
AlSi10Mg/ 3.4 vol.%TiB_2	Prox DMP 200 SLM P = 210 W v = 1000 mm/s d = 30 μm h = 100 μm E_v = 70 J/mm^3	99.975	2.08	σ_u = 522.9–529	$\varepsilon \approx$ 7.5–8.6	-	[59]
AlSi10Mg/ 1 wt.%TiB_2	SLM 150 P = 450 W v = 1600–2600 mm/s d = 50 μm h = 50 μm E_v = 69.2–112.5 J/mm^3	Up to 99.09	6.32 ± 0.07	$\sigma_y \approx$ 270 σ_u = 397	$\varepsilon \approx$ 3.6	~124 HV0.2	[73]
AlSi10Mg/ 2 wt.% TiB_2		Up to 99	2.20 ± 0.11	$\sigma_y \approx$ 283 $\sigma_u \approx$ 444	$\varepsilon \approx$ 4.2	~127 HV0.2	
AlSi10Mg/ 5 wt.% TiB_2		~96–97.8	1.55 ± 0.14	$\sigma_y \approx$ 270 σ_u = 422	$\varepsilon \approx$ 4.1	~129 HV0.2	
AlSi10Mg	Prox DMP 200, 3D Systems P = 220–280 W v = 800–2000 mm/s d = 30 μm h = 90 μm	99.56 ± 0.16	4.64	σ_y = 270.1 ± 4.3 σ_u = 430.7 ± 1.6	ε = 4.7 ± 0.4	125.9 ± 1.4 HV10	[77]
AlSi10Mg/ 0.5 wt.% TiB_2		99.82 ± 0.10	3.45	σ_y = 317.6 ± 2.1 σ_u = 484.1 ± 3.3	ε = 9.5 ± 0.3	140.5 ± 1.3 HV10	
AlSi10Mg/ 2 wt.% TiB_2		99.92 ± 0.04	2.0	σ_y = 320.1 ± 3.2 σ_u = 500.7 ± 3.5	ε = 12.7 ± 0.2	147.1 ± 1.5 HV10	
AlSi10Mg/ 5 wt.% TiB_2		99.91 ± 0.02	~2.0	σ_y = 323.7 ± 1.9 σ_u = 522.9 ± 3.6	ε = 8.7 ± 0.5	151.1 ± 2.1 HV10	
AlSi10Mg/ 8 wt.% TiB_2		99.92 ± 0.05	~2.0	σ_y = 340.8 ± 1.7 σ_u = 544.4 ± 2.6	ε = 6.2 ± 0.2	161.5 ± 2.5 HV10	
AlSi10Mg/ 6.5 wt.%TiB_2	BLT-S310 P = 260–350 W v = 900–1500 mm/s d = 30 μm h = 110–170 μm	>99.5	1.63 μm for top	σ_y = 332.3 ± 6.7 σ_u = 536.9 ± 14.4	ε = 16.5 ± 1.7	-	[79]
			1.38 μm for side	σ_y = 277.9 ± 6.9 σ_u = 517.3 ± 9.1	ε = 15.4 ± 1.6		
AlSi10Mg/ 11.6 wt.% TiB_2	House-built P = 200–300 W v = 800–2000 mm/s d = 30 μm h = 105 μm E_v = 31.7–119.0 J/mm^3	99.5	~2	σ_u = 530 ± 16	ε = 15.5 ± 1.2	191 ± 4 HV0.3	[80]
AlCu/ ~4.7 wt.% TiB_2	Renishaw AM400 P = 250–300 W v = 1125–4500 mm/s d = 30 μm h = 90 μm	Up to 99.5	0.5–2	σ_u = 391 ± 7.3 σ_y = 317.8 ± 9.3	ε = 12.5 ± 0.8	-	[50]
Al-Cu-Mg-Si/ 5 vol.% TiB_2	SLM 250 HL P = 190 W v = 165 mm/s d = 40 μm h = 80 μm E_v = 359.8 J/mm^3	>99.0	2.5 ± 0.1	σ_{yc} = 191 ± 12	$\varepsilon_c \approx$ 60	-	[81]

Table 1. *Cont.*

System	Used Device, Process Parameters	Relative Density (%)	Average Grain Size (μm)	σ_y/σ_u (MPa)	$\varepsilon/\varepsilon_c$ (%)	Hardness (HV)	N
Al-Cu/ ~4 wt.% TiB_2	Aconity LAB P = 200 W v = 1000 mm/s d = 30 μm h = 100 μm E_v = 66.67 J/mm^3	99.9 ± 0.1	0.64 ± 0.26	σ_u = 401 ± 2	ε = 17.7 ± 0.8	113 ± 2 HV10	[82]
Al-12Si	SLM 250 HL P = 320 W v = 1655 mm/s d = 50 μm h = 110 μm E_v = 35.1 J/mm^3	-	-	σ_{yc} = 211 ± 4	-	119 HV0.05	[64,83]
Al-12Si/ 2 wt.% TiB_2		≈99.1	~5.1	σ_{yc} = 225 ± 4	ε_c ≈ 30	142 ± 6 HV0.05	
AlSi10Mg	SLM125HL P = 300 W v = 1650 mm/s d = 30 μm h = 130 μm E_v = 46.6 J/mm^3 T = 200 °C	99.08 ± 0.1	6.1	σ_y = 243 ± 9 σ_u = 420 ± 9	ε_{tr} ≈ 5.5 ε_{long} ≈ 3.7	-	[84]
AlSi10Mg/ 0.05 wt.% LaB_6		99.03 ± 0.08	4.0	σ_y ≈ 242 σ_u ≈ 430	ε_{tr} ≈ 6.4 ε_{long} ≈ 4.8		
AlSi10Mg/ 0.2 wt.% LaB_6		99.17 ± 0.05	2.5	σ_y ≈ 245 σ_u ≈ 435	ε_{tr} ≈ 7 ε_{long} ≈ 6.5		
AlSi10Mg/ 0.5 wt.% LaB_6		99.46 ± 0.18	2.2	σ_y ≈ 240 σ_u ≈ 427	ε_{tr} ≈ 6.5 ε_{long} ≈ 6.9		
AlSi10Mg/ 1 wt.% LaB_6		99.49 ± 0.13	1.8	σ_y ≈ 235 σ_u ≈ 429	ε_{tr} ≈ 7.1 ε_{long} ≈ 5.8		
AlSi10Mg/ 2 wt.% LaB_6		99.48 ± 0.22	1.6	σ_y ≈ 238 σ_u ≈ 445	ε_{tr} ≈ 7.0 ε_{long} ≈ 5.6		
2024 Al alloy	Aconity LAB machine P = 200–300 W v = 600–1200 mm/s d = 30 μm h = 100 μm E_v = 56–167 J/mm^3	98.3	-	-	-	66 ± 6 HV5	[28]
2024 Al alloy/ 2 wt.% CaB_6		>99.5	0.91 ± 0.32	σ_y = 348 ± 16 σ_u = 391 ± 22	ε = 12.6 ± 0.6	132 ± 4 HV5	

E_v—laser volumetric energy density, E_l—laser linear energy density, P—laser power, v—scanning speed, h—hatching distance, d—layer thickness, σ_u—ultimate tensile strength, σ_y—yield strength, σ_{uc}—ultimate compressive strength, σ_{yc}—compressive yield strength, ε—elongation, ε_{long}—elongation at longitudinal direction, ε_{tr}—elongation at transverse direction, ε_c—compression strain, RT—room temperature, - means no data available.

Partial melting of TiB_2 was reported in Ref. [73] despite the fact that TiB_2 is considered a refractory material. Adding 5 vol.% (or 8.3 wt.%) TiB_2 to an Al–Cu alloy [81] resulted in a remarkable grain size reduction from 23 to 2.5 μm. In Ref. [82], the in situ TiB_2 (4 wt.%) reinforced Al–Cu–Ag–Mg–Ti alloy had fine equiaxed grains with ~0.64 μm average size without preferential orientation (Figure 3p). The reported grain size was smaller than that stated in Refs. [73,80]. In Refs. [64,83], the addition of 2 wt.% TiB_2 to an Al–12Si alloy produced a textureless microstructure with an average grain size of ~5 μm, meaning that in case of similar content of incorporated TiB_2, coarser grains were grown in the Al–12Si alloy than in AlSi10Mg (Figure 3m,n). For comparison, a hot-pressed sample's EBSD image is shown in Figure 3o, which, interestingly, showed a higher degree of grain refinement.

For a bare minimum boride additive range, at least 2 wt.% TiB_2 is sufficient to significantly alter the final morphology and crystallographic texture of LPBF-processed materials [64,73,77,82,83].

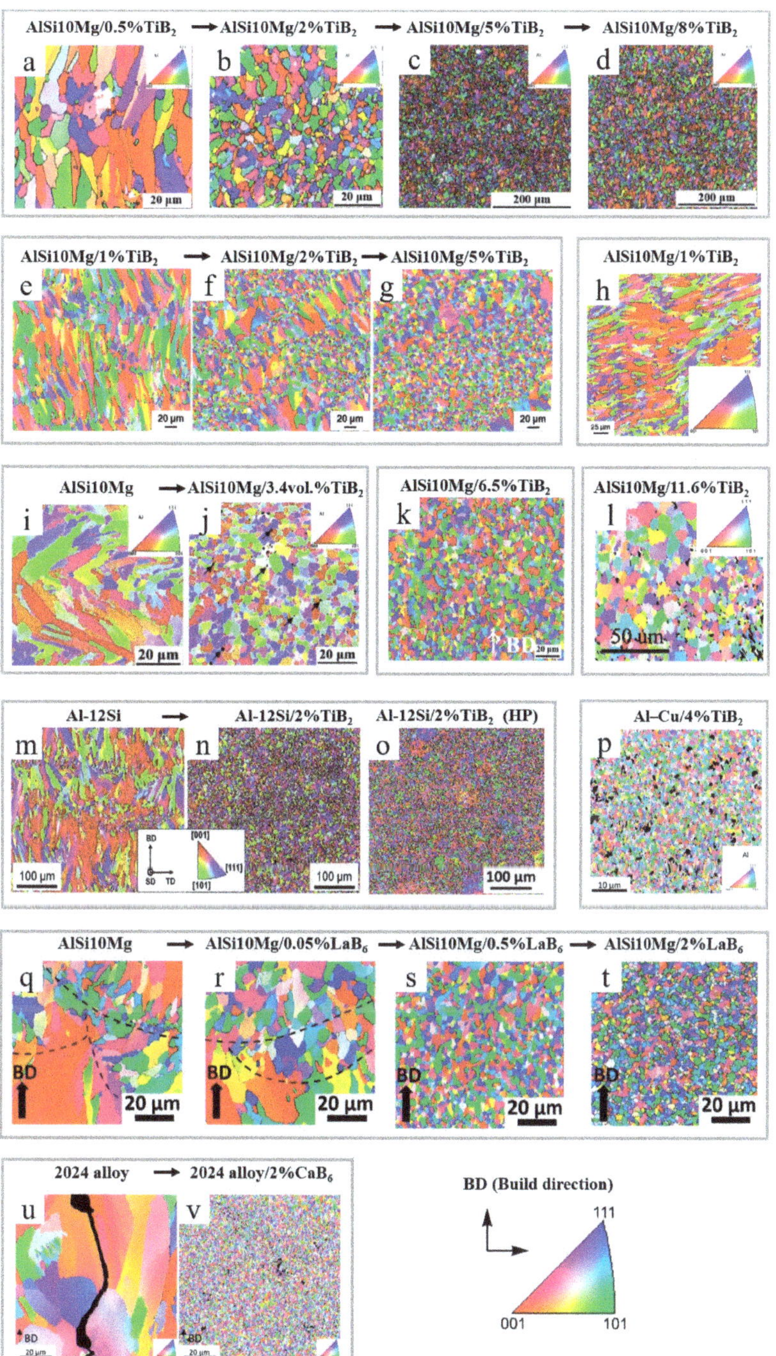

Figure 3. EBSD (electron backscatter diffraction) color maps for LPBF-prepared Al alloys and AMCs reinforced with borides ((**a–n**) and (**p–v**)) (subfigure (**o**) represents hot-pressed (HP) sample) (reproduced with permission from [28,59,64,73,77–80,83,84]).

The grain refining (columnar to equiaxed transition) effect of TiB_2 (Figure 4a,b) is ascribed to its good stability in a melt pool, supplying numerous low-energy barrier nucleation sites (crystal embryos) and a reduction in the critical amount of total undercooling required to initiate the formation of equiaxed crystals [77]. The particles pushed to the grain boundaries pin and stabilize grain boundaries and limit grain growth along the heat flux direction [59]. Furthermore, due to a lower thermal conductivity of TiB_2 (~77.8 W/mK) as compared to Al (~108 W/mK) [73], TiB_2 particles prevent heat flux at a high temperature, reducing the temperature gradient. The latter results in the formation of fine equiaxed grains, weakening the texture and anisotropy of fabricated AMCs [59]. Overall, grain refinement is justified with a combination of high cooling rates during LPBF, an increased number of nucleation sites and limitations on grain growth [73,80], which lie beneath three main mechanisms: constitutional supercooling, heterogeneous nucleation and Zener pinning. Meanwhile, random orientations of TiB_2 particles provide the randomization of Al grain orientation and texture elimination [77].

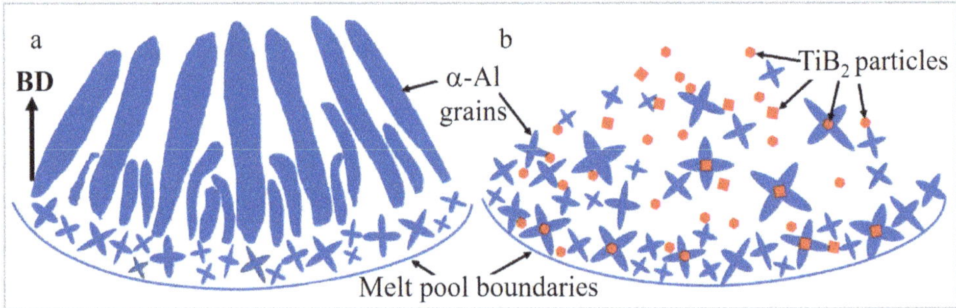

Figure 4. Graphical illustration of grain formation during solidification in a melting pool of AlSi10Mg (**a**) and AlSi10Mg-TiB_2 AMC (**b**) (reproduced with permission from [77]).

The grain refining effect of TiB_2 is also reported to be a result of the formation of Al_3Ti and the crystallographically coherent interface between Al_3Ti and TiB_2, which promotes the nucleation of Al_3Ti on the surface of TiB_2 particles in an Al melt. Without the Al_3Ti layer, TiB_2 additives are easily contaminated by impurities with a high tendency to form a eutectic microstructure with Al and, therefore, being insufficient in nucleating α-Al grains [85]. However, in Ref. [81], a preferable natural stacking sequence of Al atoms on TiB_2 and direct refining are reported. Meanwhile, in Ref. [82], it was highlighted that the absence of the Al_3Ti layer does not prove a lack of nucleation, since the Al_3Ti layer can fully transform into α-Al during the cooling process via a peritectic reaction.

Besides TiB_2, other borides, such as CaB_6 and LaB_6, had shown a promising refining capability. The addition of 0.05–2 wt.% LaB_6 to AlSi10Mg resulted in grain refinement down to 1.6 μm (Figure 3q–t). LaB_6 particles form a highly coherent interface with the Al matrix. A higher amount of LaB_6 nanoparticles (>0.5 wt.%) did not further provide grain refinement and restricted longitudinal elongation due to the weakening of melt pool boundaries by segregation of the excess LaB_6 nanoparticles [84]. The addition of 2 wt.% CaB_6 nanoparticles to the high-strength 2024 aluminum alloy resulted in an equiaxed, crack-free microstructure with an average grain size of 0.91 ± 0.32 μm and a highly coherent interface with Al (Figure 3u,v and Figure 5a,b) [28]. No decomposition of CaB_6 was observed. However, not every CaB_6 nanoparticle functions as a nucleant; a large quantity of them is acquired in the liquid phase between the growing grains, and they are forced to the grain boundaries where they stabilize the microstructure via Zener pinning.

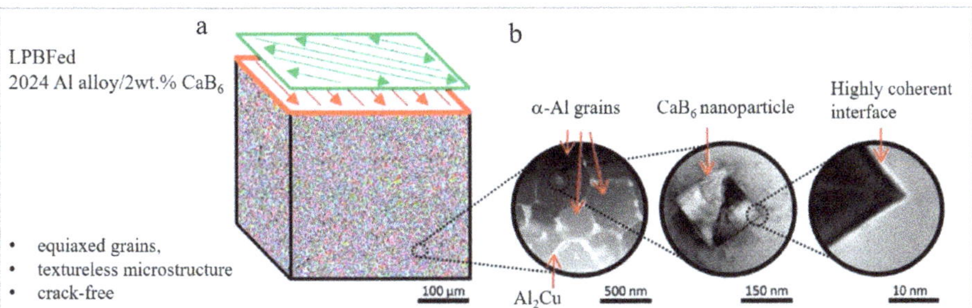

Figure 5. EBSD inverse pole figure grain orientation map of LPBFed 2024 alloy—2 wt.% CaB_6 (**a**), respective HAADF-STEM and ADF-STEM images of CaB_6 nanoparticles within α-Al grain (**b**) (HAADF-STEM stands for high-angle annular dark-field scanning transmission electron microscope, ADF for annular dark-field)(reproduced with permission from [28]).

In Ref. [77], the addition of 0.5–8 wt.% TiB_2 to AlSi10Mg resulted in increased strength (up 544 MPa) and hardness (with 20%); however, the high content of TiB_2 (>2%) resulted in a reduced ductility (6.2%), which was still higher than for a reference AlSi10Mg. Simultaneous enhancement of strength (up to 537 MPa and 530 MPa) and ductility (16.5% and 15.5%) was achieved in Refs. [79,80], respectively, when 6.5 wt.% and 11.6 wt.% TiB_2 were introduced to AlSi10Mg. The increased strength was mainly attributed to the Hall–Petch relationship, loading-bearing and Orowan strengthening mechanisms. The grain boundary modification by TiB_2 nano-particulates and the promoted dislocation plasticity by nano-Si precipitates improved ductility. LaB_6 addition resulted in a subtle improvement of strength and ductility; however, the reinforcing effect was not as pronounced, as in the case of TiB_2.

The highest elongation (~17.7%) was recorded in Ref. [82], when the Al–Cu alloy was reinforced with 4 wt.% TiB_2; however, the alloys exhibited a significantly lower strength and hardness. The addition of 2 wt.% CaB_6 [28] resulted in an increased elongation of 2024 alloy, up to 12.6%, and improved tensile and yield strength (Table 1).

2.2. *Carbides: Grain Refining and Strengthening Effect of TiC, SiC, B_4C*

2.2.1. Titanium Carbide: TiC

TiC exhibits several favorable characteristics required for Al alloys reinforcement; among them, there are moderate density (4.91 g/cm^3), high hardness (28–32 GPa) [86], high modulus of elasticity (up to 440 GPa) [87], good wettability, good laser absorptivity (higher than TiB_2) and low lattice mismatch (6.9%) with Al. TiC particle reinforced AMCs have a high strength, stiffness and modulus, good corrosion and wear performance [22,72]. However, when formed in situ in the melt pool, the TiC phase possesses unstable chemical composition (portrayed as TiC_x, where x is in 0.48–1 range) due to the generation of carbon atom vacancies. Consequently, the nucleating behavior of TiC_x for α-Al is not consistent, since the TiC_x+Al→Al_4C_3 reaction is favored, which results in weakened grain refining performance [88].

In Ref. [89], an increase in the TiC content from 1 to 10 wt.% when added to the Al–15Si alloy resulted in an increase in melt pool fluidity and a decrease in the undercooling degree, leading to significant grain coarsening (Figure 6). Ultimately, with the added threshold limit of TiC (10 wt.%), the primary Si particles precipitate out and distribute on the surface of the Al matrix (Figure 6d).

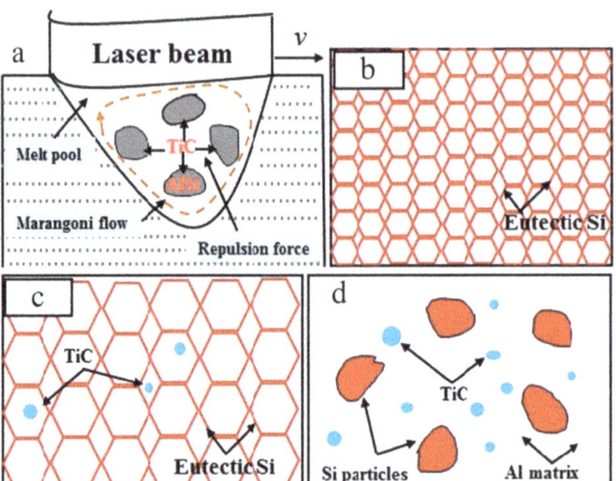

Figure 6. Microstructure evolution of the Al–15Si alloy reinforced with 1 wt.% (**a**), 2.5 wt.% (**b**), 7.5 wt.% (**c**) and 10 wt.% TiC (**d**) (reproduced with permission from [89]).

Alternatively, the fabrication of AlSi10Mg/5 wt.%-nano-TiC [70] under an increased laser energy caused the nano-TiC particles to accumulate in clusters, forming the micron-sized agglomerates. However, the dispersion of reinforcement became more uniform, as shown in Figure 7a–d.

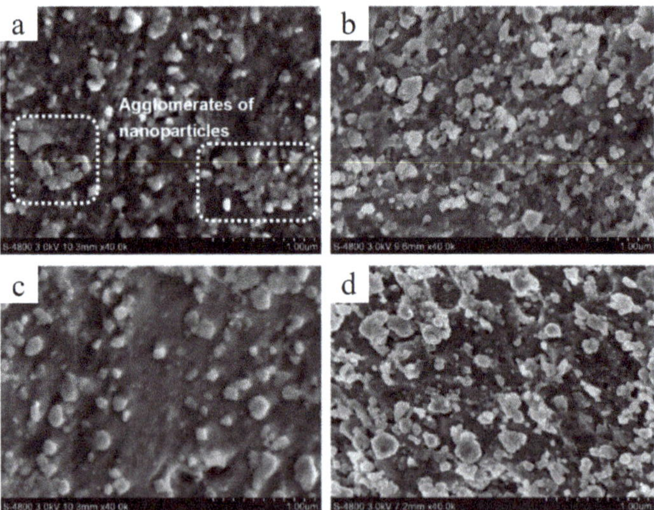

Figure 7. SEM images portraying dispersion degree of TiC and respective microstructure of fabricated AlSi10Mg/5 wt.%TiC composite processed at various E_l (E_v): 314 J/m (125.71 J/mm^3) (**a**), 440 J/m (176.0 J/mm^3) (**b**), 733 J/m (293.3 J/mm^3) (**c**) and 1100 J/m (440.0 J/mm^3) (**d**) (reproduced with permission from [70]).

An increase in energy input resulted in change in TiC appearance, from aggregate to ring (circular) structures, due to intensive Marangoni flow (Figure 8a–d) in LPBFed AlSi10Mg/3 wt.%TiC composites [71].

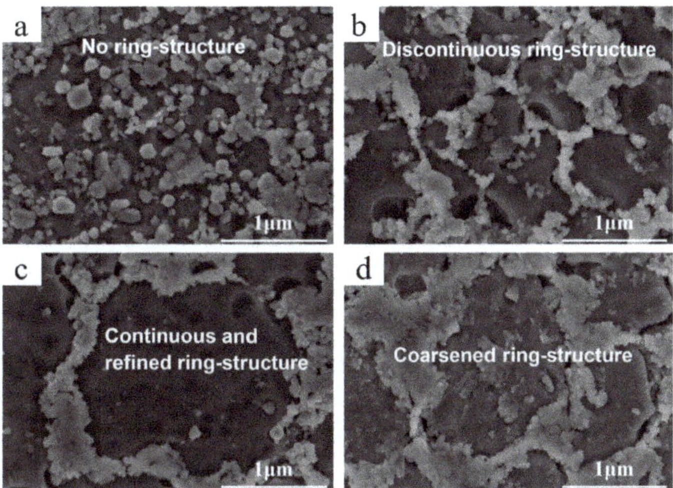

Figure 8. SEM images demonstrating the dispersion states of nano-TiC particles in LPBFed AlSi10Mg/3 wt.%TiC composites at E_v = 160 J/mm^3 (**a**), E_v = 200 J/mm^3 (**b**), E_v = 240 J/mm^3 (**c**) and E_v = 280 J/mm^3 (**d**) (reproduced with permission from [71]).

The formation of ring-structured TiC was reported in Ref. [22] as well. At 5 and 7.5 wt.% TiC addition, at elevated Marangoni force and a lower viscous drag force, the ceramic particulates are captured in the circular melt motion (Figure 9b,c) and generate distinct circular structures in solidified build (Figure 9e–g). The circular-structured TiC agglomerate formation was not found in Ref. [70] when 5 wt.% TiC was used, which can probably be justified by the application of different process parameters.

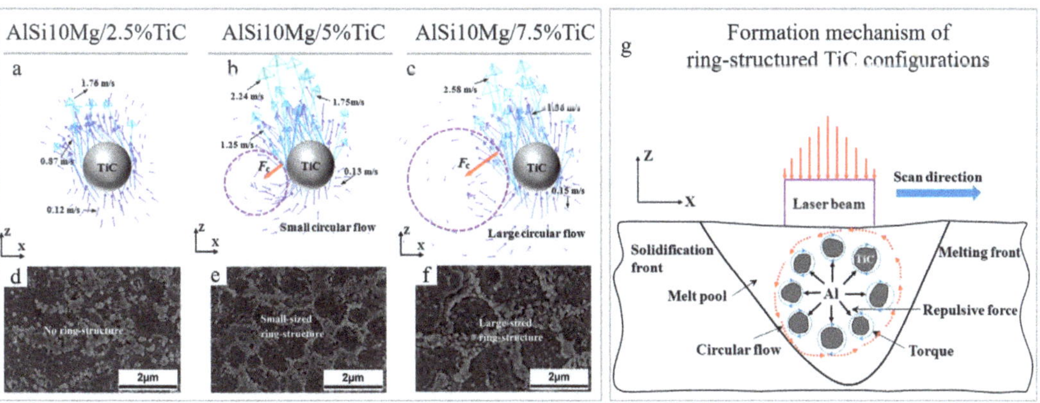

Figure 9. Velocity vector plots around a TiC reinforcing particle in the melt pool (the dashed circles highlight the circular motion) and micrographs demonstrating typical morphology of LPBF-processed AlSi10Mg/TiC nanocomposites with different TiC contents: 2.5 wt.% (**a**,**d**), 5 wt.% (**b**,**e**) and 7.5 wt.% (**c**,**f**). Schematics of the formation mechanism of novel circular TiC configurations during fusion process at fixed E_v = 571.43 J/mm^3 (**g**) (reproduced with permission from [22]).

The presence of in situ formed D0$_{22}$-Al$_3$Ti inoculants (with tetragonal structure) was revealed in Ref. [31] for the AlSi10Mg/5 wt.%TiC composite. Heterogeneous nucleation of α-Al on the D0$_{22}$-Al$_3$Ti nanoparticles (Figure 10c–f) occurred, leading to (i) columnar-to-equiaxed transition with subsequent grain refinement from ~80 μm to ~1 μm (Figure 11a,b), and (ii) the preferred orientation of the of α-Al (200) phase was removed (Figure 10a,b). In situ formed Al_3Ti served as a more effective nucleant as compared to TiC, mainly due to the small lattice mismatch between Al and Al_3Ti, which was reduced to 0.09%.

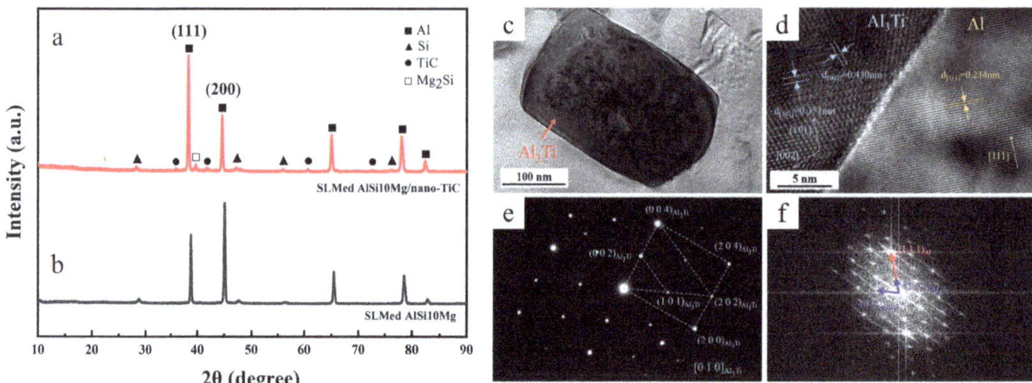

Figure 10. Diffractograms of the LPBFed AlSi10Mg (**a**) and AlSi10Mg/5 wt.%TiC (**b**) specimens, HRTEM image of the $D0_{22}$–Al_3Ti/Al matrix (**c**) and interface (**d**), SAED patterns taken at the $D0_{22}$-Al_3Ti along (010) Al_3Ti (**e**), FFT patterns of the $D0_{22}$–Al_3Ti/Al matrix interface (**f**) (SAED stands for selected area electron diffraction and FFT for fast Fourier transform, (reproduced with permission from [31]).

Yet, another variable parameter centers on powder production for the LPBF process. In Ref. [90], the LPBF of the ball-milled composite powder of AlSi10Mg/5 wt.%TiC is reported. After printing, the TiC particles maintained their nanoscale nature and were not subjected to a significant coarsening, which resulted in an increased hardness of the alloy from 140 to 185 $HV_{0.1}$ and the tensile strength from 400 to 482 MPa (Table 2). The elongation of the composite part (10.8%) was similar to the elongation measured for the pure AlSi10Mg alloy. This can be explained by various effects: (i) an increased dislocation density near reinforcement/matrix interface, (ii) TiC nanoparticles acting as a barrier for dislocation movement, (iii) delaying crack propagation, thus improving the tensile strength. Alternating the TiC concentration, laser energy density and powder processing technique yield different composite attributes, as shown in Table 2.

Table 2. Characteristics of carbide reinforced AMCs fabricated by laser powder-bed fusion.

System	Used Device, Process Parameters	Relative Density (%)	Average Grain Size (μm)	σ_y/σ_u (MPa)	$\varepsilon/\varepsilon_c$ (%)	Hardness (HV)	N
Al-15Si	SLM125 P = 360 W v = 600 mm/s d = 20 μm h = 60 μm	>98.5	-	σ_u = 398	ε = 2.6	154 HV1	[89]
Al-15Si/ 1 wt.% TiC				σ_u = 578	ε = 7.86	146 HV1	
Al-15Si/ 2.5 wt.% TiC				σ_u ≈ 450	ε ≈ 4	150 HV1	
Al-15Si/ 10 wt.% TiC				σ_u ≈ 313	ε = 2.24	177 HV1	

Table 2. *Cont.*

System	Used Device, Process Parameters	Relative Density (%)	Average Grain Size (μm)	σ_y/σ_u (MPa)	$\varepsilon/\varepsilon_c$ (%)	Hardness (HV)	N
AlSi10Mg/ 3 wt.% TiC	SLM system P = 80, 100, 120 and 140 W v = 200 mm/s d = 50 μm h = 50 μm E = 160 J/mm^3	>98.5	-	σ_u = 452	ε = 9.8	157.4 HV0.1	[71]
	E = 200 J/mm^3			-	-	≈173 HV0.1	
	E = 240 J/mm^3			σ_u = 486	ε = 10.9	188.3 HV0.1	
	E = 280 J/mm^3			-	-	180.6 HV0.1	
AlSi10Mg/ 5 wt.% TiC	SLM system P = 110 W v = 100–350 mm/s d = 50 μm h = 50 μm E_l = 1100, 733, 440, 314 J/m	>98	-	-	-	181.2 HV0.2	[70]
AlSi10Mg/ 5 wt.% TiC	EOS M290 P = 320 W v = 1100 mm/s d = 30 μm h = 130 μm	99.75	0.5–1	σ_u ≈ 456 σ_y ≈ 338	ε = 2.97	131 HV0.05	[31]
AlSi10Mg/ 5 wt.% TiC	SLM system P = 100 W v = 150 mm/s d = 50 μm h = 50 μm	Full dense	-	σ_u = 482	ε = 10.8	185 HV0.1	[90]
AlSi10Mg/ 10 wt.% Al-Ti-C-B master alloy	3D Systems ProX DMP 320 P = 300 W v = 1400 mm/s d = 30 μm h = 100 μm	-	~3	σ_u = 488 ± 6 σ_y = 287 ± 3	ε = 10.1 ± 2.2	-	[88]
2024 alloy	EOS M290 P = 200 W v = 100 mm/s d = 40 μm h = 90 μm T = 180 °C	98.2	~30	σ_u = 240 ± 10	ε = 0.3 ± 0.2	108 HV0.2	[92]
2024/ 1 wt.% TiC		98.5	-	-	-		
2024/ 1 wt.% TiH_2		95.7	-	-	-	-	
2024/ (1 wt.% TiC +1 wt.% TiH_2)		97.1	~2	σ_u = 390 ± 15	ε = 12.0 ± 0.5	120 HV0.2	
AlSi10Mg	EOS M280 P = 270 W v = 1600 mm/s d = 30 μm h = 110 μm	98.22	12.1	σ_u = 393.8 ± 14.5 σ_y = 224.2 ± 7.2	ε = 4.5 ± 0.9	127.8 ± 2.4 HV.1	[93]
ASi10Mg/ 1.5 wt.% TiC +1.5 wt.% TiB_2		99.02	1.5	σ_u = 552.4 ± 12.1 σ_y = 325 ± 10.2	ε = 12 ± 0.6	142 ± 2.9 HV0.1	
ASi10Mg/ 3 wt.% TiB_2		97.12	7.7	σ_u = 360.6 ± 8.5 σ_y = 200 ± 8.8	ε = 3.8 ± 0.2	134.4 ± 1.4 HV0.1	
ASi10Mg/ 3 wt.% TiC		98.23	1.7	σ_u = 453 ± 10 σ_y = 267.5 ± 7.8	ε = 4.8 ± 1.1	138.3 ± 1.7 HV0.1	

Table 2. *Cont.*

System	Used Device, Process Parameters	Relative Density (%)	Average Grain Size (μm)	σ_y/σ_u (MPa)	$\varepsilon/\varepsilon_c$ (%)	Hardness (HV)	N
AlSi10Mg	SLM-125HL P = 150 W v = 1200 mm/s d = 30 μm h = 105 μm T = 200 °C	At RT full dense	-	RT σ_u = 356 ± 10 σ_y = 220 ± 4	ε = 4.5 ± 0.5	-	[91]
				100 °C σ_u = 327 ± 2 σ_y = 230 ± 3	ε = 5 ± 1		
				150 °C σ_u = 282 ± 3 σ_y = 213 ± 3	ε = 11.5 ± 2.5		
				200 °C σ_u = 245 ± 8 σ_y = 194 ± 7	ε = 11 ± 1.2		
AlSi10Mg/ 2 vol.% TiCN		At RT full dense	RT <1.5	RT σ_u = 333 ± 2 σ_y = 227 ± 7	ε = 2.8 ± 0.	-	
			-	100 °C σ_u = 344 ± 2 σ_y = 245 ± 2	ε = 3.5 ± 0.2		
			-	150 °C σ_u = 308 ± 9 σ_y = 235 ± 4	ε = 4.2 ± 0.2		
			-	200 °C σ_u = 270 ± 1 σ_y = 209 ± 10	ε = 4.9 ± 0.4		
AlSi10Mg	SLM-120 P = 200 W v = 1200 mm/s d = 30 μm h = 70 μm T = 200 °C	Almost full dense	-	σ_u = 366 σ_y = 193	ε = 6.8	~141 HV0.2	[94]
AlSi10Mg/ 0.7 wt.% (B_4C+Ti)				σ_u = 417 σ_y = 234	ε = 5.2	~139 HV0.2	
AlSi10Mg/ 5.7 wt.% (B_4C+Ti)				σ_u = 307 σ_y = 126	ε = 3.6	~170 HV0.2	
AlSi10Mg/ 11.5 wt.% (B_4C+Ti)				σ_u = 218 σ_y = 117	ε = 3.4	~175 HV0.2	
AlSi10Mg/ 17.2 wt.% (B_4C+Ti)				σ_u = 165 σ_y = 72	ε = 1.7	~222 HV0.2	
AlSi7Mg	EOSINT M280 P = 350 W v = 1200 mm/s d = 40 μm h = 190 μm T = 80 °C	Porosity ≈0.59%	~4.55	σ_u = 388.3 ± 49.6	ε = 7.03 ± 1.25	≈1.85 GPa nano-hardness	[8]
AlSi7Mg/ 2 wt.% SiC		Porosity ≈0.25%	~3.14	σ_u = 502.94	ε = 10.64 ± 1.06	≈2.11 GPa nano-hardness	
AlSi10Mg/ 2 vol.% SiC (~2.4 wt.%)	SLM280HL P = 120 W v = 250 mm/s d = 30 μm h = 60 μm T = 150 °C E_v = 267 J/mm^3	~92.04	-	-	-	-	[95]

Table 2. *Cont.*

System	Used Device, Process Parameters	Relative Density (%)	Average Grain Size (μm)	σ_y/σ_u (MPa)	$\varepsilon/\varepsilon_c$ (%)	Hardness (HV)	N
AlSi10Mg/ 2 vol.% SiC (~2.4 wt.%)	P = 150 W E_v = 333 J/mm^3	98.7	4.44	σ_u = 343 ± 59	ε = 3.3 ± 1.7	134.4 ± 3.2 HV0.1	[95]
	P = 180 W E_v = 400 J/mm^3	97.69	4.96	σ_u = 377 ± 28	ε = 2.9 ± 0.95	135.6 ± 3.5 HV0.1	
	P = 210 W E_v = 467 J/mm^3	97.36	6.73	σ_u = 440 ± 17	$\varepsilon \approx 7.4$	131.7 ± 2.6 HV0.1	
	P = 240 W E_v = 533 J/mm^3	97.40	-	σ_u = 450 ± 30	ε = 4.9	129.7 ± 6.9 HV0.1	
Al–12Si/ 10 vol.% SiC (~11.8 wt.%)	ReaLizer SLM-100 P = 200 W v = 375–1500 mm/s d = 50 μm h = 100 μm $E_v \approx$ 20–80 J/mm^3	97.4 (by X-ray micro tomography (XMT))	-	-	-	-	[34]
AlSi10Mg/ 10 wt.% SiC	EOSINT M280 P = 240–320 W v = 500–1800 mm/s d = 30 μm h = 80–160 μm	-	2.35	$\sigma_u \approx 450$ $\sigma_y \approx 410$	-	208.5 HV0.1	[96]
AlSi10Mg/ 15 wt.% SiC	Self-developed NRD-SLM-III P = 340–490 W v = 600–2100 mm/s d = 40 μm h = 60–180 μm T = 200 °C	97.7	-	σ_u = 341.9	$\varepsilon \approx 3$	217.4 HV0.2	[97]
AlSi10Mg/ 15 wt.% SiCp (300 mesh)	Self-developed NRD-SLM-III P = 500 W v = 1200 mm/s d = 40 μm h = 120 μm T = 200 °C	≈97.8	-	σ_{uc} = 545.4	$\varepsilon_c \approx 4.7\%$	≈210 HV0.2	[98]
AlSi10Mg/ 15 wt.% SiCp (600 mesh)		≈98.5		σ_{uc} = 642.4	$\varepsilon_c \approx 6.1\%$	≈240 HV0.2	
AlSi10Mg/ 15 wt.% SiCp (1200 mesh)		98.9		σ_{uc} = 764.1	$\varepsilon_c \approx 7.0\%$	316.1 HV0.2	
AlSi10Mg/ 20 wt.% SiC	Self-developed P = 80–110 W N = 100 mm/s d = 50 μm h = 50 μm E_l = 800–1100 J/m	~89.2–96.1	-	-	-	214 HV0.1	[11]
AlSi10Mg/ 20 wt.% SiC $D50_{SiC}$ = 50 μm	SLM apparatus with Yb laser P = 100 W v = 100 mm/s d = 30 μm h = 50 μm	86.4	-	-	-	~127 HV0.1	[13]
AlSi10Mg/ 20 wt.% SiC $D50_{SiC}$ = 15 μm		93.7				188 HV0.1	
AlSi10Mg/ 20 wt.% SiC $D50_{SiC}$ = 5 μm		~97.2				218.5 HV0.1	

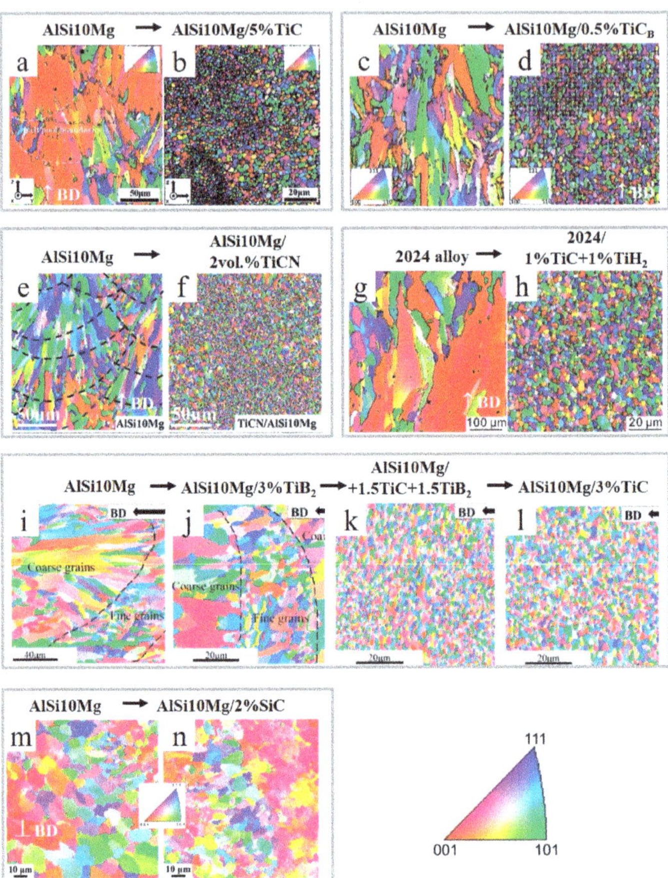

Figure 11. EBSD color maps for LPBF-prepared Al alloys and AMCs reinforced with carbides, carbonitride, carbide/hydride and carbide/boride additives (**a**–**n**) (reproduced with permission from [31,88,91–93]).

While using a single carbide reinforcement has proven to be an effective way for grain refinement, the use of a second additive was shown to complement the effects of a single species. In Ref. [92], the dual reinforcing phases were used, resulting in a crack-free sample produced from the 2024 alloy/1 wt.%TiC+1 wt.%TiH_2 powders mixture. It was shown that unreinforced alloy contained columnar microstructure (Figures 11g and 12a–c), while the 2024 alloy/1 wt.%TiC+1 wt.%TiH_2 composite was composed of superfine equiaxed grains (Figures 11h and 12d–h).

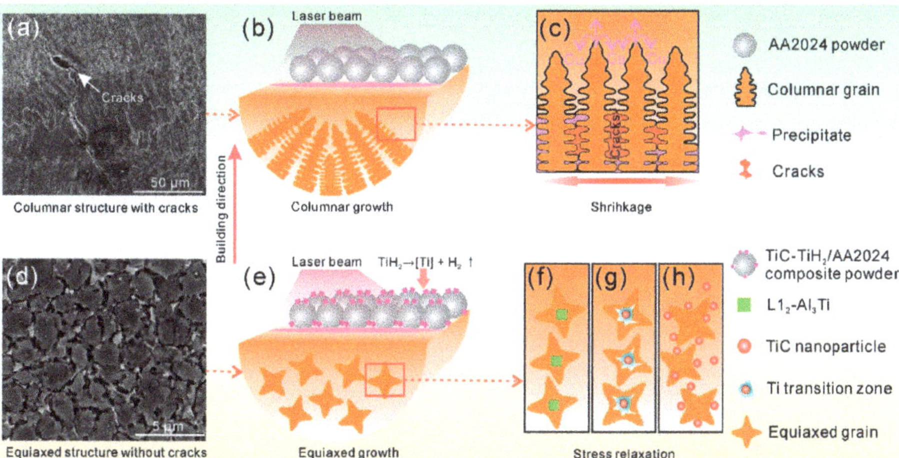

Figure 12. Schematic representation of microstructures and solidification mechanisms of LPBF-fabricated 2024 Al alloy (**a–c**) and 2024/TiC-TiH_2 composite (**d–h**) (reproduced with permission from [92]).

Ti-rich particles (TiC and Al_3Ti) with irregular or cubic shape are present in the grains exhibited in Figure 13a,b. The $L1_2$-Al_3Ti with a face-center-cubic (FCC) structure is a result of TiH_2 decomposition ($TiH_2 \rightarrow Ti + H_2$) and reaction between Ti and Al melt. It is worth mentioning that in Ref. [31], a formation of the $D0_{22}$-Al_3Ti phase with a tetragonal structure was reported. A highly coherent interface between $L1_2$-Al_3Ti and α-Al was observed (with 0.24% lattice mismatch) (Figure 13c), indicating that $L1_2$-Al_3Ti might serve as substrate for heterogeneous α-Al nucleation; however, a coherent interface was not generated between TiC and Al (Figure 13d). Following the "Ti transition zone" theory (demonstrated in Figure 12), Ti-covered TiC nanoparticles, and then TiC particles themselves, become the effective nucleation substrates for α-Al as well.

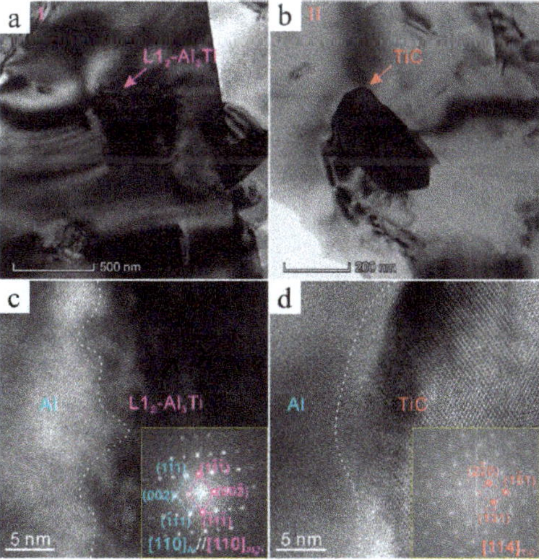

Figure 13. TEM images of $L1_2$-Al_3Ti (**a**) and TiC particles (**b**), HRTEM image and respective FFT pattern of α-Al/$L1_2$-Al_3Ti interface (**c**) and α-Al/TiC interface (**d**) (reproduced with permission from [92]).

On account of the inhibition of columnar grains, elimination of cracks, a refined microstructure and Orowan strengthening, the 2024 alloy/TiC+TiH_2 AMC showed a simultaneous enhancement of tensile strength and ductility.

Another study on the fabrication of double TiB_2-TiC reinforced AMCs [93] revealed that the addition of dual ceramic phases improved laser absorptivity by almost two-fold, substantially refining the Al grains (Figure 11i,k) and resulting in the increment in tensile strength (552 MPa) and elongation (12%) (Table 2). It was revealed that the dual reinforcement more remarkably affected the mechanical performance, improved densification and grain refinement compared to the single reinforcement with the same total content (Table 2 and Figure 11j,l).

Double or triple reinforcements formed during in situ chemical reactions generate a composite material highly coherent with the metal matrix. When 0–17.2 wt.% (Ti-B_4C) mixture was added to AlSi10Mg [94], the full densification of samples and in situ formation of ceramic phases were reported due to the combined LPBF and combustion synthesis (CS) process. Silicon atoms released from the alloy combine with Ti and C atoms, yielding the formation of transitional ternary carbide Ti_3SiC_2, while the remaining B_4C and Ti are responsible for the formation of TiB_2 and TiC particulates (Figure 14). The generation of the Ti_3SiC_2 phase resulted in a significant drop in porosity of the fabricated sample. The heat released during the combustion reaction allowed for carrying out the fabrication in low laser energy regime.

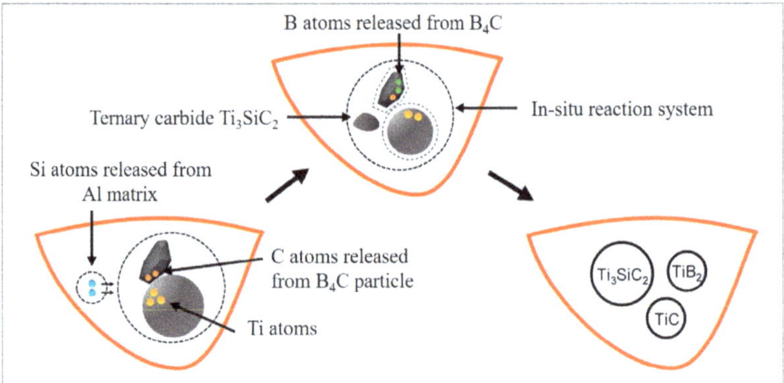

Figure 14. In situ formation mechanism of TiB_2, TiC, Ti_3SiC_2 ceramic phases in the molten pool (reproduced with permission from [94]).

2.2.2. Silicon Carbide: SiC

The SiC particle reinforced AMCs are applied in aerospace and electronic encapsulation, both in military and civilian fields, due to their high specific strength and stiffness, in addition to abrasion resistance. SiC has a much higher laser absorptivity (≈78%) than aluminum (≈7%), moderate density (3.21 g/cm^3), and it increases the laser absorptivity of the blended mixture [13,34,97,98]. During laser irradiation, SiC particles tend to heat up to extremely high temperature, leading to rapid reaction rates. Hence, the decrease in thermal conductivity results in further rise in temperature, the lifetime and fluidity of the melt pool. Meanwhile, an increase in SiC content in the initial feedstock and, hence, in the blend melt pool, increases the viscosity of a liquid melt and results in a lower fluidity. Therefore, both thermo-kinetic factors should be considered before selecting the content and size of the reinforcing SiC [11,13].

The chemical reaction between silicon carbide and aluminum melt at temperatures exceeding 940 K may result in SiC decomposition according to $4Al(l)+3SiC(s) \rightarrow Al_4C_3(s)+3Si(s)$ reaction. Al_4C_3 compound is known to be brittle and unstable, causing degradation of the mechanical properties of the AMCs. It is reactive with H_2O in humid conditions and might form amorphous aluminum hydroxide. This process is followed by a volume increase and can induce the residual stresses into the surrounding aluminum matrix. Therefore, the inhibition of the Al_4C_3 formation is a crucial issue to be overcome [11,34].

At a processing temperature above 1670 K, Al_4SiC_4 (ternary carbide) is formed following the $4Al_{(l)}+4SiC_{(S)} \rightarrow Al_4SiC_{4(S)}+3Si$ reaction [13]. Al_4SiC_4, due to its high hardness of 1200 HV, low brittleness, remarkable chemical stability in wet conditions, is a favored reinforcement for aluminum [11]. At temperatures above 2800 °C, SiC particles partially or fully decompose into silicon and carbon vapor [34,97]. The increase in applied energy results in a high degree of SiC decomposition, causing surface turbulence, melt pool instability, non-continuous scan tracks and, consequently, an uneven surface finish.

It should be noted that the size of used SiC reinforcing particles ranges from tens of micrometers down to nanoscale, and the resultant mechanical properties of AMCs are significantly affected by particle size [8,13]. In Refs. [8,34], the LPBF of AlSi7Mg/2 wt.% nano-SiCp (40 nm) and Al-12Si/10 vol.%SiC ($\approx$11.7 wt.%) (SiC $\approx$ 25 µm), respectively, were reported. Nano SiC in AlSi7Mg matrix serves as a grain refinement agent (Figure 11m,n) due to the nucleation of numerous heterogenous sites and formation of nanosized Al_4C_3 (Figure 15b,c). The use of nano-SiC yielded low porosity, near-full densification and improvement in tensile strength without sacrificing ductility. However, inferior densification was observed in Ref. [34] when a micron size reinforcement was used.

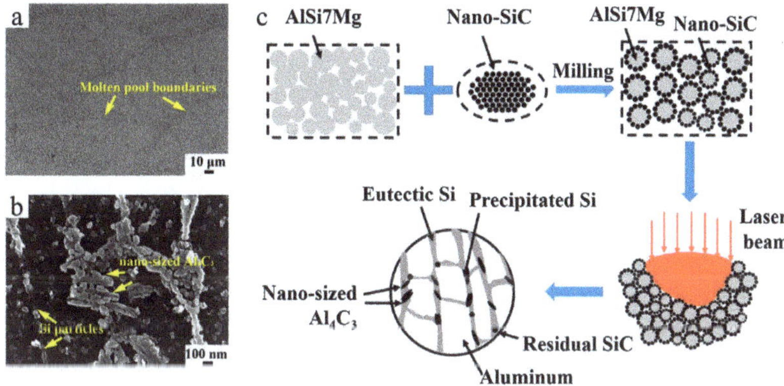

Figure 15. Cross section SEM images of the LPBF-ed AlSi7Mg/2 wt.% nano-SiC composite (**a**,**b**) and the illustration of the formation route of different phases during the LPBF process (**c**) (reproduced with permission from [8]).

The successful fabrication of AlSi10Mg/2 vol.% nano-SiC (~2.4 wt.%) composite reinforced by Al_4SiC_4 phase was reported in Ref. [95]. With an increase in laser power, the eutectic structure gradually changed from thick flakes to network shapes and then to a fine structure, as shown in Figure 16.

At low applied energy, the eutectic structure represents a collection of thick flakes. In contrast, high energy input provides sufficient wettability between SiC and Al, promoting the reaction product transformation into Al_4SiC_4 and a homogeneously dispersed eutectic structure (Figure 17), which positively affects the mechanical properties of the AMC. Despite the analogous content of nano SiC added to the Al alloy, the mechanical properties of the samples in this work are far inferior to those reported in Ref. [8].

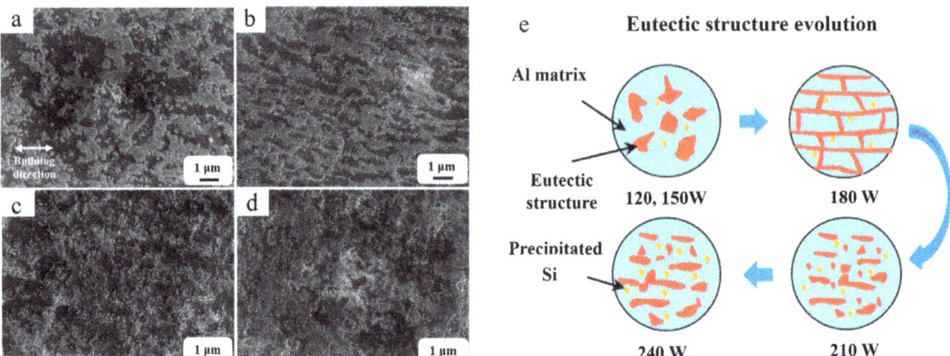

Figure 16. High-magnification SEM micrographs of as-built AlSi10Mg/SiC composites fabricated at different laser powers of 120 W (**a**), 180 W (**b**), 210 W (**c**), 240 W (**d**) and graphical illustration for development of eutectic structure (**e**) (reproduced with permission from [95]).

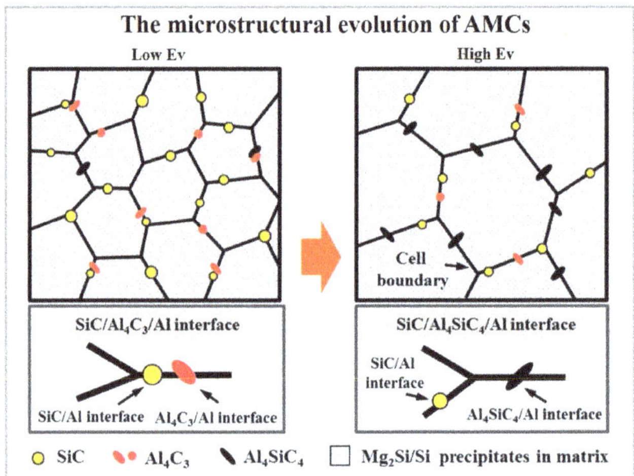

Figure 17. Microstructure changes of the composites at low to high energy application (reproduced with permission from [95]).

An increase in SiC content up to 10 wt.% resulted in increased tensile and yield strength; however, the SiC, Si and in situ formed Al_4SiC_4, reduce the elongation of the composites [96]. When comparing the properties of AlSi10Mg/15–20 wt.% SiC composites [11,13,97,98], it should be mentioned that the highest hardness (316.2$HV_{0.2}$) and density (98.9%) were achieved for AlSi10Mg/15 wt.%SiC, when the SiC particle size was 1200 mesh [98] (Table 2). The larger SiC particles reduced tensile strength as compared to a pure alloy [97]. The use of finer SiC particulates yields to a higher degree of densification, elevated microstructural uniformity and simultaneous improvement in compressive strength, hardness and strain [11,98]. In Refs. [11,13], the in situ formed Al_4SiC_4 is shown to serve as a transition zone, limiting the interaction of SiC and aluminum crystals simultaneously with reinforcing capacity for the Al.

2.3. Nitrides: Grain Refinement and Strengthening Effect

2.3.1. Titanium Nitride: TiN

Besides the favorable characteristics of ceramic materials, TiN (titanium nitride) also demonstrates excellent light absorptivity. TiN has good coherency with Al, owing to small difference (4.72%) in lattice parameters (a_{Al} = 0.4049 nm and a_{TiN} = 0.4240 nm). Meanwhile, the laser reflectivity (at 1064 nm laser wavelength) of the AlSi10Mg/TiN composite powder is around 25%, which is much lower than that of AlSi10Mg powder (62%) [99].

In Refs. [99,100], when fabricating AlSi10Mg/2 wt.%TiN composite, the mutual diffusion and in situ reaction between the TiN clusters and aluminum generates a graded interfacial layer composed of $Al_{3.21}Si_{0.47}$ and (Ti,Al)N (Figure 18).

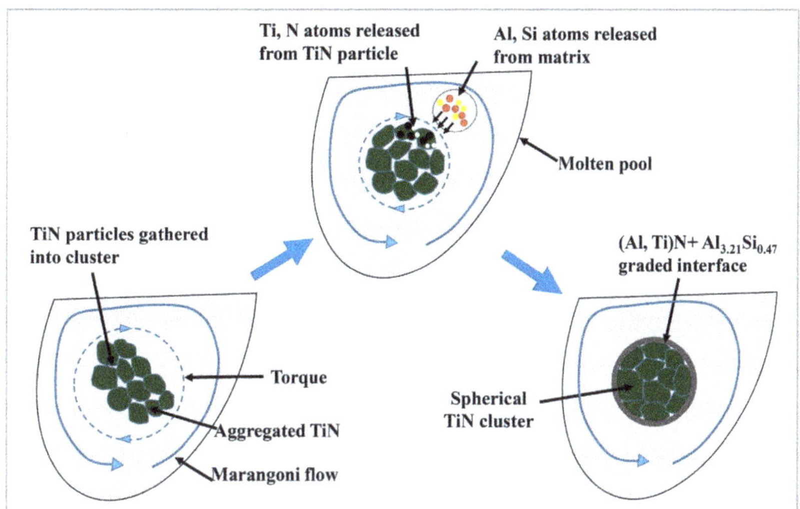

Figure 18. Graphical representation of the movements of aggregated TiN particles and the novel graded layer formation mechanism (reproduced with permission from [100]).

The formed layer is of central importance to the enhancement in microhardness due to an improved interface bonding and a precipitation of stiff (Al,Ti)N. The combined influence of superfine grains (0.284 μm), uniform particle dispersion, formed novel layer and high densification significantly improve the mechanical and wear characteristics of the fabricated AMCs. The Al matrix–Mg_2Si–TiN coherent interfaces lead to a precipitation strengthening, benefiting the enhancement in strength [100].

An increase in TiN content (0–6 wt.%) improves strength, ductility and hardness of nano-TiN particle reinforced AlSi10Mg [101]. It was shown that 4 wt.% TiN is a critical threshold to inhibit porosity. The composites had a relatively random grain orientation, and the grain size decreased from 3.86 to 1.19 μm when the content of TiN increased from 0–6 wt.% due to intensive heterogenous nucleation (Figures 19a–d and 20, Table 3).

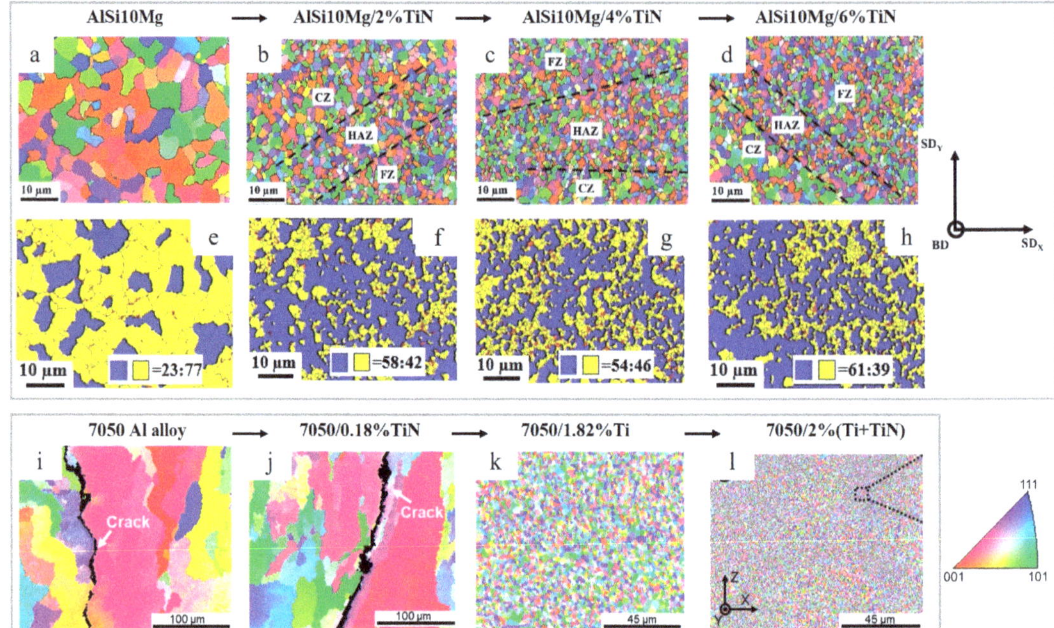

Figure 19. EBSD orientation maps from the top view and distribution of sub-structured (in yellow) and recrystallized (in blue) grains of the as-built AlSi10Mg reinforced with 0% TiN (**a**,**e**), 2%TiN (**b**,**f**), 4% TiN (**c**,**g**) and 6%TiN (**d**,**h**) EBSD color maps of 7050 Al alloy (**i**), 7050-0.18TiN (**j**), 7050-1.82Ti (**k**) and 7050-2(Ti+TiN) (**l**) (reproduced with permission from [66,101]).

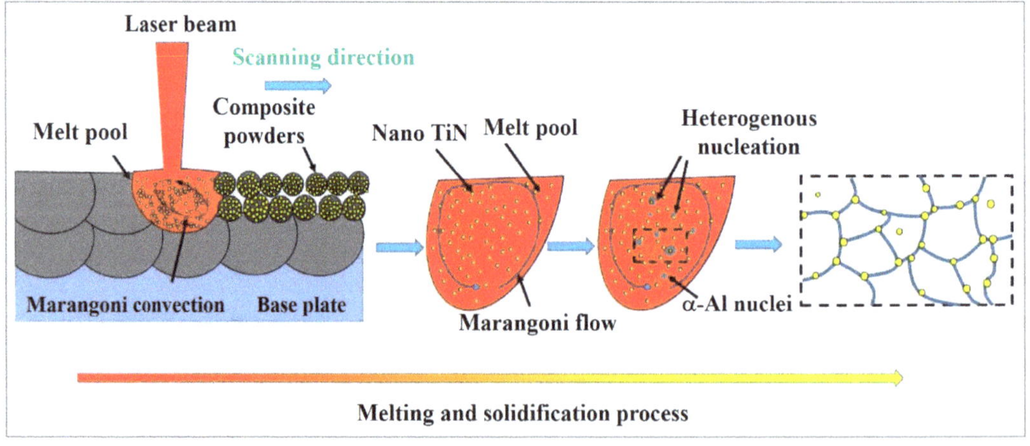

Figure 20. Graphical illustration demonstrating the morphology evolution for the TiN/AlSi10Mg AMC during LPBF (reproduced with permission from [101]).

Table 3. Characteristics of nitride reinforced AMCs fabricated by laser powder-bed fusion.

System	Used Device, Process Parameters	Relative Density (%)	Average Grain Size (μm)	σ_{y}/σ_{u} (MPa)	$\varepsilon/\varepsilon_{c}$ (%)	Hardness (HV)	N
AlSi10Mg/ 2 wt.% TiN ($D50_{TiN}$ = 80 nm)	Dimetal-80 SLM system P = 100 W v = 200–600 mm/s d = 30 μm h = 80 μm	97.6	0.284	-	-	145 ± 4.9 HV0.1	[99,100]
AlSi10Mg	SLM-280 HL P = 100 W v = 1200 mm/s d = 30 μm h = 90 μm	Porosity =0.9%	3.86	σ_u = 359.4 ± 8.5 σ_y = 264 ± 10.5	ε = 3.9 ± 0.3	134.6 ± 4.4 HV0.1	[101]
AlSi10Mg/ 2 wt.% TiN		Porosity =0.2%	1.37	σ_u = 386.1 ± 12.6 σ_y = 295.9 ± 4.6	ε = 4.4 ± 0.27	148.5 ± 4.1 HV0.1	
AlSi10Mg/ 4 wt.% TiN		Porosity =0.01%	1.24	σ_u = 491.8 ± 5.5 σ_y = 315.4 ± 5.2	ε = 7.5 ± 0.29	156.9 ± 4.9 HV0.1	
AlSi10Mg/ 6 wt.% TiN		Porosity =3.7%	1.19	σ_u = 325.1 ± 14.2 σ_y = 261.6 ± 3.5	ε = 2.9 ± 0.32	150.4 ± 3.1 HV0.1	
7050 Al alloy	SLM-280 HL P = 210 W v = 115 mm/s d = 30 μm h = 50 μm	98.5	91.8	σ_u = 75 ± 25	ε ≈ 0.6	-	[66]
7050/0.18 wt.% TiN		98.9	88	σ_u = 111 ± 3	ε = 1.1 ± 0.2		
7050/0.36 wt.% TiN		-	-	σ_u ≈140	ε ≈ 1		
7050/0.54 wt.% TiN		-	-	σ_u ≈ 60	ε ≈ 0.9		
7050/1.82 wt.% Ti		99.6	2.3	σ_u = 427 ± 12	ε = 3.9 ± 1.1		
7050/3.64 wt.% Ti		-	-	σ_u ≈ 480	ε ≈ 6.1		
7050/5.46 wt.% Ti		-	-	σ_u ≈ 350	ε ≈ 2.5		
7050/2 wt.% (TiN+Ti)		99.7	0.775	σ_u ≈ 550	ε ≈ 8.6		
7050/4 wt.% (TiN+Ti)		-	-	σ_u = 613±15	ε = 8.8 ± 0.8		
7050/6 wt.% (TiN+Ti)		-	-	σ_u ≈ 408	ε ≈ 13.2		
AlSi10Mg/ 1 wt.% AlN (50 nm)	SLM apparatus P = 200 W v = 100–300 mm/s d = 30 μm h = 60–100 μm E_v = 1100 J/mm^3	97	4.5	-	-	-	[67]
	E_v = 660 J/mm^3	60	2				
	E_v = 420 J/mm^3	Full dense	1.4				
	E_v = 220 J/mm^3	Full dense	2				
AlSi10Mg/ 2 wt.% AlN	Self-made P = 200 W v = 100 mm/s d = 30 μm h = 80 μm	-	-	-	-	77–85.3 HV0.05	[102]
AlSi10Mg	EOSINT M290 P = 380 W v = 1300 mm/s d = 30 μm h = 200 μm	Porosity =0.15%	-	σ_u ≈ 180	ε ≈ 5.6	103 HV0.2	[103]
AlSi10Mg/ 1 wt.% BN		Porosity =0.81%		σ_u = 230	ε ≈ 2.3	136 HV0.2	
AlSi10Mg	EOSINT M290 P = 180–300 W v = 300–800 mm/s d = 30 μm h = 30–70 μm T = 150 °C	-	-	σ_u = 432 ± 15 σ_y = 275 ± 13	ε = 5.12 ± 0.29	128 ± 3 HV0.2	[104]
AlSi10Mg/ 5 vol.% Si_3N_4 (~5.8 wt.%)		99.49 ± 0.17		σ_u = 447 ± 18 σ_y = 308 ± 12	ε = 3.58 ± 0.15	140 ± 7 HV0.2	
AlSi10Mg/ 10 vol.% Si_3N_4 (~11.5 wt.%)		99.18 ± 0.16		σ_u = 485 ± 12 σ_y = 362 ± 18	ε = 2.47 ± 0.23	153 ± 3 HV0.2	
AlSi10Mg/ 15 vol.% Si_3N_4 (~17.1 wt.%)		98.41 ± 0.22		σ_u = 399 ± 21	ε = 0.66 ± 0.31	187 ± 13 HV0.2	

As shown in Figure 20, only a fraction of TiN serves as heterogenous nucleation substrates, and the majority of particles are dispersed along the grain boundaries owing to the pushing effects of the solidification front.

It was found that all the specimens were dominated by high-angle grain boundaries (HAGBs), and with an increase in TiN content, the volume of low-energy HAGBs increased. TiN nanoparticles also promote recrystallization and possesses a crucial role in recrystallized nucleation during the LPBF process, as shown in Figure 19e–h.

The use of hybrid Ti–TiN reinforcements for 7050 Al alloy was reported in Ref. [66], exhibiting significant synergistic grain refinement and a higher strengthening as compared to pure 7050 Al alloy and a single reinforced 7050-TiN and 7050-Ti. Although both single-Ti-reinforced and hybrid-reinforced alloy possessed a crack-free microstructure (Figure 21g–l), the hybrid reinforcement provided greater grain refinement (Figure 19k,l).

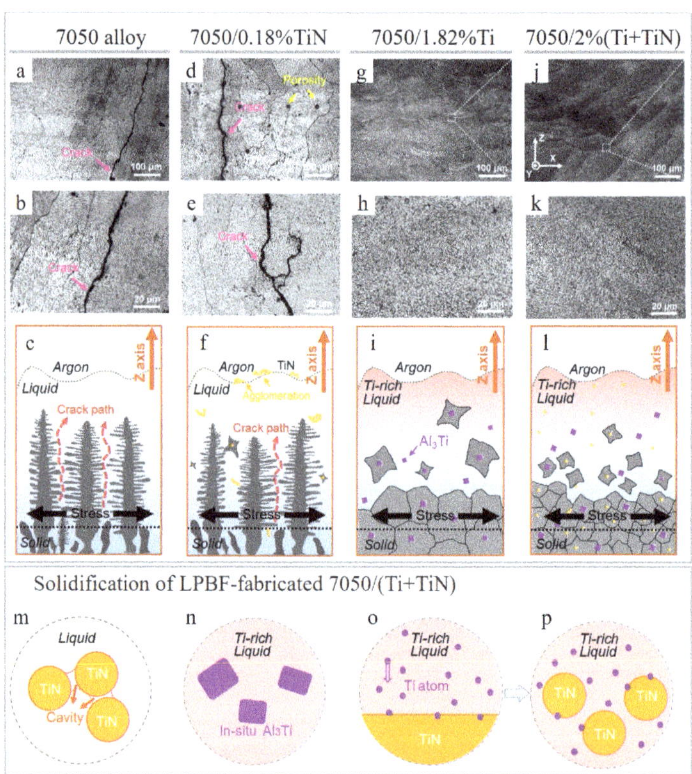

Figure 21. SEM images of LPBF-fabricated 7050 alloy (**a**,**b**), 7050/0.18%TiN (**d**,**e**), 7050/1.82%Ti (**g**,**h**) and 7050/2%(Ti+TiN) (**j**,**k**) samples after etching. Schematic diagram of solidification, columnar and equiaxed grain formation of fabricated 7050 (**c**), 7050-TiN (**f**), 7050-Ti (**i**) and 7050/(Ti+TiN) (**l**). Solidification of 7050/(Ti+TiN): agglomeration of TiN particles in high-temperature liquid Al (**m**), in situ Al_3Ti in Ti-rich liquid Al (**n**), Ti absorption at the interface between TiN and liquid Al (**o**), dispersion of TiN in Ti-rich liquid Al (**p**) (reproduced with permission from [66]).

Meanwhile, the 7050 and 7050-0.18%TiN specimens are prone to cracking, consist of columnar grains and possess relatively high porosity (Figure 19i,j and Figure 21a–f). The reason for grain refinement, when Ti is added to pure alloy and to 7050-TiN, is the formation of $L1_2$ structured Al_3Ti, which promotes heterogeneous nucleation and contributes to the rapid formation of constitutional supercooling zones (Figure 21n). Besides Al_3Ti, fine $MgZn_2$ phase was formed with coherent interface with Al; however, the in situ formed Al_2CuMg showed non-coherent interface with Al. Ultrafine grains (775 nm) were reported in the LPBF-prepared 7050-2 wt.%(Ti+TiN) composite, vastly benefiting from the Ti/TiN synergism.

It can be concluded that the addition of 2–4 wt.% TiN-Ti hybrid additives notably improved the quality of LPBF-fabricated AMCs.

2.3.2. Aluminum Nitride: AlN

AlN is one of the favorable reinforcing candidates for aluminum alloys due to its superior combination of high thermal conductivity (~250 W/mK) [105] and high hardness (~12 GPa) [106]. AlN shows high chemical stability, good compatibility with Al alloy combined with a good interfacial adherence without any interfacial reaction [107]. Besides, due to a low thermal expansion coefficient (similar to Si), AlN has been broadly employed in the aviation and transportation and is shown to be an appropriate reinforcement for aluminum alloys [102].

In a series of works [67,107,108], it was observed that the applied energy had a dramatic effect on the AlN particle distribution. At low energy, random AlN distribution occurred due to the relatively consistent pressure around the introduced particles (Figure 22a,c); and at high laser energy, a circular-structured AlN distribution was compelled by the centripetal force (Figure 22b,d).

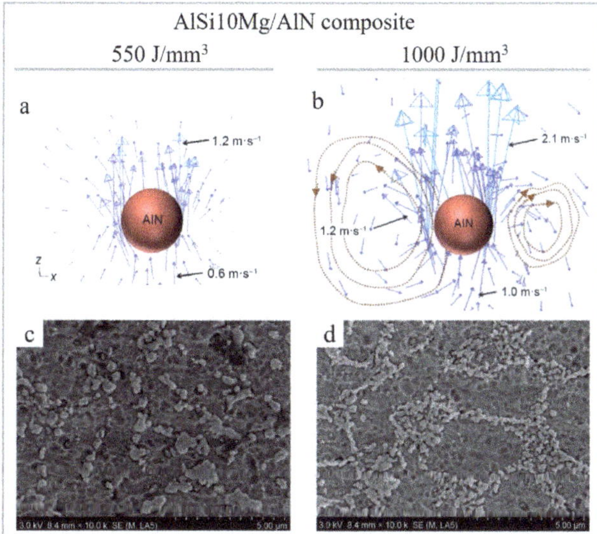

Figure 22. Characteristics of velocity vector obtained around AlN reinforcing particles and their respective distribution state in the solidified Al matrix at E_v = 550 J/mm^3 (**a**,**c**) and E_v = 1000 J/mm^3 (**b**,**d**) (reproduced with permission from [108]).

However, excessive energy results in particles coarsening and a deconstruction of the circular-structured AlN. In Ref. [58], the preparation of an almost fully densified composite with 1 wt.% AlN and refined grains of increased wear resistance has been reported. In Ref. [102], it was shown that during LPBF of AlSi10Mg-2 wt.%AlN powders mixture, the solidified material undergoes various microstructural transformations from the first to the fourth layer (directional columnar microstructure to coarse cellular microstructure), affirming the importance of added particles, solidification rate, the lifespan of the melt pools and subsequent crystal growth rate.

2.3.3. Boron Nitride: BN

The high tensile strength and low density (2.1 g/cm^3, which is close to that of pristine Al), makes hexagonal boron nitride (h-BN) an effective reinforcing agent for the AMCs [109]. It was revealed that even 1 wt.% addition of BN micro-flakes to AlSi10Mg increased the tensile strength and hardness as compared to a pure alloy due to the formation of AlN and AlB_2 phases via solid-state Al–BN reaction [103].

2.3.4. Silicon Nitride: Si_3N_4

A whole basket of favorable properties of Si_3N_4 (silicon nitride), including remarkable strength, high hardness, high elastic modulus, lower CTE, superior hardness compared to other ceramics, [110–112], similar density with aluminum, which will ensure homogeneous dispersion, and high wettability with the aluminum matrix [104] makes it a promising reinforcing agent. The enhanced strength and elastic modulus of the LPBF-prepared AlSi10Mg-Si_3N_4 composite, owing to the impeded dislocation motion during deformation and load-bearing effect of added reinforcing Si_3N_4, are achieved. The mutual diffusion of Al and Si atoms and the absence of in situ formed brittle phases increased the Al matrix-Si_3N_4 particles bonding strength [104]. The addition of Si_3N_4 to the Al alloy, however, reduces process stability and thus narrows the optimal range of process parameters [104].

3. Comparison of Ceramic Reinforcements' Influence on LPBF Process and the Properties of the AMCs

As shown above, even small portions of ceramic or hybrid additives (metal–ceramic), such as 0.5–0.7 wt.%, are able to dramatically improve the performance of the AMCs. Accordingly, matching ceramic additives with an optimized fraction and particle size provides good wettability, compatible interfaces and a strong bonding between the constituents, which hinder crack propagation and contribute to a hardening and strengthening of AMCs.

The addition of TiB_2 to the AlSi10Mg alloy results in fully dense samples with significantly refined grains (down to 0.5 μm), randomized crystallographic orientation, increased hardness up to 191 HV, tensile strength up to 540 MPa and elongation to 17.7% (Figures 23–26). Similarly, high tensile strength is observed for the TiC/Al-15Si, double-reinforced TiC-TiB_2/AlSi10Mg and hybrid TiN-Ti/7050 AMCs, however, with lower elongation (Figure 23a,b).

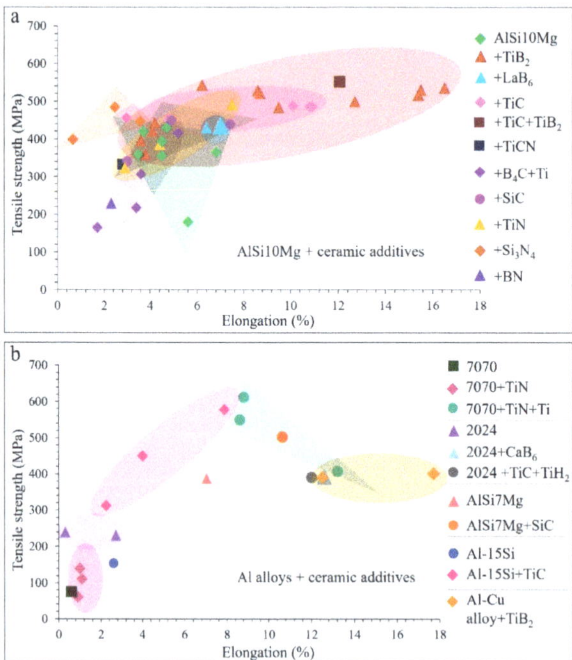

Figure 23. Tensile strength and elongation results of LPBF prepared ceramic particulate reinforced AlSi10Mg (**a**) and other Al alloys (**b**) (data from [8,28,31,50,59,66,71,73,77,79,80,82,84,88–98,101,103,104]).

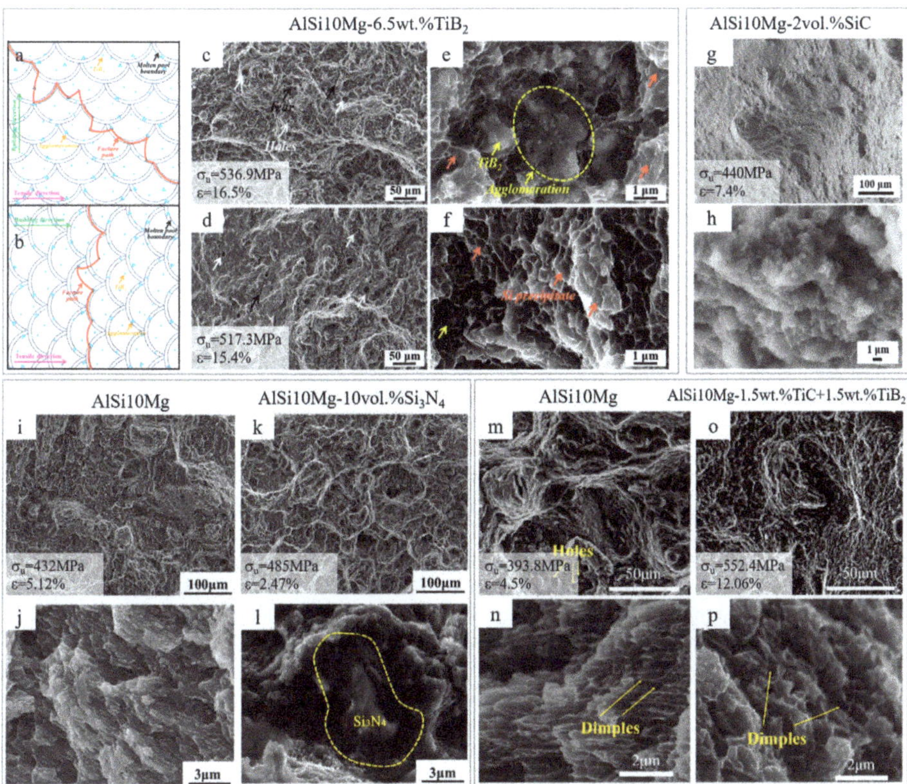

Figure 24. The schematic diagram of probable crack propagation path (**a**,**b**) and tensile fracture morphology for horizontal (**c**,**e**) and vertical samples (**d**,**f**) of the AlSi10Mg/6.5 wt.%TiB_2 composite. Fracture SEM images of AlS10Mg/2 vol.%SiC (**g**,**h**), AlSi10Mg (**i**,**j**) and AlSi10Mg/10 vol.%Si_3N_4 (**k**,**l**), AlSi10Mg (**m**,**n**) and AlSi10Mg/1.5 wt.%TiC+1.5 wt.%TiB_2 (**o**,**p**) (reproduced with permission from [79,93,95,101]).

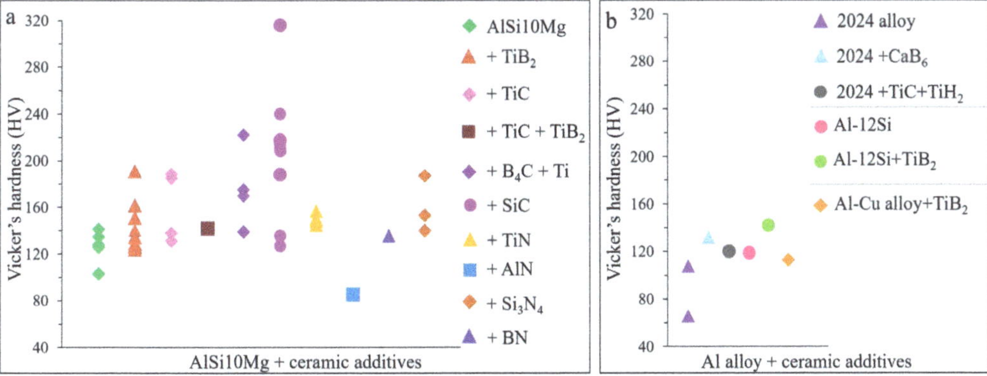

Figure 25. Hardness results of LPBF prepared ceramic particulate reinforced AlSi10Mg (**a**) and other Al alloys (**b**) (data from [11,13,28,31,64,70,71,73,77,78,80,82,83,89,90,92–96,98–105]).

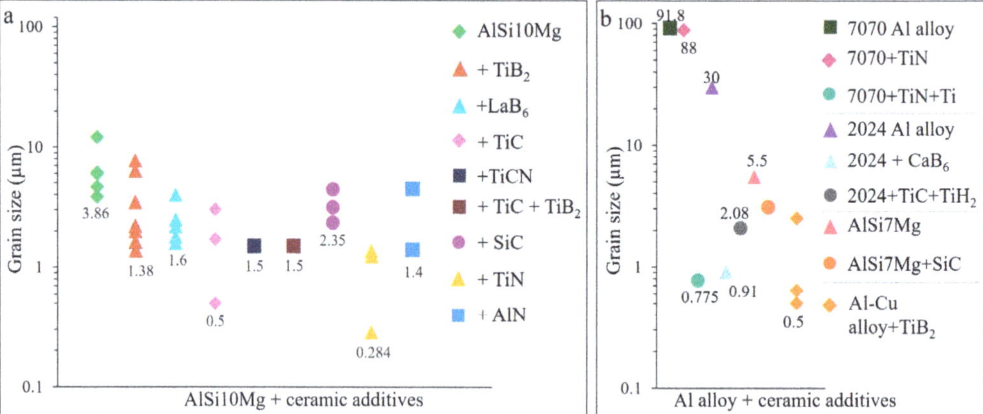

Figure 26. Average grain size of LPBF prepared ceramic particulate reinforced AlSi10Mg (**a**) and other Al alloys (**b**) (data from [8,28,31,50,59,66,67,73,77–84,88,91–93,96,99–101]).

The tensile fracture of the AlSi10Mg-6.5 wt.%TiB_2 composite showed that the fracture path of the AMC is not flat, as in the case of AlSi10Mg, but rather random for both horizontal and vertical samples (Figure 24a,b) [79]. Generally, the reinforced composites with refined microstructure have high ductility due to less stress concentration. Based on the fine-sized equiaxed dimples (Figure 24e,f), the failure mode of the AMC is a ductile fracture, stating improved ductility. However, the holes and the tears on the fracture surface might have led to premature failure of the AMC (Figure 24c,d). Similarly, in the AlSi10Mg-0.2 wt.%LaB_6 composite, cracking predominantly occurred within the melt pool boundaries, and the LaB_6 nanoparticles led to more ductile fracture of the composite, owing to fine equiaxed dimples [84]. Ductile-type failure was reported for AlSi10Mg with homogeneously dispersed circular-structured TiC (3 wt.%). The latter contributed to the improvement of tensile strength without sacrificing ductility [71]. The dual TiB_2 and TiC reinforced AMC's tensile fracture (Figure 24m,n) possesses fewer pores and deeper dimples as compared to AlSi10Mg (Figure 24o,p) and shows mixed ductile and brittle fracture mode. The relatively hard intragranular TiB_2 and TiC particles accommodate the dislocations in the grains, contributing to strain hardening and uniform elongation [93]. Both brittle and ductile fractures were observed in the case of 0.7 wt.% hybrid Ti-B_4C addition. However, the further increase in additive content led to fracture changes from ductile to brittle [94].

When analyzing SiC reinforced AlSi10Mg, huge attention was given to applied energy, as under low energy, brittle Al_4C_3 is formed. However, higher energy promotes the formation of Al_4SiC_4, along with a well-dispersed eutectic structure, hence prohibiting the premature failure of the composite [95]. Similar to SiC (Figure 24g,h), in Si_3N_4 reinforced AMC (Figure 24k,l), the nature of the fracture is ductile brittle, dominated by brittle, whereas pure AlSi10Mg (Figure 24i,j) shows a ductile-brittle composite fracture dominated by ductile. Due to Si_3N_4, crack propagation is suppressed when the tip meets the Si_3N_4–AlSi0Mg interface. However, because of the irregular distribution of Si_3N_4 and the changes in propagation path of the connected cracks, more cleavage steps were formed [104]. When TiN nanoparticles are added to AlSi10Mg, the fracture behavior of the alloy remains in mixed failure mode; however, large-size agglomerates formed during excess addition of TiN, decreasing both strength and ductility [101].

Analyses show that the highest hardness was shown by 15 wt.%SiC reinforced AMCs, followed by the 17.2 wt.% hybrid B_4C-Ti and 11.6 wt.%TiB_2 reinforced materials (Figure 25a). Hardness values of TiC and Si_3N_4 reinforced AMCs are comparable with TiB_2. Meanwhile, ceramic reinforced 2024, Al–12Si and Al–Cu alloys show inferior hardness compared to AlS10Mg with similar additives (Figure 25b).

The AMCs reinforced with TiB_2, TiC, hybrid TiN-Ti and TiC-TiH_2 additives are subjected to in situ formation of L1$_2$-Al_3Ti or D0$_{22}$-Al_3Ti (Table 4), which serve as active nucleation sites and promote grain refinement in the 0.5–2 μm range (Figure 26a,b). The substantial grain refinement, down to submicron level, is achieved by the incorporation of TiN and CaB_6 into AMCs, resulting in both significantly enhanced hardness and tensile strength (Figure 26a,b).

Table 4. The effect of reinforcing compounds on the fabrication and properties of AMCs and their optimal content limit.

Reinforcing Compound	Influence on the LPBF Process and the Properties of the Al Alloys	Minimum Optimal Limit
TiB_2	Exhibits good wettability, interfacial compatibility with Al. Increases densification level, serves as grain refiner along with in situ formed Al_3Ti, stabilizes grain boundaries, leads to randomized crystallographic orientation, dramatically improves strength, hardness and ductility.	2–6.5 wt.%
LaB_6	Forms highly coherent interface with Al, leads to significant grain refinement, microstructural homogeneity, isotropic mechanical properties, does not have huge effect on strength enhancement, but improves ductility.	Up to 0.5 wt.%
CaB_6	Serves as excellent grain refiner, microstructure stabilizer at the grain boundaries, forms highly coherent interface with Al, improves hardness, tensile strength, without sacrificing ductility.	Up to 2 wt.%
TiC	Using fine TiC particles leads to fully dense part fabrication with improved strength, ductility and hardness. The in situ formed D0$_{22}$-Al_3Ti inoculants provide heterogeneous nucleation of α-Al, leading to grain refinement, and remove the preferred orientation of the α-Al (200) phase. Depending on the TiC content and process parameters, novel circular (ring) structures are formed within the matrix, enhancing the mechanical performance of AMCs.	Up to 5 wt.%
TiC_B	The gas-atomized powders release enormous TiC_B particles during LPBF process, largely promoting the nucleation of Al grains, grain refinement and resulting in weak crystallographic texture of AMCs. TiC_B particles along with precipitated Si enhance the yield strength, tensile strength and elongation.	~0.5 wt.%
TiCN	The addition of TiCN significantly reduces the average grain size, improves yield strength and ductility over native LPBF AlSi10Mg and rarely induces the formation of brittle Al_4C_3.	2 wt.%
TiC+TiH_2	Due to decomposition of TiH_2 and reaction of Al with Ti, a well-bonded interface between L1$_2$-Al_3Ti and α-Al was observed acting as substrate for α-Al heterogeneous nucleation. Meanwhile, the presence of Ti creates "Ti transition zone" between TiC and matrix, creating potent nucleation sites for α-Al as well. Owing to restriction of columnar grain growth, the joint effect of refinement strengthening, the reinforced AMCs exhibit enhanced mechanical performance, tensile strength and ductility.	1 wt.%TiC 1 wt.%TiH_2
TiC+TiB_2	Dual TiB_2+TiC particles induce heterogeneous nucleation of Al and significantly refine the grains of the Al matrix. Double reinforcement results in simultaneous enhancement in strength, ductility and hardness, acting more efficiently than single species.	1.5 wt.%TiC 1.5 wt.%TiH_2
SiC	Use of fine (nanosized or few-micron-sized) SiC results in grain refinement, decrease in porosity, enhancement of hardness, tensile strength and ductility but, depending on the process parameters, can cause in situ formation of Al_4C_3 or Al_4SiC_4 phases.	Up to 2 wt.%

Table 4. *Cont.*

Reinforcing Compound	Influence on the LPBF Process and the Properties of the Al Alloys	Minimum Optimal Limit
$Ti+B_4C$	In situ formed TiC, TiB_2 and Ti_3SiC_2 serve as nucleants and reinforcements. The $Ti+B_4C$ content increase results in improvement in hardness, however much lower elongation and tensile strength. The released heat during the combustion reaction allows for fabricating the materials at low applied laser energy.	0.7 wt.%
Al_4C_3	Al_4C_3 itself is a brittle and unstable phase and is best avoided. However, small amounts of formed nanosized Al_4C_3 can enhance the mechanical properties of AMCs.	-
Al_4SiC_4	Al_4SiC_4 along with intermetallic Mg_2Si increase reinforcement/matrix wettability and the resultant interfacial bonding coherence. Al_4SiC_4 serves as the transition zone, which hinders the direct contact of SiC and aluminum crystals. Ultrafine Al_4SiC_4 has a reinforcing effect, improving the mechanical properties of SiC reinforced AMCs.	-
TiN	TiN particles refine the α-Al grains due to intensive heterogeneous nucleation and increase the fraction of low-energy high-angle grain boundaries, enhancing the hardness and strength. Due to the Al+TiN reaction, $Al_{3.21}Si_{0.47}$ and a (Ti,Al)N graded layer is formed, which significantly enhances the hardness due to improving interface bonding strength. The coherent interfaces between the matrix, Mg_2Si and TiN particles lead to precipitation strengthening, which contributes to the overall strength increase.	4 wt.%
TiN+Ti	Provides crack-free microstructure and significant grain refinement due to formation of Al_3Ti phase and different precipitates, improves the hardness and tensile strength.	4 wt.%
AlN	The AlN particles show high chemical stability and good compatibility with Al alloy. They promote densification, refine the α-Al grains, create strain-hardened tribo-layer, enhancing the wear resistance and stabilizing the coefficient of friction.	1 wt.%
BN	The formation of AlN and AlB_2 phases during the solid-state reaction of Al+BN results in increased tensile strength and hardness, though at the expense of porosity increase. However, increase in BN content and particle size decreases wettability and prevents uniform metal spreading.	1 wt.%
Si_3N_4	Si_3N_4 particles increase the melt pool's viscosity and disturb the stability, suggesting a much narrower window for LPBF process parameters. Owing to hindered dislocation motion during deformation (because of difference of Al and Si_3N_4) and the load-bearing effect of Si_3N_4 particles, the AMCs possess improved strength and elastic modulus.	10 vol.%

The degree of improvement depends on additive content and composition of the Al alloy. Table 4 briefly summarizes the influence of the reported ceramic additives on the LPBF process and their content limitation.

4. Summary and Outlook

LPBF technologies are now commercially available and attract a huge deal of attention in research community. Although the number of aluminum alloys suitable for AM through LPBF is quite limited, the process keeps evolving, and, in the nearest future, a widespread application of AM of high-strength aluminum alloys is expected to occur in the aerospace market.

The cost of industrial metal printers remains the chief capital expenditure of AM parts to achieve economies-of-scale cost reduction. Although the industry has suffered due to COVID-19, the reverse has now begun. In light of current metal printers' high prices, they are mostly used in high-value industries, such as aerospace, defense and medical.

Other fields, such as energy, are starting to show interest in powder bed fusion technology, although developing economically viable applications requires sufficient time.

A 2.6 percent annual growth rate is predicted for aluminum consumption globally up to 2029. In 2021, global aluminum consumption is projected at 64.2 million metric tons alone (Figure 27).

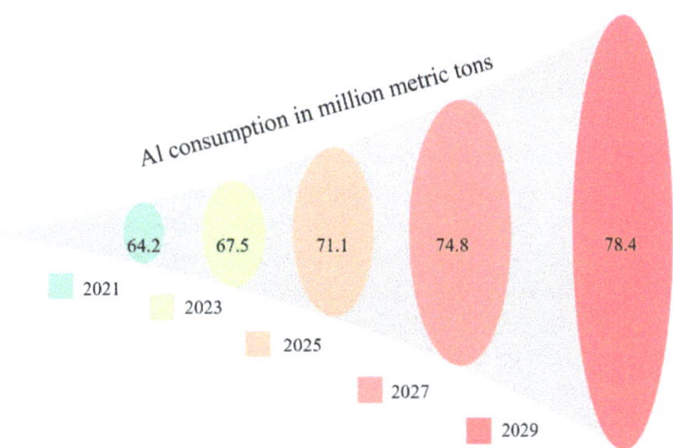

Figure 27. Calculated aluminum consumption up to 2029 (adapted from Ref. [113]).

However, fuel efficiency and low carbon emission are the mantra for new-era airliners, which have groundbreaking design equipped with composite materials comprising 50 percent of the primary structure, hence eliminating the use of numerous aluminum parts [114]. In addition, the world's biggest aluminum producers are limiting the production of Al, planning to reduce energy consumption and encourage the producers to develop green and low-carbon technologies and produce high-quality, high-strength and long-life aluminum products through innovations [115]. This means that there is a need for revolutionary actions to keep additive manufacturing of aluminum alloys on track.

Over the next decade, the development of new 3D printable Al alloys is expected to bring down the cost and enlarge the materials' capacity and portfolio. For example, the lightweight aluminum–lithium alloys could contribute to reducing aircraft weight, also benefiting from excellent fatigue resistance and cryogenic toughness in addition to light weight and high specific modulus.

As numerous reinforcements are used to further enhance the properties of Al alloys, one big step ahead will be using different reinforcing particles (ceramics) and covering them with compatible coatings to provide suitable wettability and interface, or incorporating the reinforcing particles into Al alloy particles to provide a homogeneous distribution. Another main challenge is the recycling of the used feedstock and the utilization of the spattered debris to prepare new powders for further use.

As the design of new alloys applicable for the LPBF process is time and cost consuming, a high-throughput and reliable technique is needed to experimentally validate the custom alloys and effectively introduce them into the market. Therefore, a deep understanding of the impact of the alloying constituents on the processability of the feedstock by LPBF and, ultimately, the properties of the produced items in application, is of a crucial importance.

In this review paper, the effect of non-oxide ceramic (borides, nitrides, carbides) and hybrid reinforcing additives on the densification, grain refinement and respective mechanical characteristics of LPBF-fabricated AMCs was discussed. A comprehensive analysis of research studies on densification, compositional and microstructural characteristics of the in situ and ex situ reinforced aluminum alloys produced by LPBF method was accomplished to demonstrate the capability of different ceramic additives to tailor the mechanical properties with application to a wide variety of process parameters.

- Generally, an incorporation of the ceramic particles into Al alloys results in a significant improvement in strength, ductility and hardness of the fabricated parts accompanied

by a refined microstructure and with randomization of crystallographic orientation of reinforced AMCs.
- Most of the AMCs can be densified to over 99% relative density; moreover, non-oxide ceramic additives significantly improve laser absorptivity of a powder feedstock.
- The addition of ceramic particulates shifts the process window to a higher energy regime; however, an applied excess energy may result in the evaporation or decomposition of ceramics particles (mainly SiC).
- The application of a laser re-melting strategy can further increase the densification degree and the surface quality of AMCs; however, it also can cause the evaporation and loss of ceramic particles.
- Hybrid reinforcements are proven to be the effective additives, providing the formation of a wide variety of reinforcing phases with a coherent interface with matrices.
- The use of ceramics with a fine-particle size results in an increased degree of densification, microstructural and compositional uniformity, as well as an apparent grain refinement.
- The addition of TiB_2, CaB_6, TiC, TiN to Al alloys leads to a considerable grain refinement, down to the submicron level, due to the intensive heterogeneous nucleation and grain growth inhibition.
- An addition of matching ceramics prevents the hot tearing and gives the prospect to consolidate crack-susceptible Al alloys by a laser powder-bed fusion technique.
- The highest elongation of 17.7% is demonstrated by the AlSi10Mg/TiB_2 composite; however, the highest strength of 613 MPa is recorded for the hybrid TiN-Ti reinforced AMCs.
- The highest hardness of 316 HV is estimated for SiC reinforced AMCs, which possess a relatively high strength and moderate ductility.

Author Contributions: Conceptualization, T.M. and I.H.; data curation, I.H.; funding acquisition, I.H.; investigation, T.M.; methodology, I.H. and T.M.; resources, I.H.; supervision, I.H.; visualization, T.M.; writing—original draft, T.M.; writing—review and editing, I.H. and T.M. All authors have read and agreed to the published version of the manuscript.

Funding: This work was supported by the Estonian Research Council (ETAG), Estonia, under the grants PRG643 (I. Hussainova) and PSG220 (S. Aydinyan).

Institutional Review Board Statement: Not applicable.

Informed Consent Statement: Not applicable.

Data Availability Statement: The data supporting the findings of this study is available within the article.

Conflicts of Interest: The authors declare no conflict of interest.

References

1. Spierings, A.B.; Dawson, K.; Uggowitzer, P.J.; Wegener, K. Influence of SLM scan-speed on microstructure, precipitation of Al3Sc particles and mechanical properties in Sc- and Zr-modified Al-Mg alloys. *Mater. Des.* **2018**, *140*, 134–143. [CrossRef]
2. Otani, Y.; Sasaki, S. Effects of the addition of silicon to 7075 aluminum alloy on microstructure, mechanical properties, and selective laser melting processability. *Mater. Sci. Eng. A* **2020**, *777*, 139079. [CrossRef]
3. Muhammad, M.; Nezhadfar, P.; Thompson, S.; Saharan, A.; Phan, N.; Shamsaei, N. A comparative investigation on the microstructure and mechanical properties of additively manufactured aluminum alloys. *Int. J. Fatigue* **2021**, *146*, 106165. [CrossRef]
4. Li, P.; Li, R.; Yang, H.; Yuan, T.; Niu, P.; Wang, M.; Li, L.; Chen, C. Selective laser melting of Al-3.48Cu-2.03Si-0.48Sc-0.28Zr alloy: Microstructure evolution, properties and metallurgical defects. *Intermetallics* **2020**, *129*, 107008. [CrossRef]
5. Qian, W.; Zhao, Y.; Kai, X.; Yan, Y.; Gao, X.; Jin, L. Microstructure and properties of 6111Al matrix composites reinforced by the cooperation of in situ ZrB2 particles and Y. *J. Alloys Compd.* **2020**, *829*, 154624. [CrossRef]
6. Qbau, N.; Nam, N.; Hien, N.; Ca, N. Development of light weight high strength aluminum alloy for selective laser melting. *J. Mater. Res. Technol.* **2020**, *9*, 14075–14081. [CrossRef]
7. Totten, G.E.; Tiryakioğlu, M.; Kessler, O. (Eds.) *Encyclopedia of Aluminum and Its Alloys*; CRC Press: Boca Raton, FL, USA, 2018. [CrossRef]
8. Wang, M.; Song, B.; Wei, Q.; Shi, Y. Improved mechanical properties of AlSi7Mg/nano-SiCp composites fabricated by selective laser melting. *J. Alloys Compd.* **2019**, *810*, 151926. [CrossRef]

9. Tan, Q.; Fan, Z.; Tang, X.; Yin, Y.; Li, G.; Huang, D.; Zhang, J.; Liu, Y.; Wang, F.; Wu, T.; et al. A novel strategy to additively manufacture 7075 aluminium alloy with selective laser melting. *Mater. Sci. Eng. A* **2021**, *821*, 141638. [CrossRef]
10. Zhang, J.; Song, B.; Wei, Q.; Bourell, D.; Shi, Y. A review of selective laser melting of aluminum alloys: Processing, microstructure, property and developing trends. *J. Mater. Sci. Technol.* **2018**, *35*, 270–284. [CrossRef]
11. Gu, D.; Chang, F.; Dai, D. Selective Laser Melting Additive Manufacturing of Novel Aluminum Based Composites With Multiple Reinforcing Phases. *J. Manuf. Sci. Eng.* **2015**, *137*, 021010. [CrossRef]
12. Famodimu, O.H.; Stanford, M.; Oduoza, C.F.; Zhang, L. Effect of process parameters on the density and porosity of laser melted AlSi10Mg/SiC metal matrix composite. *Front. Mech. Eng.* **2018**, *13*, 520–527. [CrossRef]
13. Chang, F.; Gu, D.; Dai, D.; Yuan, P. Selective laser melting of in-situ Al_4SiC_4 + SiC hybrid reinforced Al matrix composites: Influence of starting SiC particle size. *Surf. Coatings Technol.* **2015**, *272*, 15–24. [CrossRef]
14. Zhou, L.; Huynh, T.; Park, S.; Hyer, H.; Mehta, A.; Song, S.; Bai, Y.; McWilliams, B.; Cho, K.; Sohn, Y. Laser powder bed fusion of Al–10 wt% Ce alloys: Microstructure and tensile property. *J. Mater. Sci.* **2020**, *55*, 14611–14625. [CrossRef]
15. Wallis, C.; Buchmayr, B.; Bermejo, R.; Supancic, P. Fabrication of 3D metal-ceramic (Al-AlN) architectures using laser-powder bed fusion process. *Addit. Manuf.* **2020**, *38*, 101799. [CrossRef]
16. Minasyan, T.; Aghayan, M.; Liu, L.; Aydinyan, S.; Kollo, L.; Hussainova, I.; Rodríguez, M.A. Combustion synthesis of MoSi2 based composite and selective laser sintering thereof. *J. Eur. Ceram. Soc.* **2018**, *38*, 3814–3821. [CrossRef]
17. Minasyan, T.; Ivanov, R.; Toyserkani, E.; Hussainova, I. Laser powder-bed fusion of $Mo(Si,Al)_2$—Based composite for elevated temperature applications. *J. Alloys Compd.* **2021**, *884*, 161034. [CrossRef]
18. Wang, J.; Liu, T.; Luo, L.; Cai, X.; Wang, B.; Zhao, J.; Cheng, Z.; Wang, L.; Su, Y.; Xue, X.; et al. Selective laser melting of high-strength TiB2/AlMgScZr composites: Microstructure, tensile deformation behavior, and mechanical properties. *J. Mater. Res. Technol.* **2021**, *16*, 786–800. [CrossRef]
19. Minasyan, T.; Ivanov, R.; Toyserkani, E.; Hussainova, I. $Mo(Si,Al)_2$ by laser powder bed fusion of AlSi10Mg and combustion synthesized MoSi2. *Mater. Lett.* **2021**, *307*, 131041. [CrossRef]
20. Minasyan, T.; Aydinyan, S.; Toyserkani, E.; Hussainova, I. Parametric Study on In Situ Laser Powder Bed Fusion of $Mo(Si_{1-x},Al_x)_2$. *Materials* **2020**, *13*, 4849. [CrossRef]
21. Kuai, Z.; Li, Z.; Liu, B.; Liu, W.; Yang, S. Effects of remelting on the surface morphology, microstructure and mechanical properties of AlSi10Mg alloy fabricated by selective laser melting. *Mater. Chem. Phys.* **2022**, 125901. [CrossRef]
22. Gu, D.; Yuan, P. Thermal evolution behavior and fluid dynamics during laser additive manufacturing of Al-based nanocomposites: Underlying role of reinforcement weight fraction. *J. Appl. Phys.* **2015**, *118*, 233109. [CrossRef]
23. Yu, W.; Sing, S.; Chua, C.; Kuo, C.; Tian, X. Particle-reinforced metal matrix nanocomposites fabricated by selective laser melting: A state of the art review. *Prog. Mater. Sci.* **2019**, *104*, 330–379. [CrossRef]
24. Kumar, M.B.; Sathiya, P. Methods and materials for additive manufacturing: A critical review on advancements and challenges. *Thin-Walled Struct.* **2020**, *159*, 107228. [CrossRef]
25. Bayat, M.; Nadimpalli, V.K.; Pedersen, D.B.; Hattel, J.H. A fundamental investigation of thermo-capillarity in laser powder bed fusion of metals and alloys. *Int. J. Heat Mass Transf.* **2020**, *166*, 120766. [CrossRef]
26. Martin, J.H.; Yahata, B.D.; Hundley, J.M.; Mayer, J.A.; Schaedler, T.A.; Pollock, T.M. 3D printing of high-strength aluminium alloys. *Nature* **2017**, *549*, 365–369. [CrossRef]
27. Griffiths, S.; Rossell, M.D.; Croteau, J.; Vo, N.Q.; Dunand, D.C.; Leinenbach, C. Effect of laser rescanning on the grain microstructure of a selective laser melted Al-Mg-Zr alloy. *Mater. Charact.* **2018**, *143*, 34–42. [CrossRef]
28. Mair, P.; Goettgens, V.S.; Rainer, T.; Weinberger, N.; Letofsky-Papst, I.; Mitsche, S.; Leichtfried, G. Laser powder bed fusion of nano-CaB6 decorated 2024 aluminum alloy. *J. Alloys Compd.* **2021**, *863*, 158714. [CrossRef]
29. Zhou, L.; Hyer, H.; Chang, J.; Mehta, A.; Huynh, T.; Yang, Y.; Sohn, Y. Microstructure, mechanical performance, and corrosion behavior of additively manufactured aluminum alloy 5083 with 0.7 and 1.0 wt% Zr addition. *Mater. Sci. Eng. A* **2021**, *823*, 141679. [CrossRef]
30. Plotkowski, A.; Sisco, K.; Bahl, S.; Shyam, A.; Yang, Y.; Allard, L.; Nandwana, P.; Rossy, A.M.; Dehoff, R. Microstructure and properties of a high temperature Al–Ce–Mn alloy produced by additive manufacturing. *Acta Mater.* **2020**, *196*, 595–608. [CrossRef]
31. Fan, Z.; Yan, X.; Fu, Z.; Niu, B.; Chen, J.; Hu, Y.; Chang, C.; Yi, J. In situ formation of D022-Al3Ti during selective laser melting of nano-TiC/AlSi10Mg alloy prepared by electrostatic self-assembly. *Vacuum* **2021**, *188*, 110179. [CrossRef]
32. Kaufman, J.G. Introduction to Aluminum Alloys and Tempers. Available online: https://books.google.ee/books?hl=en&lr=&id=idmZIDcwCykC&oi=fnd&pg=PR7&dq=Introduction+to+Aluminum+Alloys+and+Tempers&ots=YF2Do8uYO4&sig=lVfaG-D2QRHLKNTb8Nivh0VpmmA&redir_esc=y#v=onepage&q=Introduction%20to%20Aluminum%20Alloys%20and%20Tempers&f=false (accessed on 25 August 2021).
33. Montero-Sistiaga, M.L.; Mertens, R.; Vrancken, B.; Wang, X.; Van Hooreweder, B.; Kruth, J.-P.; Van Humbeeck, J. Changing the alloy composition of Al7075 for better processability by selective laser melting. *J. Mater. Process. Technol.* **2016**, *238*, 437–445. [CrossRef]
34. Astfalck, L.; Kelly, G.K.; Li, X.; Sercombe, T.B. On the Breakdown of SiC during the Selective Laser Melting of Aluminum Matrix Composites. *Adv. Eng. Mater.* **2017**, *19*, 1600835. [CrossRef]

35. Li, X.; Wang, X.; Saunders, M.; Suvorova, A.; Zhang, L.; Liu, Y.; Fang, M.; Huang, Z.; Sercombe, T. A selective laser melting and solution heat treatment refined Al–12Si alloy with a controllable ultrafine eutectic microstructure and 25% tensile ductility. *Acta Mater.* **2015**, *95*, 74–82. [CrossRef]
36. Zhuo, L.; Wang, Z.; Zhang, H.; Yin, E.; Wang, Y.; Xu, T.; Li, C. Effect of post-process heat treatment on microstructure and properties of selective laser melted AlSi10Mg alloy. *Mater. Lett.* **2018**, *234*, 196–200. [CrossRef]
37. Wang, M.; Song, B.; Wei, Q.; Zhang, Y.; Shi, Y. Effects of annealing on the microstructure and mechanical properties of selective laser melted AlSi7Mg alloy. *Mater. Sci. Eng. A* **2018**, *739*, 463–472. [CrossRef]
38. Zhang, C.; Zhu, H.; Hu, Z.; Zhang, L.; Zeng, X. A comparative study on single-laser and multi-laser selective laser melting AlSi10Mg: Defects, microstructure and mechanical properties. *Mater. Sci. Eng. A* **2019**, *746*, 416–423. [CrossRef]
39. Li, W.; Li, S.; Liu, J.; Zhang, A.; Zhou, Y.; Wei, Q.; Yan, C.; Shi, Y. Effect of heat treatment on AlSi10Mg alloy fabricated by selective laser melting: Microstructure evolution, mechanical properties and fracture mechanism. *Mater. Sci. Eng. A* **2016**, *663*, 116–125. [CrossRef]
40. Ji, Y.; Dong, C.; Kong, D.; Li, X. Design materials based on simulation results of silicon induced segregation at AlSi10Mg interface fabricated by selective laser melting. *J. Mater. Sci. Technol.* **2020**, *46*, 145–155. [CrossRef]
41. Wang, Y.; Shi, J. Effect of hot isostatic pressing on nanoparticles reinforced AlSi10Mg produced by selective laser melting. *Mater. Sci. Eng. A* **2020**, *788*, 139570. [CrossRef]
42. Bi, J.; Lei, Z.; Chen, Y.; Chen, X.; Tian, Z.; Lu, N.; Qin, X.; Liang, J. Microstructure, tensile properties and thermal stability of AlMgSiScZr alloy printed by laser powder bed fusion. *J. Mater. Sci. Technol.* **2020**, *69*, 200–211. [CrossRef]
43. Thapliyal, S.; Shukla, S.; Zhou, L.; Hyer, H.; Agrawal, P.; Agrawal, P.; Komarasamy, M.; Sohn, Y.; Mishra, R.S. Design of heterogeneous structured Al alloys with wide processing window for laser-powder bed fusion additive manufacturing. *Addit. Manuf.* **2021**, *42*, 102002. [CrossRef]
44. Lu, J.; Lin, X.; Kang, N.; Cao, Y.; Wang, Q.; Huang, W. Keyhole mode induced simultaneous improvement in strength and ductility of Sc modified Al–Mn alloy manufactured by selective laser melting. *Mater. Sci. Eng. A* **2021**, *811*, 141089. [CrossRef]
45. Thapliyal, S.; Komarasamy, M.; Shukla, S.; Zhou, L.; Hyer, H.; Park, S.; Sohn, Y.; Mishra, R.S. An integrated computational materials engineering-anchored closed-loop method for design of aluminum alloys for additive manufacturing. *Materialia* **2019**, *9*, 100574. [CrossRef]
46. Yang, K.; Shi, Y.; Palm, F.; Wu, X.; Rometsch, P. Columnar to equiaxed transition in Al-Mg(-Sc)-Zr alloys produced by selective laser melting. *Scr. Mater.* **2018**, *145*, 113–117. [CrossRef]
47. Kurnsteiner, P.; Bajaj, P.; Gupta, A.; Benjamin, W.; Weisheit, A.; Li, X.; Leinebach, C.; Gault, B.; Jagle, E.; Raabe, D. Control of thermally stable core-shell nano-precipitates in additively manufactured Al-Sc-Zr alloys. *Addit. Manuf.* **2020**, *32*, 100910. [CrossRef]
48. Zhou, L.; Hyer, H.; Thapliyal, S.; Mishra, R.S.; McWilliams, B.; Cho, K.; Sohn, Y. Process-Dependent Composition, Microstructure, and Printability of Al-Zn-Mg and Al-Zn-Mg-Sc-Zr Alloys Manufactured by Laser Powder Bed Fusion. *Met. Mater. Trans. A* **2020**, *51*, 3215–3227. [CrossRef]
49. Zhou, S.; Su, Y.; Wang, H.; Enz, J.; Ebel, T.; Yan, M. Selective laser melting additive manufacturing of 7xxx series Al-Zn-Mg-Cu alloy: Cracking elimination by co-incorporation of Si and TiB_2. *Addit. Manuf.* **2020**, *36*, 101458. [CrossRef]
50. Biffi, C.A.; Bassani, P.; Fiocchi, J.; Albu, M.; Tuissi, A. Selective laser melting of AlCu-TiB2 alloy using pulsed wave laser emission mode: Processability, microstructure and mechanical properties. *Mater. Des.* **2021**, *204*, 109628. [CrossRef]
51. Jia, Q.; Rometsch, P.; Kürnsteiner, P.; Chao, Q.; Huang, A.; Weyland, M.; Bourgeois, L.; Wu, X. Selective laser melting of a high strength Al Mn Sc alloy: Alloy design and strengthening mechanisms. *Acta Mater.* **2019**, *171*, 108–118. [CrossRef]
52. Kang, N.; El Mansori, M.; Lin, X.; Guittonneau, F.; Liao, H.; Huang, W.; Coddet, C. In-situ synthesis of aluminum/nano-quasicrystalline Al-Fe-Cr composite by using selective laser melting. *Compos. Part B Eng.* **2018**, *155*, 382–390. [CrossRef]
53. Kang, N.; Fu, Y.; Coddet, P.; Guelorget, B.; Liao, H.; Coddet, C. On the microstructure, hardness and wear behavior of Al-Fe-Cr quasicrystal reinforced Al matrix composite prepared by selective laser melting. *Mater. Des.* **2017**, *132*, 105–111. [CrossRef]
54. Demir, A.G.; Previtali, B. Multi-material selective laser melting of Fe/Al-12Si components. *Manuf. Lett.* **2017**, *11*, 8–11. [CrossRef]
55. Aboulkhair, N.T.; Simonelli, M.; Parry, L.; Ashcroft, I.; Tuck, C.; Hague, R. 3D printing of Aluminium alloys: Additive Manufacturing of Aluminium alloys using selective laser melting. *Prog. Mater. Sci.* **2019**, *106*, 100578. [CrossRef]
56. Mair, P.; Braun, J.; Kaserer, L.; March, L.; Schimbäck, D.; Letofsky-Papst, I.; Leichtfried, G. Unique microstructure evolution of a novel Ti-modified Al-Cu alloy processed using laser powder bed fusion. *Mater. Today Commun.* **2022**, *31*, 103353. [CrossRef]
57. Wang, Z.; Wang, X.; Chen, X.; Qiu, C. Complete columnar-to-equiaxed transition and significant grain refinement in an aluminium alloy by adding Nb particles through laser powder bed fusion. *Addit. Manuf.* **2022**, *51*, 102615. [CrossRef]
58. Tan, Q.; Zhang, J.; Mo, N.; Fan, Z.; Yin, Y.; Bermingham, M.; Liu, Y.; Huang, H.; Zhang, M.-X. A novel method to 3D-print fine-grained AlSi10Mg alloy with isotropic properties via inoculation with LaB6 nanoparticles. *Addit. Manuf.* **2020**, *32*, 101034. [CrossRef]
59. Xiao, Y.; Bian, Z.; Wu, Y.; Ji, G.; Li, Y.; Li, M.; Lian, Q.; Chen, Z.; Addad, A.; Wang, H. Effect of nano-TiB2 particles on the anisotropy in an AlSi10Mg alloy processed by selective laser melting. *J. Alloys Compd.* **2019**, *798*, 644–655. [CrossRef]
60. Kotadia, H.; Gibbons, G.; Das, A.; Howes, P. A review of Laser Powder Bed Fusion Additive Manufacturing of aluminium alloys: Microstructure and properties. *Addit. Manuf.* **2021**, *46*, 102155. [CrossRef]

61. Wang, L.; Jue, J.; Xia, M.; Guo, L.; Yan, B.; Gu, D. Effect of the Thermodynamic Behavior of Selective Laser Melting on the Formation of In situ Oxide Dispersion-Strengthened Aluminum-Based Composites. *Metals* **2016**, *6*, 286. [CrossRef]
62. Minasyan, T.; Aydinyan, S.; Liu, L.; Volubujeva, O.; Toyserkani, E.; Hussainova, I. Mo(Si_{1-x},Al_x)$_2$-based composite by reactive laser powder-bed fusion. *Mater. Lett.* **2020**, *281*, 128776. [CrossRef]
63. Dadbakhsh, S.; Mertens, R.; Hao, L.; Van Humbeeck, J.; Kruth, J. Selective Laser Melting to Manufacture "In Situ" Metal Matrix Composites: A Review. *Adv. Eng. Mater.* **2018**, *21*, 1801244. [CrossRef]
64. Xi, L.; Wang, P.; Prashanth, K.; Li, H.; Prykhodko, H.; Scudino, S.; Kaban, I. Effect of TiB2 particles on microstructure and crystallographic texture of Al-12Si fabricated by selective laser melting. *J. Alloys Compd.* **2019**, *786*, 551–556. [CrossRef]
65. Macías, J.G.S.; Douillard, T.; Zhao, L.; Maire, E.; Pyka, G.; Simar, A. Influence on microstructure, strength and ductility of build platform temperature during laser powder bed fusion of AlSi10Mg. *Acta Mater.* **2020**, *201*, 231–243. [CrossRef]
66. Li, X.; Li, G.; Zhang, M.-X.; Zhu, Q. Novel approach to additively manufacture high-strength Al alloys by laser powder bed fusion through addition of hybrid grain refiners. *Addit. Manuf.* **2021**, *48*, 102400. [CrossRef]
67. Dai, D.; Gu, D.; Xia, M.; Ma, C.; Chen, H.; Zhao, T.; Hong, C.; Gasser, A.; Poprawe, R. Melt spreading behavior, microstructure evolution and wear resistance of selective laser melting additive manufactured AlN/AlSi10Mg nanocomposite. *Surf. Coatings Technol.* **2018**, *349*, 279–288. [CrossRef]
68. Wang, P.; Eckert, J.; Prashanth, K.-G.; Wu, M.-W.; Kaban, I.; Xi, L.-X.; Scudino, S. A review of particulate-reinforced aluminum matrix composites fabricated by selective laser melting. *Trans. Nonferrous Met. Soc. China* **2020**, *30*, 2001–2034. [CrossRef]
69. Tjong, S.C. Novel Nanoparticle-Reinforced Metal Matrix Composites with Enhanced Mechanical Properties. *Adv. Eng. Mater.* **2007**, *9*, 639–652. [CrossRef]
70. Gu, D.; Wang, H.; Chang, F.; Dai, D.; Yuan, P.; Hagedorn, Y.-C.; Meiners, W. Selective Laser Melting Additive Manufacturing of TiC/AlSi10Mg Bulk-form Nanocomposites with Tailored Microstructures and Properties. *Phys. Procedia* **2014**, *56*, 108–116. [CrossRef]
71. Gu, D.; Wang, H.; Dai, D.; Yuan, P.; Meiners, W.; Poprawe, R. Rapid fabrication of Al-based bulk-form nanocomposites with novel reinforcement and enhanced performance by selective laser melting. *Scr. Mater.* **2015**, *96*, 25–28. [CrossRef]
72. Gu, D.; Wang, H.; Dai, D.; Chang, F.; Meiners, W.; Hagedorn, Y.-C.; Wissenbach, K.; Kelbassa, I.; Poprawe, R. Densification behavior, microstructure evolution, and wear property of TiC nanoparticle reinforced AlSi10Mg bulk-form nanocomposites prepared by selective laser melting. *J. Laser Appl.* **2015**, *27*, S17003. [CrossRef]
73. Xi, L.; Gu, D.; Guo, S.; Wang, R.; Ding, K.; Prashanth, K.G. Grain refinement in laser manufactured Al-based composites with TiB2 ceramic. *J. Mater. Res. Technol.* **2020**, *9*, 2611–2622. [CrossRef]
74. Kusoglu, I.M.; Gökce, B.; Barcikowski, S. Use of (nano-)additives in Laser Powder Bed Fusion of Al powder feedstocks: Research directions within the last decade. *Procedia CIRP* **2020**, *94*, 11–16. [CrossRef]
75. Liu, L.; Minasyan, T.; Ivanov, R.; Aydinyan, S.; Hussainova, I. Selective laser melting of TiB2-Ti composite with high content of ceramic phase. *Ceram. Int.* **2020**, *46*, 21128–21135. [CrossRef]
76. Gu, D.; Yang, Y.; Xi, L.; Yang, J.; Xia, M. Laser absorption behavior of randomly packed powder-bed during selective laser melting of SiC and TiB_2 reinforced Al matrix composites. *Opt. Laser Technol.* **2019**, *119*, 105600. [CrossRef]
77. Xiao, Y.; Chen, H.; Bian, Z.; Sun, T.; Ding, H.; Yang, Q.; Wu, Y.; Lian, Q.; Chen, Z.; Wang, H. Enhancing strength and ductility of AlSi10Mg fabricated by selective laser melting by TiB_2 nanoparticles. *J. Mater. Sci. Technol.* **2021**, *109*, 254–266. [CrossRef]
78. Xi, L.; Guo, S.; Gu, D.; Guo, M.; Lin, K. Microstructure development, tribological property and underlying mechanism of laser additive manufactured submicro-TiB_2 reinforced Al-based composites. *J. Alloys Compd.* **2019**, *819*, 152980. [CrossRef]
79. Feng, Z.; Tan, H.; Fang, Y.; Lin, X.; Huang, W. Selective laser melting of TiB_2/AlSi10Mg composite: Processability, microstructure and fracture behavior. *J. Mater. Process. Technol.* **2021**, *299*, 117386. [CrossRef]
80. Li, X.; Ji, G.; Chen, Z.; Addad, A.; Wu, Y.; Wang, H.; Vleugels, J.; Van Humbeeck, J.; Kruth, J. Selective laser melting of nano-TiB_2 decorated AlSi10Mg alloy with high fracture strength and ductility. *Acta Mater.* **2017**, *129*, 183–193. [CrossRef]
81. Wang, P.; Gammer, C.; Brenne, F.; Niendorf, T.; Eckert, J.; Scudino, S. A heat treatable TiB_2/Al-3.5Cu-1.5Mg-1Si composite fabricated by selective laser melting: Microstructure, heat treatment and mechanical properties. *Compos. Part B Eng.* **2018**, *147*, 162–168. [CrossRef]
82. Mair, P.; Kaserer, L.; Braun, J.; Weinberger, N.; Letofsky-Papst, I.; Leichtfried, G. Microstructure and mechanical properties of a TiB2-modified Al–Cu alloy processed by laser powder-bed fusion. *Mater. Sci. Eng. A* **2020**, *799*, 140209. [CrossRef]
83. Xi, L.X.; Zhang, H.; Wang, P.; Li, H.C.; Prashanth, K.G.; Lin, K.J.; Kaban, I.; Gu, D.D. Comparative investigation of microstructure, mechanical properties and strengthening mechanisms of Al-12Si/TiB2 fabricated by selective laser melting and hot pressing. *Ceram. Int.* **2018**, *44*, 17635–17642. [CrossRef]
84. Tan, Q.; Yin, Y.; Fan, Z.; Zhang, J.; Liu, Y.; Zhang, M.-X. Uncovering the roles of LaB6-nanoparticle inoculant in the AlSi10Mg alloy fabricated via selective laser melting. *Mater. Sci. Eng. A* **2020**, *800*, 140365. [CrossRef]
85. Wearing, D.; Horsfield, A.P.; Xu, W.; Lee, P.D. Which wets TiB_2 inoculant particles: Al or Al_3Ti? *J. Alloys Compd.* **2016**, *664*, 460–468. [CrossRef]
86. Savalani, M.; Ng, C.; Li, Q.; Man, H. In situ formation of titanium carbide using titanium and carbon-nanotube powders by laser cladding. *Appl. Surf. Sci.* **2012**, *258*, 3173–3177. [CrossRef]
87. Masanta, M.; Shariff, S.; Choudhury, A.R. Evaluation of modulus of elasticity, nano-hardness and fracture toughness of TiB_2–TiC–Al_2O_3 composite coating developed by SHS and laser cladding. *Mater. Sci. Eng. A* **2011**, *528*, 5327–5335. [CrossRef]

88. Gao, T.; Zhang, S.; Liu, G.; Sun, Q.; Liu, J.; Sun, Q.; Sun, J.; Wang, Z.; Liu, X.; Wang, X. A high strength AlSi10Mg alloy fabricated by laser powder bed fusion with addition of Al Ti C B master alloy powders. *Materialia* **2021**, *16*, 101103. [CrossRef]
89. Zhou, Y.; Wen, S.; Wang, C.; Duan, L.; Wei, Q.; Shi, Y. Effect of TiC content on the Al-15Si alloy processed by selective laser melting: Microstructure and mechanical properties. *Opt. Laser Technol.* **2019**, *120*, 105719. [CrossRef]
90. Wang, H.; Gu, D. Nanometric TiC reinforced AlSi10Mg nanocomposites: Powder preparation by high-energy ball milling and consolidation by selective laser melting. *J. Compos. Mater.* **2014**, *49*, 1639–1651. [CrossRef]
91. He, P.; Kong, H.; Liu, Q.; Ferry, M.; Kruzic, J.J.; Li, X. Elevated temperature mechanical properties of TiCN reinforced AlSi10Mg fabricated by laser powder bed fusion additive manufacturing. *Mater. Sci. Eng. A* **2021**, *811*, 141025. [CrossRef]
92. Liu, X.; Liu, Y.; Zhou, Z.; Wang, K.; Zhan, Q.; Xiao, X. Grain refinement and crack inhibition of selective laser melted AA2024 aluminum alloy via inoculation with $TiC–TiH_2$. *Mater. Sci. Eng. A* **2021**, *813*, 141171. [CrossRef]
93. Cheng, W.; Liu, Y.; Xiao, X.; Huang, B.; Zhou, Z.; Liu, X. Microstructure and mechanical properties of a novel $(TiB_2+TiC)/AlSi_{10}Mg$ composite prepared by selective laser melting. *Mater. Sci. Eng. A* **2021**, *834*, 142435. [CrossRef]
94. Yi, J.; Zhang, X.; Rao, J.H.; Xiao, J.; Jiang, Y. In-situ chemical reaction mechanism and non-equilibrium microstructural evolution of $(TiB_2$ + TiC)/AlSi10Mg composites prepared by SLM-CS processing. *J. Alloys Compd.* **2020**, *857*, 157553. [CrossRef]
95. Wang, Z.; Zhuo, L.; Yin, E.; Zhao, Z. Microstructure evolution and properties of nanoparticulate SiC modified AlSi10Mg alloys. *Mater. Sci. Eng. A* **2021**, *808*, 140864. [CrossRef]
96. Zhang, D.; Yi, D.; Wu, X.; Liu, Z.; Wang, W.; Poprawe, R.; Schleifenbaumc, J.H.; Zieglerd, S. SiC reinforced AlSi10Mg composites fabricated by selective laser melting. *J. Alloys Compd.* **2021**, *894*, 162365. [CrossRef]
97. Xue, G.; Ke, L.; Zhu, H.; Liao, H.; Zhu, J.; Zeng, X. Influence of processing parameters on selective laser melted SiCp/AlSi10Mg composites: Densification, microstructure and mechanical properties. *Mater. Sci. Eng. A* **2019**, *764*, 138155. [CrossRef]
98. Xue, G.; Ke, L.; Liao, H.; Chen, C.; Zhu, H. Effect of SiC particle size on densification behavior and mechanical properties of SiCp/AlSi10Mg composites fabricated by laser powder bed fusion. *J. Alloys Compd.* **2020**, *845*, 156260. [CrossRef]
99. Gao, C.; Xiao, Z.; Liu, Z.; Zhu, Q.; Zhang, W. Selective laser melting of nano-TiN modified AlSi10Mg composite powder with low laser reflectivity. *Mater. Lett.* **2018**, *236*, 362–365. [CrossRef]
100. Gao, C.; Wang, Z.; Xiao, Z.; You, D.; Wong, K.; Akbarzadeh, A. Selective laser melting of TiN nanoparticle-reinforced AlSi10Mg composite: Microstructural, interfacial, and mechanical properties. *J. Mater. Process. Technol.* **2020**, *281*, 116618. [CrossRef]
101. Gao, C.; Wu, W.; Shi, J.; Xiao, Z.; Akbarzadeh, A. Simultaneous enhancement of strength, ductility, and hardness of TiN/AlSi10Mg nanocomposites via selective laser melting. *Addit. Manuf.* **2020**, *34*, 101378. [CrossRef]
102. Dai, D.; Gu, D.; Poprawe, R.; Xia, M. Influence of additive multilayer feature on thermodynamics, stress and microstructure development during laser 3D printing of aluminum-based material. *Sci. Bull.* **2017**, *62*, 779–787. [CrossRef]
103. Konopatsky, A.S.; Kvashnin, D.G.; Corthay, S.; Boyarintsev, I.; Firestein, K.L.; Orekhov, A.; Arkharova, N.; Golberg, D.V.; Shtansky, D.V. Microstructure evolution during AlSi10Mg molten alloy/BN microflake interactions in metal matrix composites obtained through 3D printing. *J. Alloys Compd.* **2021**, *859*, 157765. [CrossRef]
104. Miao, K.; Zhou, H.; Gao, Y.; Deng, X.; Lu, Z.; Li, D. Laser powder-bed-fusion of Si3N4 reinforced AlSi10Mg composites: Processing, mechanical properties and strengthening mechanisms. *Mater. Sci. Eng. A* **2021**, *825*, 141874. [CrossRef]
105. Rauchenecker, J.; Rabitsch, J.; Schwentenwein, M.; Konegger, T. Additive manufacturing of aluminum nitride ceramics with high thermal conductivity via digital light processing. *Open Ceram.* **2021**, *9*, 100215. [CrossRef]
106. Li, Q.; Wang, Z.; Wu, C.; Cheng, X. Microstructure and mechanical properties of aluminum nitride co-doped with cerium oxide via hot-pressing sintering. *J. Alloys Compd.* **2015**, *640*, 275–279. [CrossRef]
107. Dai, D.; Gu, D. Influence of thermodynamics within molten pool on migration and distribution state of reinforcement during selective laser melting of AlN/AlSi10Mg composites. *Int. J. Mach. Tools Manuf.* **2016**, *100*, 14–24. [CrossRef]
108. Gu, D.; Ma, C.; Xia, M.; Dai, D.; Shi, Q. A Multiscale Understanding of the Thermodynamic and Kinetic Mechanisms of Laser Additive Manufacturing. *Engineering* **2017**, *3*, 675–684. [CrossRef]
109. Xue, Y.; Jiang, X.; Bourgeois, L.; Dai, P.; Mitome, M.; Zhang, C.; Yamaguchi, M.; Matveev, A.; Tang, C.; Bando, Y.; et al. Aluminum matrix composites reinforced with multi-walled boron nitride nanotubes fabricated by a high-pressure torsion technique. *Mater. Des.* **2015**, *88*, 451–460. [CrossRef]
110. Minasyan, T.; Liu, L.; Aghayan, M.; Kollo, L.; Kamboj, N.; Aydinyan, S.; Hussainova, I. A novel approach to fabricate Si_3N_4 by selective laser melting. *Ceram. Int.* **2018**, *44*, 13689–13694. [CrossRef]
111. Minasyan, T.; Liu, L.; Aghayan, M.; Rodriguez, M.A.; Aydinyan, S.; Hussainova, I. Mesoporous fibrous silicon nitride by catalytic nitridation of silicon. *Prog. Nat. Sci.* **2019**, *29*, 190–197. [CrossRef]
112. Minasyan, T.; Liu, L.; Holovenko, Y.; Aydinyan, S.; Hussainova, I. Additively manufactured mesostructured $MoSi_2$-Si_3N_4 ceramic lattice. *Ceram. Int.* **2019**, *45*, 9926–9933. [CrossRef]
113. Worldwide Aluminum Consumption Forecast 2029 | Statista. Available online: https://www.statista.com/statistics/863681/global-aluminum-consumption/ (accessed on 31 January 2022).
114. Boeing: First of Many: 787 Dreamliner Celebrates 10 Years since First Flight. Available online: https://www.boeing.com/features/2019/12/787-1st-flight-anniversary-12-19.page (accessed on 31 January 2022).
115. China Sets Carbon Reduction Plans for Steel, Aluminium | Argus Media. Available online: https://www.argusmedia.com/en/news/2266222-china-sets-carbon-reduction-plans-for-steel-aluminium (accessed on 31 January 2022).

MDPI

Review

Online Monitoring Technology of Metal Powder Bed Fusion Processes: A Review

Zhuo-Jun Hou [1], Qing Wang [1], Chen-Guang Zhao [1], Jun Zheng [1,2,*], Ju-Mei Tian [3,*], Xiao-Hong Ge [1] and Yuan-Gang Liu [4,*]

1 School of Materials Science and Engineering, Xiamen University of Technology, Xiamen 361021, China
2 3D METALWERKS Co., Ltd., Xiamen 361021, China
3 Engineering Research Center of Fujian University for Stomatological Biomaterials, Xiamen Medical College, Xiamen 361023, China
4 College of Chemical Engineering, Huaqiao University, Xiamen 361021, China
* Correspondence: jason@3dmetalwerks.com (J.Z.); tjm@xmmc.edu.cn (J.-M.T.); ygliu@hqu.edu.cn (Y.-G.L.)

Abstract: Metal powder bed fusion (PBF) is an advanced metal additive manufacturing (AM) technology. Compared with traditional manufacturing techniques, PBF has a higher degree of design freedom. Currently, although PBF has received extensive attention in fields with high–quality standards such as aerospace and automotive, there are some disadvantages, namely poor process quality and insufficient stability, which make it difficult to apply the technology to the manufacture of critical components. In order to surmount these limitations, it is necessary to monitor the process. Online monitoring technology can detect defects in time and provide certain feedback control, so it can greatly enhance the stability of the process, thereby ensuring its quality of the process. This paper presents the current status of online monitoring technology of the metal PBF process from the aspects of powder recoating monitoring, powder bed inspection, building process monitoring, and melt layer detection. Some of the current limitations and future trends are then highlighted. The combination of these four–part monitoring methods can make the quality of PBF parts highly assured. We unanimously believe that this article can be helpful for future research on PBF process monitoring.

Keywords: additive manufacturing; powder bed fusion; online monitoring; selective laser melting; electron beam melting

Citation: Hou, Z.-J.; Wang, Q.; Zhao, C.-G.; Zheng, J.; Tian, J.-M.; Ge, X.-H.; Liu, Y.-G. Online Monitoring Technology of Metal Powder Bed Fusion Processes: A Review. *Materials* **2022**, *15*, 7598. https://doi.org/10.3390/ma15217598

Academic Editors: Swee Leong Sing and Wai Yee Yeong

Received: 29 August 2022
Accepted: 24 October 2022
Published: 28 October 2022

1. Introduction

AM (Figure 1) is a layered manufacturing technique. Compared with subtractive manufacturing technology, it can produce complex geometric shapes, lower the use of raw materials, and greatly reduce the cost. Currently, AM is widely used in aerospace, automobile manufacturing, biomedical, and other fields [1–3]. According to different feeding methods, metal AM is divided into two types: direct energy deposition and powder bed fusion [4]. The former adopts the method of synchronous powder feeding or wire feeding, and fills melt pool area with the raw material while scanning the high–energy beam; the latter adopts the method of laying the powder bed in the forming area in advance.

Metal PBF processes (Figure 2) comprise selective laser sintering (SLS), selective laser melting (SLM), and electron beam melting (EBM) [5]. In SLS, the process uses a laser beam as a heat source with high forming accuracy and surface finish. It has a wide range of materials that can form almost any geometrical part, especially for parts with complex internal structures. In EBM, the process uses a high–energy electron beam as a heat source, has faster forming speed and low forming thermal residual stresses, which can form high melting point material and brittle material.

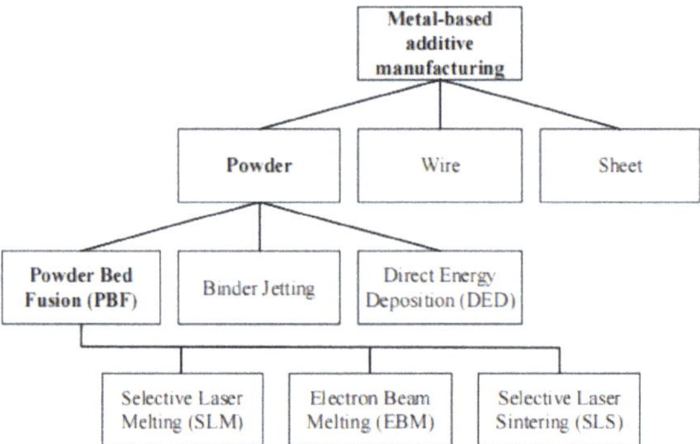

Figure 1. Overview metal–based additive manufacturing.

Melting and depositing a powder bed by laser or electron beam is a highly dynamic and complex process with multiple physical phenomena and transformations. Metallic AM parts are prone to macro defects such as warping deformation [6,7], spheroidization [8], cracking [9], and internal defects such as porosity [10–14], incomplete fusion [15], and inclusions [16]. The timely detection and suppression of the defects in the formed parts can greatly improve the forming quality of the metal powder bed and eliminate the limitation of the technical instability of the process development.

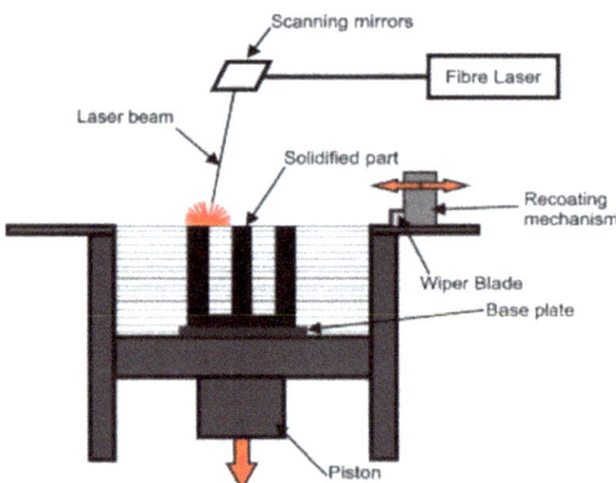

Figure 2. Schematic of the PBF process [17].

Online monitoring is a method of detecting defects in a timely manner to ensure the quality of parts manufactured by the PBF process. On the one hand, it can provide researchers with records and optimize process parameters; On the other hand, it can conduct real–time monitoring and data analysis on the process for online diagnosis, real–time defect repair, and provide key data for process records. Currently, AM is widely used in several fields. Compared with conventional manufacturing processes, parts formed by PBF are smaller in size and therefore more sensitive to defects. In recent years, many research teams have obtained a variety of online monitoring methods for PBF process. These monitoring methods have shown irreplaceable technical potential, but they are still in the research and development stage.

This article builds on previous studies and reviews methods for monitoring defects during PBF. Subsequentially, the monitoring results are analyzed, and the reliability, limitations, and improvement trends of the monitoring method are described. The paper is divided into six sections, including the present one. Section 2 surveys the monitoring of powder recoating, which is the first step in the PBF process and one of the key steps in stable forming. Section 3 discusses the powder bed inspection. Section 4 reveals the monitoring of building process. With a particular emphasis on monitoring the melt pool and temperature. Section 5 presents melt layer detection, namely the monitoring of temperature and surface topography. Section 6 summarizes existing work and identifies future research trends.

2. Powder Recoating Monitoring

Powder coating is the process of spreading powder on the forming area by a recoating device to form a powder bed [18]. It is the first step of PBF and one of the key steps for stable forming. For the faults that are prone to the coating process and abnormal damage to the coating machine, Berumen et al. [19] used a digital camera to monitor powder feed and coater defects during powder coating. Although this method is low cost, there are two major disadvantages. Firstly, it is necessary to correct the images taken, which often decreases the accuracy of PBF defect monitoring. Secondly, there is a trade−off between spatial resolution and the field of view. Reinarz and Witt [20] installed a piezoelectric accelerometer on coater of the SLM equipment (Figure 3) and monitored the powder coating process by measuring speed change signal of coater. When coater collides with previously deposited layer, it will cause large vibration or even jam. The signal of the accelerometer can reflect the smoothness of coating process and unevenness of previously deposited layer. Kleszczynski et al. [21] validated the reliability of the monitoring system by detecting defects with a high–resolution CCD camera. Experiments have shown that, with the help of image processing, the system can detect the defects such as powder deficiency, poor support, and damage to the coater. Process stability is monitored by studying critical process parameters and critical geometries. In the future, this approach can be used to build a knowledge base for specific materials, which can clearly understand the causes of defects and propose appropriate solutions.

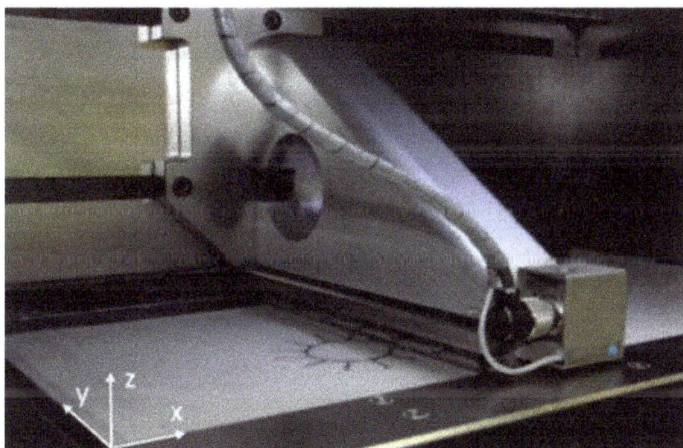

Figure 3. Integration of acceleration sensor at the recoating mechanism [20].

Liu et al. [22] proposed an on–site quantitative detection technology suitable for EBM and SLM. By evaluating the entire powder bed rear rake using edge projection profilometry, defects such as part thermal expansion, powder overfeed, and powder shortage can be inspected. The technique relies on a surface–fitting algorithm to calibrate the phase error during tracking to ensure the reliability of the tracking method. The results of the experiments show that the method can effectively inspect powder defects and use the results as feedback during the building process. The method has the advantages of low cost, fast acquisition time, and a no–vacuum environment. In the future, it can be considered for improving the processing speed and intelligent measurement of the calibration algorithm and realizing the automatic classification and feedback of defects. Seita et al. [23] revealed that powder layer defects in PBF systems (SLS, SLM, and EBM) and binder jetting can be detected with a high spatial resolution of ~5 μm using a powder scanner (Figure 4). By installing a line scanner on the coating machine, the two move synchronously to acquire the image of the powder bed, so this inspection technology has a lower cost and higher time utilization. Using an improved Laplacian algorithm to quantify out–of–focus areas in the image, the automated detection of powder bed defects (such as powder unevenness, ultra–high edges, and grooves) can be achieved. It is important to note that a slight difference between the movement speed of the coating machine and the sampling rate of the contact image sensor (CIS) in the powder scanner can cause image distortion.

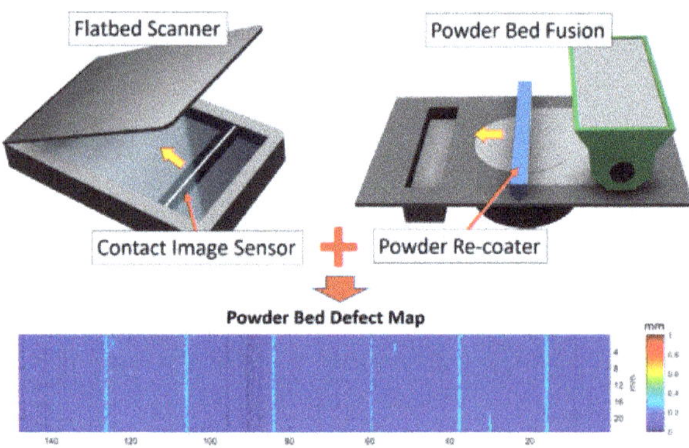

Figure 4. Components of powder scanner [23].

3. Powder Bed Inspection

A powder bed formed after powder coating is the basis for scanning melted powder with an electron or laser beam. If the powder bed is not flat, it will cause the melt pool to become unstable during the scanning process and cause abnormal defects such as protrusions or voids that may affect subsequent forming.

There are many reasons for the unevenness of a powder bed: the wear on and damage to the recoater, causing the powder bed to produce gullies or ridges distributed along the direction of the paving; recoater streaking occurs when the recoater blade drags a piece of debris or a tuft of powder onto the powder bed; the melt layer is raised, and the scraper is forced to jump at the bulge, resulting in vertical direction gully or bulge perpendicular to the direction of the paving; and an insufficient amount of powder or no powder at the end of the powder bed. Therefore, the appearance of the powder bed can not only reflect the working state of the powder coating device but also reflect the quality of the upper melt layer. The following are the methods available for monitoring powder bed defects in existing research.

Figure 5 is a schematic diagram of the system for the visible light detection of a powder bed in the SLM process, including an optical camera and several flash sources. The system uses an off–axis arrangement where the camera aligns the powder bed laterally, while the flash source is tilted at different angles. The purpose of the flash source is to provide the appropriate background light to capture a clear, high−contrast image of the powder bed, simplifying the subsequent defect identification process. Craeghs et al. [24] used a visible light inspection system to detect powder bed defects due to damage or wear of the coater. Some grayscale distributions perpendicular to the powder coating direction were extracted from the grayscale powder bed images, and the average distribution was compared with a reasonable gray level to effectively identify errors and material discontinuities at the powder bed level (Figure 6). The coating is linear in this method, so it is not possible to accurately identify a defect at a certain point. Kleszczynski et al. [21] and Jacobsmühlen et al. [25] performed threshold processing on grayscale images based on the characteristics of bright areas generated by bulges in the melt layer to achieve the effective extraction of protrusions. In addition to powder bed defect detection, the method can also be used for parameter optimization (material identification). Jacobsmühlen et al. [26] provided an image−based two−dimensional acceleration method for investigating the influence of the angle of the suspended structure and the parameters of the supporting structure on the bulge melt layer. The method can rank the stability of different components and determine the best parameters for guaranteeing their stability. In the future, the approach could be used to guide the building design and validate the boundaries of the established parameters. Abdelrahman et al. [27] extracted the area corresponding to the section of the part from the powder bed image and superimposed it to form a three−dimensional powder bed model, which can well monitor the precise location of the powder bed anomaly corresponding to the part.

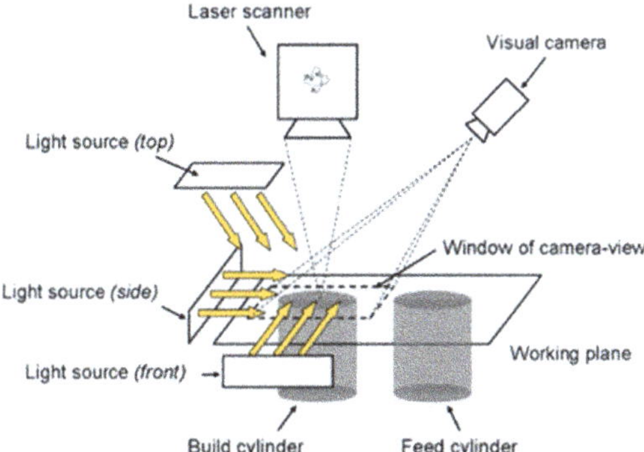

Figure 5. Visible light inspection system for powder bed [21].

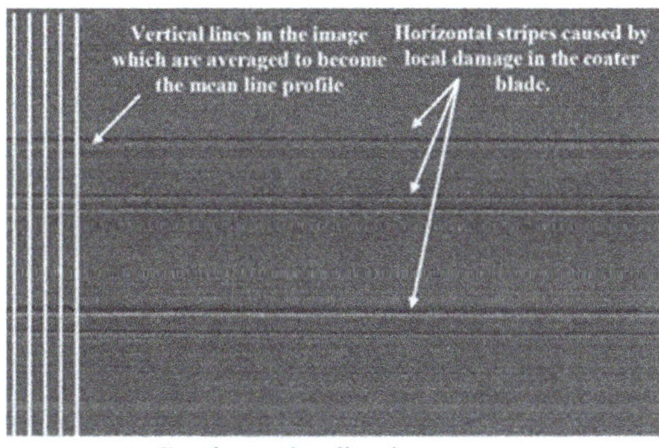

Figure 6. Image of deposited powder bed with worn coater blade [21].

Neef et al. [28] proposed the use of low−coherence interferometry to detect the flatness of a powder bed in the SLM processes. The principle of low–coherence interference imaging is to scan the powder bed by measuring the laser beam, to measure the difference in an optical path between the reflected light and the reference light through the spectrometer, and to compensate for the deviation caused by the angle deflection to obtain the height distribution of the different scanning points. It can be seen from Figure 7 that low–coherence interference technology can effectively detect the height and low fluctuation of the powder bed and can identify the groove of 50 μm depth on the powder bed. Fleming et al. [29] revealed inline coherence imaging (ICI) tracking topography, providing the instant inspection of surface roughness, damage to the recoater blade, and powder–packing density. The level of energy during 3D build affects surface roughness, which can be corrected based on ICI measurements. Moreover, the method successfully realizes manual closed–loop control and full feedback control. Boschetto et al. [30] proposed the use of digital image processing to monitor the defects of powder beds during SLM. In this study, thousands of images of powder beds taken by CCD cameras were analyzed using 2D and 3D analysis to identify single−layer defects and defects between powder beds, respectively. Using this method, the location and size of defects can be found accurately and quickly and at a low cost. However, the drawback of this method is that the image may be distorted and must be calibrated.

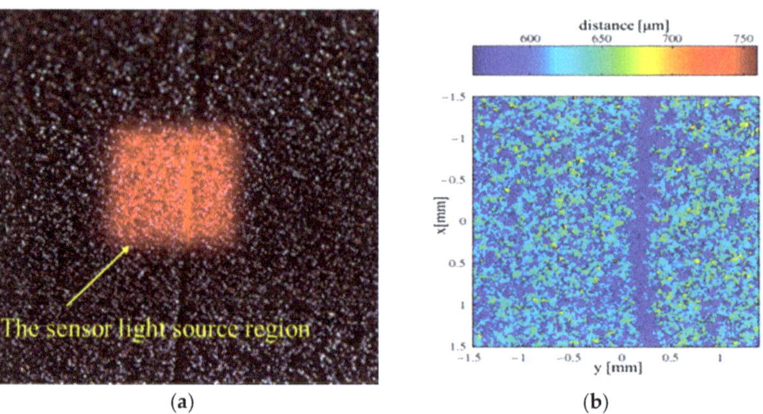

Figure 7. Powder bed inspection using low–coherence interference imaging in SLM. (**a**) Optical morphology of powder bed surfaces; (**b**) Profile scanning of powder materials [28].

The above optical detection has many limitations in the EBM process, so they are mainly aimed at the SLM process. These limitations can be seen from the following: The detection method based on optical imaging has strict requirements for the location of sensors and light sources, so the molding equipment needs to be modified accordingly, thereby increasing the difficulty of system integration. The inert gas in the forming chamber of SLM can inhibit the evaporation of metal, providing good heat dissipation conditions, so that the sensor can be directly put into the forming chamber, which makes the system integration simple. However, the EBM process provides a vacuum environment, and the metal evaporation in EBM cannot be solved well, so the ambient temperature is very high, and the radiation is very strong.

Li et al. [31] revealed an enhanced phase−measurement profiling technique (EPMP) to monitor defects such as inhomogeneities in powder beds. This method can significantly improve inspection efficiency while having high accuracy compared with the conventional phase−measurement profiling technique. In addition, this method can be used to monitor fusion area defects. Currently, the technique does not allow for real–time closed–loop control and automatic defect identification and classification.

Grasso [32] investigated a method suitable for monitoring powder defects in the EBM process. This method acquires images by layering them with a camera. Image processing is then used to identify inhomogeneities in the powder layers. It is experimentally demonstrated that the technique can currently detect three defect anomalies, namely, powder deficiency, powder overload, and powder bed contamination. Compared with the method of streak projection field measurement in the abovementioned article, this method has a higher data collection rate because of the layered image acquisition. This method also has a high practicality and can be readily used in industrial−grade EBM devices. Currently, the method cannot be used to monitor surface powder oxidation.

The monitoring methods used above usually only target specific defects in the powder bed and require the use of high–resolution cameras. Therefore, the entire powder bed is usually not monitored, which has limitations. In order to improve detection efficiency, machine learning (ML) algorithms have attracted people's attention. To verify that ML can properly and accurately track defects in AM processes, numerical simulation is necessary because of the high cost of AM technology. Jahan et al. [33] modeled the LPBF process by computer hydrodynamics and generated a large number of simulated values. The simulated values are entered as input data into a graphics−based neural network machine learning algorithm to monitor defects in LPBF processes. Experimental results show that this machine learning algorithm can predict the defects caused by thermal anomalies in PBF processes. In addition, feedback control is carried out by optimizing process parameters to reduce the formation of defects.

Xiao et al. [34] proposed a method for monitoring powder bed defects using the time−spatial convolutional neural network (TSCNN) model, which is suitable for different systems of PBF (SLS, SLM, and EBM). Start by taking an image of the powder bed with a digital camera. The image is then split into three monochromatic channels corresponding to the RBG. Finally, TSCNN and the region proposal network are utilized to detect powder bed defects in selective laser sintering. Compared with other methods for detecting defects, this approach has higher accuracy and efficiency while resisting geometric distortion and image blurring. In the future, surface defects can be detected for other related applications and can also be used to set machining parameters in real time.

Scrime et al. [35] presented a multiscale convolutional neural network (MSCNN) based on a machine learning algorithm for the automatic, high−accuracy classification of anomalies detected during powder bed monitoring (coater jumps, recoat streaks, and debris). Figure 8 shows the powder bed image taken on the 2709 floor. L. Scrime and J. Beuth used MSCNN technology to analyze the anomalies detected in the 2709 layer and found that the recoil jumper (blue−green) and partially damaged (magenta) and incompletely diffused (yellow) fragments were captured, as shown in Figure 9. In the future, anomaly classification accuracy can be further improved by incorporating additional data in the neural network. In addition, the real–time analysis of high–resolution images is performed by designing a neural network structure.

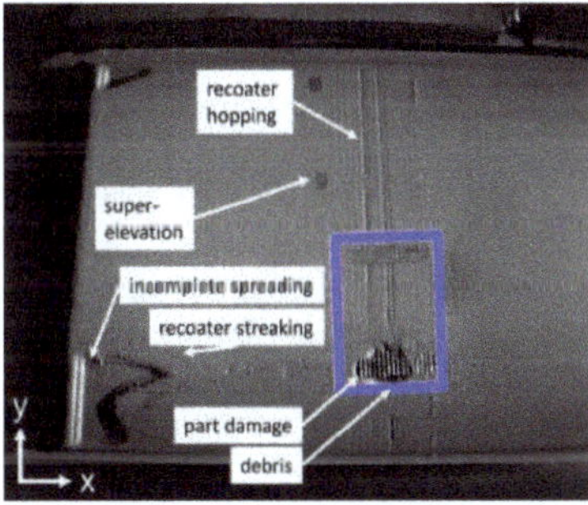

Figure 8. Anomaly diagram of powder bed [35].

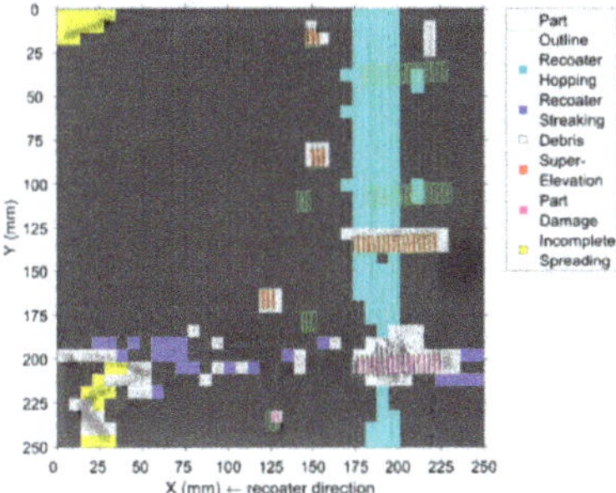

Figure 9. MSCNN anomaly analysis diagram [35].

Scrime et al. [36] proposed to monitor and classify defects in powder layers using ML algorithms. The powder layer image captured by the digital single−lens reflex (DSLR) camera was fed into the ML model. An algorithm is used to classify images so that similar images are grouped. In this experiment, the authors detected six different powder layer anomalies: recoater jumps, recoater streaks, debris, superelevation, part failure, and incomplete spreading. Although this method reduces the amount of computation, it is less accurate for monitoring recoating streaks due to the small amount of training data.

4. Building Process Monitoring

During melting, a high−energy electron beam or laser beam melts powder to form a melt pool that is solidified and deposited to form a solid cross section. Therefore, the fused deposition process directly determines the quality of the final melt layer. Currently, the monitoring of the melt process is primarily directed at the melting pool temperature and the entire forming area.

4.1. Melt Pool Monitoring

In the PBF processes, the melt pool is formed by laser beam or electron beam scanning of the powder bed and has the characteristics of small size and fast–moving speed. The quality of the castings is greatly influenced by the shape, size, and temperature of the melt pool. Melt pool monitoring is the real−time measurement of radiation intensity and shape characteristics during laser or electron beam scanning and the real–time analysis of measurement data to identify spheroidization and warpage.

Melt pool monitoring usually uses a coaxial layout, including that of Berumen et al. [37]. As shown in Figure 10, the sensing channel overlaps with the shaped laser beam path in order to acquire the melt pool signal in real time without adding a complicated melt pool tracking system. The high–power forming laser beam enters the scanning system after being reflected on the surface of the 45° half mirror, and the melt pool radiation signal propagates in the opposite direction. After passing through the half mirror, the filter is used to filter out the specific band signal into the sensor or through splitting. The mirror splits light into two beams for sensor acquisition. Compared with traditional cameras, the method has high timeliness and high local spatial resolution. Moreover, the method enables the real–time control and recording of the manufacturing process. The research groups Kruth et al. [38], Clijsters et al. [39], and Craeghs et al. [40] used coaxial sensing to screen the 740~950 nm radiation wave and split it. One beam is used for the photodiode to collect the pool intensity signal; the other beam is imaged by a high–speed CMOS camera to extract geometric information such as melt pool area, length, and width. The light intensity of the pool collected by the photodiode is proportional to the size of the pool. In addition, the size of the melt pool can be kept constant by controlling the laser power. Both of them can effectively detect process problems such as spheroidization during the forming of suspended structures, convex hulls at corners during U–shaped scanning, and powder coating failures. The research groups used laser power as the control object and used the output voltage of the photodiode and the pixel area of the melt pool as feedback variables to establish a closed–loop control system for SLM. The research results show that the above two closed–loop control systems can effectively improve the accuracy of forming surface structures formation and realize the adaptation of process parameters. Kanko et al. [41] applied a coaxial optical path to low–coherence interferometry to measure the height of a melt pool in real time during the SLM forming process. Figure 11 shows that when the laser beam is swept over the suspended area, the melt pool overheats due to poor heat dissipation conditions, which causes severe fluctuations. Low–coherence interferometry can quickly capture changes in melting pool height and identify globular defects caused by overheating.

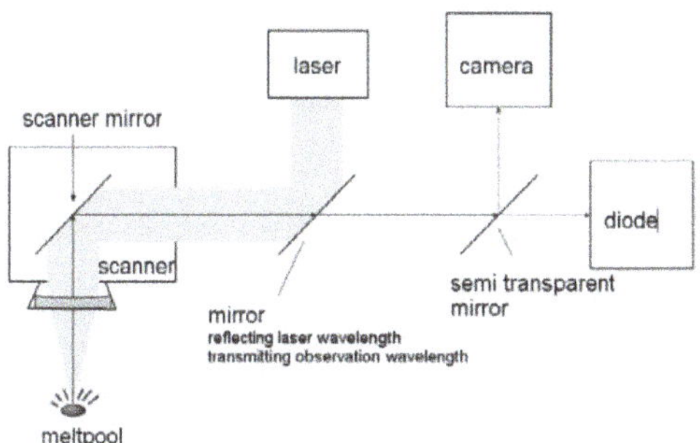

Figure 10. Schematic assembly of online process control [37].

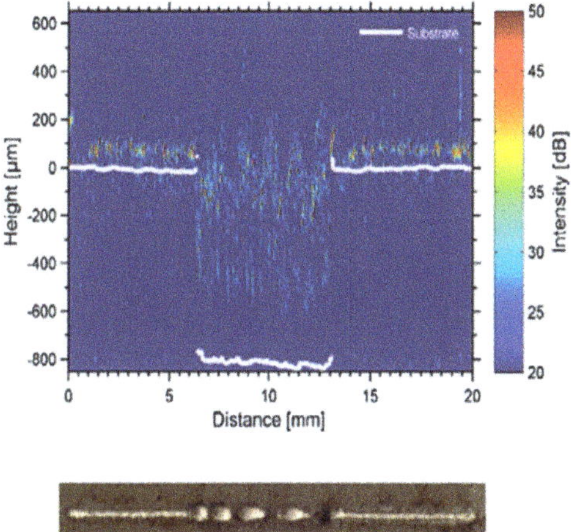

Figure 11. Single−channel SLM pool monitoring based on low coherence interference [41].

In order to facilitate defect identification and positioning, Clijsters et al. [39] implemented location–based visual pore detection. The monitoring system uses photodiodes to monitor the strength of the molten pool and uses a near–infrared thermal CMOS camera to monitor the area of the molten pool. Data monitored by both are visualized through a mapping algorithm. The pixel area of the melting pool is arranged into two dimensions according to the scanning position, and the time series signal is transformed into a spatial distribution image. The geometric parameters of the steady−state melt pool during a filling scan and contour scan were obtained experimentally and used as reference data. In addition, by obtaining the position−dependent melt pool area distribution layer by layer, the three−dimensional spatial positioning of the internal pore defects of the formed part is realized. The experimental results show that the defect identification and localization results are in good agreement with the computed tomography results. This method has a high sampling rate and can be monitored in real time, but no feedback control is currently implemented. In the future, automated detection can be achieved by adding algorithms. Krauss et al. [42] proposed the use of thermal imaging to monitor defects (pores) and irregularities (irregularities close to overhang structures) in SLM processes due to insufficient heat dissipation using an off−axis sensor arrangement to align the infrared camera from the front observation window of the SLM device to the forming area and take a thermal image during the laser beam scanning. By processing the infrared thermal image, geometric parameters such as the area of the melt pool, aspect ratio, and roundness are extracted. The effects of process parameters such as scanning speed, laser power, hatch spacing, hatch length, and powder layer thickness on the geometric parameters of melt pool were also studied. Since the reference value has a certain relationship with the moving direction of the current process area, there is a parameter deviation in this method. This technique will be explored in the future for monitoring coater wear and powder layer thickness. Feedback control could not be achieved with the above methods. Le et al. [43] revealed a method for monitoring the melting pool scale in PBF processes using CMOS cameras. In this method, the experimental measurement data and the simulation results are compared, and the results show that the numerical simulation can predict the size of the melting pool with a small error. This provides great help to the feedback control of parameters in the process of melting pool monitoring to avoid defects to the greatest extent possible.

The abovementioned method for monitoring melt pools basically uses a coaxial system in SLM processes since coaxial monitoring can track melt pools well, and the output signal is simple. Real–time process monitoring and feedback control have been implemented. In the case of an on−axis setup, the laser can be affected by lens characteristics in the Lagrangian reference frame. In the EBM process, due to structural limitations such as the electromagnetic deflection system, only the off−axis arrangement can be used, and the rapid tracking of the melt pool becomes a problem. Coupled with severe evaporation effects, the real–time monitoring of the melt pool is difficult.

4.2. Temperature Monitoring

PBF belongs to a kind of thermal processing. Recording and analyzing its temperature change process is of great value for understanding the inherent mechanism of the process and verifying simulation models. Price et al. [44] used near–infrared thermal imaging equipment to study the temperature distribution in the forming area during preheat scanning, contour scanning, and filling scanning in the EBM process. The use of a thermal imaging system in the above method does not give an accurate actual temperature. In order to make the temperature more accurate, Pavlov et al. [45] believed that the temperature of the laser impact zone could detect the changes in the SLM parameters. The method adopts a coaxial arrangement scheme, using a two–color pyrometer to measure the temperature signal of the melt pool during the laser scanning in real time. It is found that the measured values of the two–wavelength pyrometer are very sensitive to the process parameters such as filling interval, the thickness of powder layer, and filling strategy, but they cannot evaluate the integrity of the part.

Cheng et al. [46] and Price et al. [47] investigated the effects of process parameters such as the speed of scanning, electron beam current, and beam spot diameter over the longitudinal melt pool distribution (along scanning direction) as well as the melt pool size. The measured temperature distribution and the size of the melt pool were used to verify simulation model. In addition, Price et al. [48] and Gong et al. [49] explored the influence of forming height on the longitudinal temperature distribution of a melt pool and of the suspended surface at different distances from the centerline of the melt pool. It was found that longitudinal temperature distribution of the melt pool during filling scanning is repeatable and very sensitive to heat dissipation conditions. Finally, it is verified that the defect identification of temperature spatial distribution is feasible.

Internal void defects will weaken the heat dissipation ability of the local area and change the distribution and evolution characteristics of the surrounding temperature. Accordingly, Krauss, Eschey, and Zaeh [42] proposed a method for detecting internal void defects using temperature detection. When the laser beam sweeps through a predetermined defect area, the temperature distribution curve along the longitudinal direction of the melt pool is extracted. In comparison to the defect–free temperature distribution curve, the temperature distribution curve at the back end of the melt pool is found to have significant differences in defect position and size. Therefore, the dynamic data on the temperature of the powder bed can not only identify the pore defects but also obtain information such as the size of defects.

Moreover, Krauss et al. [50,51] attempted to identify defects based on the time–domain evolution of the temperature. Relevant key indicators were extracted from the dynamic temperature evolution data, including high temperature holding time, equivalent thermal diffusion coefficient, maximum temperature, and splash amount. The equivalent thermal diffusion coefficient is the cooling rate defined by a one–dimensional downward thermal diffusion simplified model. The index extracted by each layer of temperature evolution can form a frame index map. After the forming is completed, distribution maps of the layers are stacked to form a three–dimensional index distribution model. This method has a layered nature. In the subsequent process, all the information needed for part quality can be obtained by monitoring each layer. Then, the quality index diagram of each layer is derived. Finally, a 3D quality report similar to the tomography method is combined.

In terms of microstructure prediction, Price et al. [47] extracted the average cooling rate of shaped cross sections at different scanning speeds based on the temperature–time evolution curve and found that fast scanning produces higher cooling rates and smaller β columnar crystals. Raplee et al. [52] used different scanning strategies (point scan and line scan, as shown in Figure 12) during the EBM process. By analyzing these data, it is found that the thermal gradient and the velocity of the solid–liquid interface are approximately the same, so they are related to some changes in microstructure. The thermographic data were analyzed to determine the transition of material from metallic powder to a solid as–printed part. Compared with the experiment, it is found that the line scanning strategy will form a higher temperature gradient and a lower solid–liquid interface velocity, tending to form columnar crystals. The point scanning strategy will form a lower temperature gradient and a higher solid–liquid interface velocity, which is beneficial to equiaxed crystal forming (Figure 13). The above research shows that using thermal imaging to predict the microstructure of forming parts is helpful for achieving flexible preparation and effective control of forming parts in the process of adding materials.

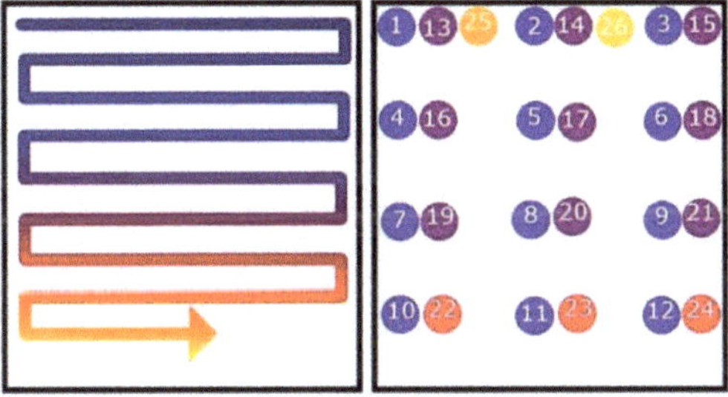

Figure 12. Line scan and point scan schematics [52].

Williams et al. [53] used a wide–field in situ infrared imaging system to monitor the powder surface temperature of the entire powder bed. This system studies the influence of interlayer cooling time by constructing cylindrical scenes with different heights. In the whole construction process, the in situ surface temperature data were obtained and compared with the results of porosity, microstructure, and mechanical properties. The research shows that using thermal imaging technology to predict the microstructure of the formed parts is helpful for controlling the part structure in the manufacturing process.

Although the real–time acquisition and analysis of dynamic temperature in the PBF process have made great progress in internal defect detection and tissue prediction, there are still many deficiencies. Due to the insufficient time resolution and spatial resolution of these thermal imaging devices, the accuracy and sensitivity of current defect detection are insufficient. Moreover, the EBM process produces a large amount of metal vapor, which makes it difficult to conduct continuous dynamic monitoring of the temperature. Although scholars have studied anti–vapor deposition systems and transmittance compensation methods, they still cannot completely eliminate the effects of evaporation [54,55]. In addition, in order to convert the radiation intensity output by the camera into an absolute temperature, it is necessary to accurately determine the parameters such as the infrared emissivity, window transmittance, and ambient temperature of the material, which also brings difficulties and challenges to real–time temperature measurement [56,57].

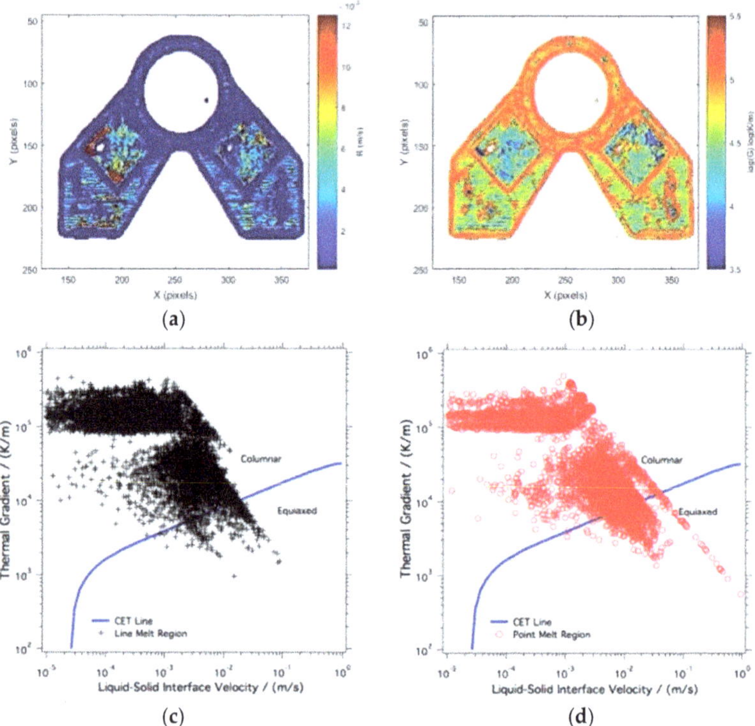

Figure 13. Influence of scan strategies on grain morphology in EBM [52]. (**a**) The layer thermal gradients; (**b**) Interface velocity; (**c**) Grain morphology distribution under line scan strategy; (**d**) Grain morphology distribution under point scanning strategy.

5. Melt Layer Detection

When a powder bed is melt–deposited by electron beam or laser beam to form a melt layer, the state of the melt layer not only reflects the quality of melt deposition and the matching of process parameters but also affects the subsequent coating and melting forming processes. Therefore, melt layer detection is a very important part of online monitoring. It can detect cross–section profiles, geometric parameters, and surface defects. At the same time, it can record the forming results of each layer and provide basic data for final quality traceability. At present, the main objects of melt layer detection are temperature and surface topography morphology.

5.1. Temperature Detection

The temperature detection of the melt layer is similar to the temperature detection in the fused deposition process. Both of them use a near–infrared/infrared thermal imager. The difference is that the temperature of the melt layer changes slowly. Usually, temperature is taken only once to extract and identify part contour defects in a single frame image. Schwerdtfeger et al. [58] used different focus bias parameters in the EBM process as a control. The thermal image of the melt layer was compared with the metallographic diagram. The results show that thermal imaging can effectively reveal unmelted material and defects within the cambium. Dinwiddie, Dehoff, Lloyd, Lowe, and Ulrich [54] used infrared thermal imaging to study the influence of EBM focusing parameters on the porosity and evolution of subsequent melt layers on suspended surface. Rodriguez et al. [56] used temperature distribution histograms to quantify the nonuniformity of the temperature distribution of the melt layer and found that temperature distribution of the superheated region is wider. It was considered that histogram analysis was a method for effectively identifying local overheating defects. Ridwan et al. [59] realized the extraction of workpiece sections through image processing and calculated the porosity of the melt layer to reflect quality. Mireles et al. [60] verified the feasibility of online defect repair by infrared thermal imaging monitoring. Compared with the thermal images before and after the remelting of melt layer, remelting effectively reduced pore defects. Based on this, the authors also proposed a closed–loop control method for the online repair of local defects.

The above research demonstrates the feasibility of using near–infrared or infrared thermal imaging technology to detect melt layer defects. Real–time melt layer detection based on this technology needs further research.

5.2. Surface Topography Detection

5.2.1. Optical Inspection

The aforementioned powder bed detection techniques, such as visible light imaging and low coherence interference, can also be applied to detect the morphology of a melt layer. Foster et al. [6] extracted contours from melt layer optical images and stacked them to form a three–dimensional solid model. The three–dimensional model not only contains size information of the formed part but also visually shows the problem of the uneven powder bed (Figure 14). Abdelrahman et al. [27] took five images of a melt layer under different lighting conditions after each layer scan. First, the melt layer contours are extracted according to a CAD model of the part. The front and back three–layer contours are averaged, and these images are segmented to obtain cross sections. The cross sections are stacked layer by layer to obtain a three–dimensional solid model. Then, an abnormality occurring in the same position on at least two adjacent melt layers is regarded as a real defect. Finally, the internal parts such as unfused material and voids were identified and located. DePond et al. [13] studied the influence of melt layer surface roughness under different filling strategies. Low–coherence interferometry was used to monitor change roughness when forming a suspended structure. Compared with the height distribution map, it was found that the melt layer has a greater roughness and a clear directionality under round–trip scanning without rotation between layers. When the uneven layers are accumulated layer by layer, the entire part will eventually be deformed. In addition, Erler et al. [61] proposed a monitoring method for measuring height distribution surfaces using 3D mapping technology and studied the influence of coating parameters and laser power on uneven melt layers. With this method, the uniformity, thickness, and layer defects of powder layers and sintered layers can be monitored. To avoid false detections during the detection process, direct process control should be added to the detection method. Imani et al. [62] used X–ray computed tomography to identify pores and obtained layered optical images of powder layers during part manufacturing. Then, spectral theory and multifractal features were extracted from the layered images of each test part. Finally, the machine learning method was used to link these features with the process parameters (Figure 15).

By measuring the laser beam to scan melt layer line by line, the laser displacement sensor receives reflected signals and calculates height at different positions. The result is a standard for calculating the height distribution of a melt layer, and the quality is judged according to this standard. Monitoring the melt layer using visible light imaging technology is difficult due to the analysis and processing of the grayscale images. Currently, contour extraction and defect recognition have been implemented, but most of this technology uses offline processing; the output of low–coherence interference imaging and 3D topography mapping is the distribution of melt layer heights, which reduces data analysis and processing, but its problem is that it must be scanned point by point or progressively, which increases the time cost. Moreover, due to the complex detection system, there are not many detection studies that directly measure height distribution.

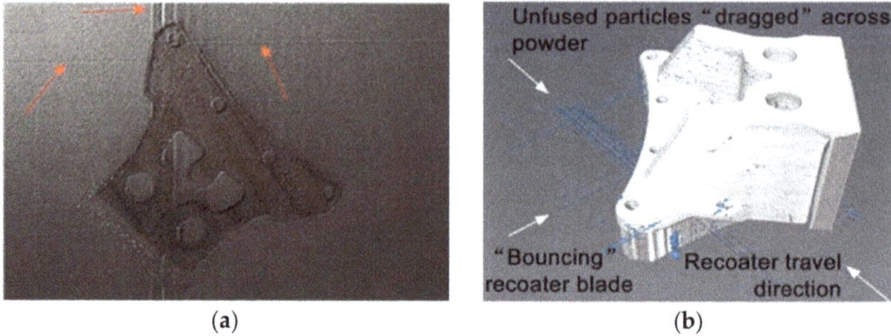

(a) (b)

Figure 14. Melt layer inspection based on visual imaging. (**a**) Melt layer image; (**b**) 3D reconstruction model and powder bed anomaly [6].

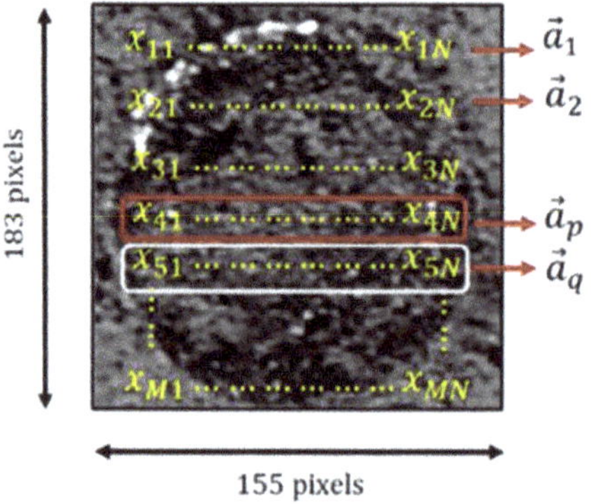

Figure 15. Diagram of linking features and process parameters using machine learning methods [62].

5.2.2. Electro–Optical Inspection

Electro–optical inspection is a unique technology for monitoring the morphology of the melt layer in EBM. Because electron imaging and scanning electron microscopy (SEM) have the same advantages, they can also be used in the current EBM detection technology. As electronic imaging is an image generation mechanism, it is not affected by the above problems such as thermal cameras and optical cameras. Watt [63] showed that after scanning layer by layer, a small electron beam was used to scan the molten layer point by point. The secondary electron and backscatter electron signals were collected and arranged into two–dimensional grayscale images according to scanning point order. Figure 16 is a schematic of the principle of electron microscopy. Electron detectors in existing studies are usually placed below the electron beam, above the building platform, and coaxial with the electron beam. Metal plates are often used as detectors due to the adverse environmental effects of high evaporation, high radiation, and high temperature.

The use of electro–optical imaging to monitor the surface topography of a melt layer can greatly reduce difficulty in extracting the contours of the melt layer and identifying defects. Due to differences in the morphologies of the powder region and the solid region, the two can be easily separated to extract cross sections, measure contour size, and create 3D reconstructions based on the electron–optical image; Furthermore, the number of backscattered or secondary electrons emitted by the sub–micron pore area is small, and they appear as dark spots on the electron optical image, so it is easy to identify and locate pores. Arnold et al. [64] found that electro–optical monitoring can effectively identify pore defects and gives information about the quality of the resulting components. Wong et al. [65] separated the powder layer and the molten layer using electronic images and used specially designed hardware to detect the interaction between the electron beam from the machine and the treatment area, thereby generating a digital electronic image. This method is simple and reliable, so it can be a very good system for establishing rapid process optimization and timely feedback. Wong et al. [66] proposed an index to estimate the spatial resolution including the information depth of backscattered electrons (BSE), and they estimated the spatial resolution that can be achieved by Arcam A1 EBM electronic imaging. The experimental results show that the spatial resolution is 0.3 to 0.4 mm at room temperature. This study is helpful for the quality evaluation of on–site monitoring EBM process.

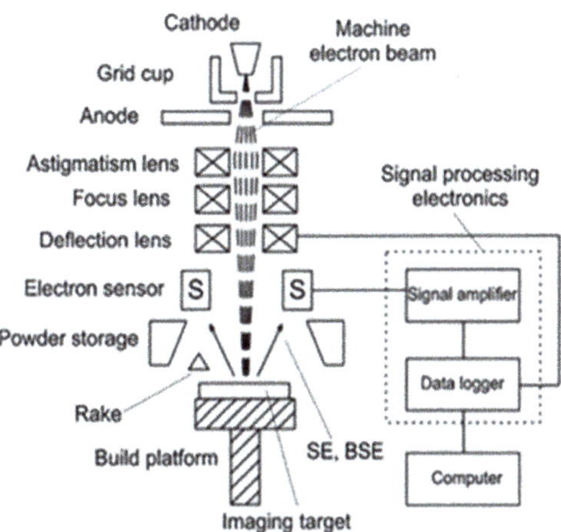

Figure 16. Electronic microscope schematic [63].

Electro–optical monitoring effectively overcomes the difficulties of high temperature, high evaporation, and strong radiation in the EBM process, greatly simplifying the difficulty of online monitoring and feedback control. Although electron optical monitoring has initially achieved defect recognition and feedback control, much research work is still needed for the analysis and interpretation of electro–optical images. In addition, in the existing research, electro–optical images are insensitive to macroscopic topography information, such as undulations and roughness, and it is impossible to extract effective macroscopic morphological information from them. Further research is needed in the detection of macromorphologies.

5.2.3. Acoustic Inspection

In the AM process (PBF), as with optics and thermals, acoustic sensing is also regarded as a key technology. Rieder et al. [67] used ultrasonic testing (UT) technology to monitor porosity. In the experiment, changes in laser power induce the formation of the porous layer. When ultrasonic waves enter the porous layer, the detector detects the reflection and scattering of the wave, thereby generating porosity. Compared with X–ray computed tomography (CT), UT can correlate online and offline data. Figure 17 shows the ultrasound scan and X–ray CT images. The technique has certain limitations and is currently limited to parts with simple geometries. In the future, the method will be used to classify defects.

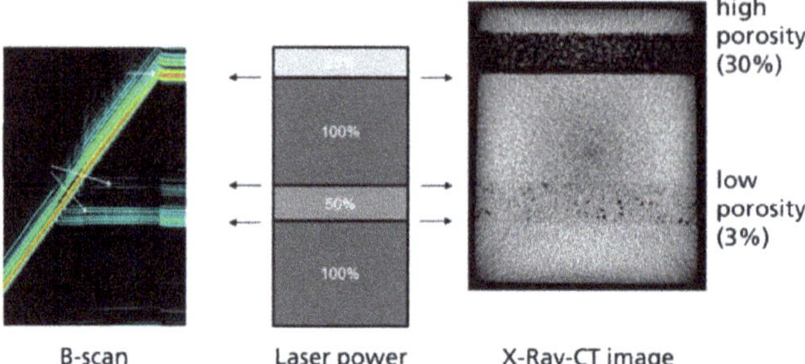

Figure 17. Ultrasound scan and corresponding X–ray CT scan [67].

Ye et al. [68] showed that there is some connection between the acoustic signal and the laser power. Using a deep belief network (DBN) to simplify the monitoring steps of the support vector machine and monitoring the PDF process by acoustic emission spectroscopy found that five different defect states could be detected, namely spheroidization, mild spheroidization, normal, mild overheating, and overheating. Smith et al. [69] presented spatially resolved acoustic spectroscopy (SRAS) to monitor porosity. However, because the signals generated in the PBF process are very complex, machine learning needs to be used to analyze these signals. Shevchik et al. [70] revealed that acoustic emission has a faster processing speed than imaging and tomography. The method combines acoustic emission spectroscopy and convolutional neural networks to pinpoint the locations of defects and monitor the porosity of parts. Furthermore, samples can be classified by porosity grade. In this study, the SLM process was monitored, and its feasibility was verified. In the future, this technology can be migrated to other PBF process monitoring.

6. Perspective

Laser selective melting and electron beam selective melting have significant differences in their monitoring. Optical monitoring technology has developed rapidly in the former for melt pool dynamic monitoring, powder bed inspection, and melt layer detection, and melt pool dynamic monitoring has been applied to process feedback control. In the field of electron beam selective melting process monitoring, optical monitoring technology is severely restricted, but electron optical imaging has become an effective means of reliably monitoring the forming quality of the electron beam selective melting process.

The following is a summary of the various parts of this paper. Table 1 summarizes the powder recoating monitoring process. Table 2 provides an overview of powder bed inspection. Table 3 reveals building process monitoring. Table 4 presents melt layer detection. Analyzing Tables 1–4, the monitoring methods summarized in this paper can detect some defects in the PBF process in time, but there are some common weaknesses:

(1) The monitoring method in this paper detects only superficial defects.

(2) Much of the data processing is offline, and the analysis of monitoring results relies on empirical data.

(3) All are measured by a single monitoring method, which may have an impact on monitoring stability.

(4) Conventional online monitoring techniques for powder–bed melting, such as those based on optical imaging, are limited by a number of factors.

(5) Some machine–based monitoring methods are less reliable.

(6) The monitoring method in this paper is only used to monitor defects in operation and does not involve the correction of defects detected.

Table 1. Summarizes the powder recoating monitoring process.

Monitoring Process	Monitoring Method	Monitored Defects	Advantages	Disadvantages
Powder recoating monitoring	Digital camera	Coater problems, low or excessive powder feed	Low cost	Precision errors, requiring a trade–off between field of view and spatial resolution
	Piezoelectric accelerometer	Smoothness of coating, unevenness of previous deposited layer	-	-
	High–resolution CCD camera	Poor supports, coater damage, insufficient powder	Easy to operate	CCD cameras have a single use
	Edge projection profilometry	Powder overfeed, powder shortage, part thermal expansion	Low cost, no vacuum environment, fast acquisition time,	No automatic feedback, no intelligent measurement
	Powder scanner	Grooves, ultra–high edges, powder unevenness	High spatial resolution, automation	To ensure the synchronization of the recoater module movement and the CIS sample rate
The coating process is the key first step, but there is little information about the formation of defects.				

Table 2. An overview of powder bed inspection.

Monitoring Process	Monitoring Method	Monitored Defects	Advantages	Disadvantages
Powder bed inspection	Visible light imaging	Damage to the coater	-	The exact location of the defect could not be obtained.
	Methods for thresholding grayscale images	Topological defects, powder bed defects	Parameter optimization (material identification)	The monitoring method based on optical imaging has higher requirements regarding the relative position of the sensor and the light source.
	Image–based two–dimensional acceleration	Influence of the angle of the overhang structure and the parameters of the support structure on the expanded melt layer	Sorting the stability of different components	
	Low coherence interferometry	Powder bed flatness	-	
	Inline coherence imaging	Surface roughness, recoater blade damage, powder packing density	Correction of surface roughness based on ICI measurements, closed loop control, full feedback control	
	3D indexing	3D locations of powder bed anomalies	Accurate location of defects	-
	Digital image processing	Single–layer defects and defects between layers	Accurate, fast, and low cost	Image distortion
	EPMP	Inhomogeneities in powder beds, irregular surface of fusion area	Reliable, high precision, high efficiency	Failure to implement real–time closed–loop control or automatic defect identification and classification
	Camera layered acquisition combined with image processing	Powder deficiency, powder overload, powder bed contamination	Highly usable for industrial EBM	Cannot be used to monitor oxidation of surface powders

Table 2. *Cont.*

Monitoring Process	Monitoring Method	Monitored Defects	Advantages	Disadvantages
Powder bed inspection	Numerical simulation combined with GBNN	Thermal anomalies	Feedback control	-
	TS-CNN model	Warpage, short feed, part shifting	High precision, high efficiency, anti–geometric distortion	No real–time control
	MSCNN model	Debris, coater jumps, recoat streaks,	High anomaly classification accuracy	Lack of real time
	ML and DSLR camera	Recoater hopping, recoater streaking, debris, superelevation, part failure, incomplete	Less amount of calculation	Less accurate monitoring of repaint streaks
The powder bed is the basis of the melting process and effectively reflects the quality of current layer. Limited to monitoring the surface state of the current layer.				

Table 3. Reveals building process monitoring.

Monitoring Process		Monitoring Method	Monitored Defects	Advantages	Disadvantages
Building process monitoring	Melt pool monitoring	Coaxial sensor	Warpage, spheroidization,	High local spatial resolution, high timeliness	With an on–axis setup, the laser can be affected by lens characteristics in the Lagrangian reference frame.
		Photodiode and CMOS camera	Spheroidization, convex hulls at corners, powder coating failures	Keeping the molten pool size constant by controlling the laser power, improving the forming accuracy of suspended surface structures	
		Low–coherence interferometry	Globular defects	High speed, real time	
		Location–based visual pore detection	Pore defects	Real time, high sampling rate	No feedback controls
		Thermal imaging and off–axis sensor	Pore defects, irregularities close to overhang structures	Widely used	Parameter deviation, no real–time control
		Numerical simulation combined with CMOS camera	Melting pool size	Low cost, feedback control	-
	Temperature monitoring	Near–infrared thermal imaging	Pore defects	Visibility	Insufficient temporal and spatial resolution, inaccurate temperature
		Two–wavelength pyrometer	Filling interval, filling strategy, thickness of powder layer	Sensitive to parameter changes	-
		Based on longitudinal temperature distribution	Internal void defects	Sensitive to heat dissipation conditions, defect size information can be obtained.	Evaporation, no real–time control, insufficient accuracy and sensitivity of defect detection

Table 3. *Cont.*

Monitoring Process	Monitoring Method	Monitored Defects	Advantages	Disadvantages
Building process monitoring	Wide–field in situ infrared imaging	Pore defects	Error calibration, prediction of the microstructure of formed parts	-
Effectively reflecting the internal formation process. Contributing to real–time defect repair and organization control.				

Table 4. Presents melt layer detection.

<table>
<tr><th colspan="2">Monitoring Process</th><th>Monitoring Method</th><th>Monitored Defects</th><th>Advantages</th><th>Disadvantages</th></tr>
<tr><td rowspan="10">Melt layer inspection</td><td>Temperature detection</td><td>Infrared or near infrared</td><td>Pore defects, Unmelted material within the cambium, non–uniformity of temperature distribution</td><td>-</td><td>Lack of real–time control</td></tr>
<tr><td rowspan="9">Surface topography detection</td><td>Visible light imaging</td><td>Inhomogeneous powder bed, internally not fused</td><td>Contour extraction, defect recognition</td><td>Analysis and processing of grayscale images, offline processing</td></tr>
<tr><td>Low–coherence interferometry</td><td>Melt layer surface rough, suspended structure rough</td><td rowspan="2">Less analysis data</td><td rowspan="2">Long time, complex system, point–by–point scan</td></tr>
<tr><td>3D mapping technology</td><td>Uniformity, thickness, layer defects</td></tr>
<tr><td>Electro–optical inspection</td><td>Pore defects, surface defects</td><td>Suitable for EBM, online monitoring, feedback control</td><td>Optical image research, roughness cannot be monitored</td></tr>
<tr><td>Ultrasonic testing</td><td>Porosity</td><td>Correlated online and offline data</td><td>This method cannot be used for complex geometry.</td></tr>
<tr><td>Acoustic emission spectroscopy</td><td>Spheroidization, slight spheroidization, slight overheating, overheating</td><td>-</td><td>-</td></tr>
<tr><td>Spatially resolved acoustic spectroscopy</td><td>Porosity</td><td>-</td><td>PBF process signals are complex and must be integrated with ML.</td></tr>
<tr><td>Acoustic emission spectroscopy and convolutional neural network</td><td>Porosity</td><td>Fast, efficient, positioning defects</td><td>-</td></tr>
<tr><td colspan="4">Most intuitively reflecting the quality of melt layer.</td></tr>
</table>

Additionally, the authors believe that the online monitoring of the PBF process has the following development trends:

(1) Gradually from monitoring surface state to monitoring internal defects and grain morphology. The online monitoring technology of early powder bed melting intends to monitor the macroscopic morphology of parts by measuring the radiation intensity of the melting pool. With the development of online monitoring technology, the online monitoring of defects in parts has become a popular topic and involves the monitoring of

grain structure, which will lay the foundation for the realization of real–time defect repair and organization control.

(2) More automated and intelligent. Most of the above online monitoring studies use offline data processing methods, and the analysis of monitoring results depends on empirical data. With the deepening of research, through the introduction of computer vision, artificial intelligence, data mining, and other technical means, defects and pores can be monitored more accurately and efficiently to further promote the development and application of online monitoring technology.

(3) Multi–information fusion monitoring. Use multiple monitoring methods to monitor different stages of the process. Comprehensively judge the stability and defect information of the forming process according to different monitoring data. This monitoring technique not only makes up for the deficiencies of a single measurement method but also avoids the uncertainty of a single–signal indicator. By realizing the comprehensive processing and judgment of multi–sensing signals and multi–physical information of the process, the exactitude and reliability of the monitoring system are improved.

(4) Active online monitoring. The conventional online monitoring of PBF processes, especially the monitoring technology based on optical imaging, is subject to many restrictions, namely lighting, metal evaporation, and temperature. Active monitoring technologies (such as low–coherence interference and electro–optical imaging) actively emit measurement beams or electron beams, which not only reduce the requirements on the working environment but also improve the sensitivity of the measurement system, the ability to counteract interference, and the adaptability to the working environment. Active monitoring is a more promising direction in the future online monitoring of PBF.

(5) Real–time and interactive data. A large amount of data will be generated in the abovementioned monitoring method based on machine algorithms. Due to the closed–loop control system, feedback from the online control process is usually not realized. Therefore, some monitoring methods based on machine algorithms have poor credibility. Improve data interaction by establishing a unified data communication protocol. The online monitoring technology based on machine algorithms obtains real–time feedback during the monitoring process, which improves the accuracy and efficiency of the monitoring system.

(6) Correction of observed deficiencies. These monitoring methods monitored deficiencies at different stages of the process, but they did not correct the deficiencies observed. According to the defect information provided by the monitoring here and some process parameters, there will be a good trend in the future research on defect correction methods.

Funding: This research was funded by Engineering Research Center of Fujian University for Stomatological Biomaterials (Xiamen Medical College) grant number No. XMMC-KQ202101.

Institutional Review Board Statement: Not applicable.

Informed Consent Statement: Not applicable.

Data Availability Statement: The raw data supporting the conclusions of this article will be made available by the authors, without undue reservation.

Conflicts of Interest: The authors declare no conflict of interest.

References

1. Wohlers, T.; Caffrey, T. *Additive Manufacturing and 3D Printing State of the Industry: Annual Worldwide Progress Report*; Wohlers Associates Inc.: Fort Collins, CO, USA, 2011.
2. Mueller, B. Additive manufacturing technologies–Rapid prototyping to direct digital manufacturing. *Assem. Autom.* **2012**, 32. [CrossRef]
3. Patterson, A.E.; Messimer, S.L.; Farrington, P.A. Overhanging features and the SLM/DMLS residual stresses problem: Review and future research need. *Technologies* **2017**, *5*, 15. [CrossRef]
4. Zhangong, Y.; Feng, L.; Yongnian, Y. Overview of Direct Metal Rapid Prototyping and Manufacturing Technologies. *Chin. J. Mech. Eng.* **2005**, *41*, 1–7.
5. Grasso, M.; Colosimo, B.M. Process defects and in situ monitoring methods in metal powder bed fusion: A review. *Meas. Sci. Technol.* **2017**, *28*, 044005. [CrossRef]

6. Foster, B.; Reutzel, E.; Nassar, A.; Hall, B.; Brown, S.; Dickman, C. Optical, layerwise monitoring of powder bed fusion. In *2014 Solid Freeform Fabrication Symposium*; University of Texas at Austin: Austin, TX, USA, 2015; pp. 10–12.
7. Grasso, M.; Laguzza, V.; Semeraro, Q.; Colosimo, B.M. In-process monitoring of selective laser melting: Spatial detection of defects via image data analysis. *J. Manuf. Sci. Eng.* **2017**, *139*, 051001. [CrossRef]
8. Zäh, M.F.; Lutzmann, S. Modelling and simulation of electron beam melting. *Prod. Eng.* **2010**, *4*, 15–23. [CrossRef]
9. Kahnert, M.; Lutzmann, S.; Zaeh, M. Layer formations in electron beam sintering. In Proceedings of the Solid Freeform Fabrication Symposium, Austin, TX, USA, 7–9 August 2017; pp. 88–99.
10. Thijs, L.; Verhaeghe, F.; Craeghs, T.; Van Humbeeck, J.; Kruth, J.-P. A study of the microstructural evolution during selective laser melting of Ti–6Al–4V. *Acta Mater.* **2010**, *58*, 3303–3312. [CrossRef]
11. Murr, L.E.; Martinez, E.; Gaytan, S.; Ramirez, D.; Machado, B.; Shindo, P.; Martinez, J.; Medina, F.; Wooten, J.; Ciscel, D. Microstructural architecture, microstructures, and mechanical properties for a nickel-base superalloy fabricated by electron beam melting. *Metall. Mater. Trans. A* **2011**, *42*, 3491–3508. [CrossRef]
12. Cunningham, R.; Narra, S.P.; Ozturk, T.; Beuth, J.; Rollett, A. Evaluating the effect of processing parameters on porosity in electron beam melted Ti-6Al-4V via synchrotron X-ray microtomography. *Jom* **2016**, *68*, 765–771. [CrossRef]
13. De Pond, P.J.; Guss, G.; Ly, S.; Calta, N.P.; Deane, D.; Khairallah, S.; Matthews, M.J. In situ measurements of layer roughness during laser powder bed fusion additive manufacturing using low coherence scanning interferometry. *Mater. Des.* **2018**, *154*, 347–359. [CrossRef]
14. Everton, S.K.; Dickens, P.; Tuck, C.; Dutton, B. Identification of Sub-Surface Defects in Parts Produced by Additive Manufacturing, Using Laser Generated Ultrasound. *Mater. Sci. Technol.* **2016**, *1*, 141–148.
15. Casati, R.; Lemke, J.; Vedani, M. Microstructure and fracture behavior of 316L austenitic stainless steel produced by selective laser melting. *J. Mater. Sci. Technol.* **2016**, *32*, 738–744. [CrossRef]
16. Puebla, K.; Murr, L.E.; Gaytan, S.M.; Martinez, E.; Medina, F.; Wicker, R.B. Effect of melt scan rate on microstructure and macrostructure for electron beam melting of Ti-6Al-4V. *Mater. Sci. Appl.* **2012**, *3*, 259–264. [CrossRef]
17. McCann, R.; Obeidi, M.A.; Hughes, C.; McCarthy, É.; Egan, D.S.; Vijayaraghavan, R.K.; Brabazon, D. In-situ sensing, process monitoring and machine control in Laser Powder Bed Fusion: A review. *Addit. Manuf.* **2021**, *45*, 102058. [CrossRef]
18. Mani, M.; Lane, B.M.; Donmez, M.A.; Feng, S.C.; Moylan, S.P. A review on measurement science needs for real-time control of additive manufacturing metal powder bed fusion processes. *Int. J. Prod. Res.* **2017**, *55*, 1400–1418. [CrossRef]
19. Berumen, S.; Bechmann, F.; Craeghs, T. Quality control system for the coating process in laser-and powder bed-based additive manufacturing technologies. In Proceedings of the Direct Digital Manufacturing Conference, Berlin, Germany, 18–19 March 2020.
20. Reinarz, B.; Witt, G. Process monitoring in the laser beam melting process-Reduction of process breakdowns and defective parts. *Proc. Mater. Sci. Technol.* **2012**, *2012*, 9–15.
21. Kleszczynski, S.; zur Jacobsmühlen, J.; Reinarz, B.; Sehrt, J.T.; Witt, G.; Merhof, D. Improving process stability of laser beam melting systems. In Proceedings of the Fraunhofer Direct Digital Manufacturing Conference, Berlin, Germany, 18–19 March 2020.
22. Liu, Y.; Blunt, L.; Zhang, Z.; Rahman, H.A.; Gao, F.; Jiang, X. In-situ areal inspection of powder bed for electron beam fusion system based on fringe projection profilometry. *Addit. Manuf.* **2020**, *31*, 100940. [CrossRef]
23. Seita, M. A high-resolution and large field-of-view scanner for in-line characterization of powder bed defects during additive manufacturing. *Mater. Des.* **2019**, *164*, 107562.
24. Craeghs, T.; Clijsters, S.; Yasa, E.; Kruth, J.-P. Online quality control of selective laser melting. In Proceedings of the 20th Solid Freeform Fabrication (SFF) Symposium, Austin, TX, USA, 8–10 August 2016; pp. 212–226.
25. zur Jacobsmühlen, J.; Kleszczynski, S.; Schneider, D.; Witt, G. High resolution imaging for inspection of laser beam melting systems. In Proceedings of the 2013 IEEE International Instrumentation and Measurement Technology Conference (I2MTC), Minneapolis, MN, USA, 6–9 May 2013; pp. 707–712.
26. zur Jacobsmühlen, J.; Kleszczynski, S.; Witt, G.; Merhof, D. Elevated region area measurement for quantitative analysis of laser beam melting process stability. In Proceedings of the 26th International Solid Freeform Fabrication Symposium, Austin, TX, USA, 10–12 August 2015; pp. 549–559.
27. Abdelrahman, M.; Reutzel, E.W.; Nassar, A.R.; Starr, T.L. Flaw detection in powder bed fusion using optical imaging. *Addit. Manuf.* **2017**, *15*, 1–11. [CrossRef]
28. Neef, A.; Seyda, V.; Herzog, D.; Emmelmann, C.; Schönleber, M.; Kogel-Hollacher, M. Low coherence interferometry in selective laser melting. *Phys. Procedia* **2014**, *56*, 82–89. [CrossRef]
29. Fleming, T.G.; Nestor, S.G.; Allen, T.R.; Boukhaled, M.A.; Smith, N.J.; Fraser, J.M. Tracking and Controlling the Morphology Evolution of 3D Powder-Bed Fusion in situ using Inline Coherent Imaging. *Addit. Manuf.* **2020**, *32*, 100978. [CrossRef]
30. Boschetto, A.; Bottini, L.; Vatanparast, S.; Veniali, F. Part defects identification in selective laser melting via digital image processing of powder bed anomalies. *Prod. Eng.* **2022**, *16*, 691–704. [CrossRef]
31. Li, Z.; Liu, X.; Wen, S.; He, P.; Zhong, K.; Wei, Q.; Shi, Y.; Liu, S. In situ 3D monitoring of geometric signatures in the powder-bed-fusion additive manufacturing process via vision sensing methods. *Sensors* **2018**, *18*, 1180. [CrossRef] [PubMed]
32. Grasso, M. In Situ Monitoring of Powder Bed Fusion Homogeneity in Electron Beam Melting. *Materials* **2021**, *14*, 7015. [CrossRef]
33. Jahan, S.A.; Al Hasan, M.; El-Mounayri, H. A framework for graph-base neural network using numerical simulation of metal powder bed fusion for correlating process parameters and defect generation. *Manuf. Lett.* **2022**, *33*, 765–775. [CrossRef]

34. Xiao, L.; Lu, M.; Huang, H. Detection of powder bed defects in selective laser sintering using convolutional neural network. *Int. J. Adv. Manuf. Technol.* **2020**, *107*, 2485–2496. [CrossRef]
35. Scime, L.; Beuth, J. A multi-scale convolutional neural network for autonomous anomaly detection and classification in a laser powder bed fusion additive manufacturing process. *Addit. Manuf.* **2018**, *24*, 273–286. [CrossRef]
36. Scime, L.; Beuth, J. Anomaly detection and classification in a laser powder bed additive manufacturing process using a trained computer vision algorithm. *Addit. Manuf.* **2018**, *19*, 114–126. [CrossRef]
37. Berumen, S.; Bechmann, F.; Lindner, S.; Kruth, J.-P.; Craeghs, T. Quality control of laser-and powder bed-based Additive Manufacturing (AM) technologies. *Phys. Procedia* **2010**, *5*, 617–622. [CrossRef]
38. Kruth, J.-P.; Mercelis, P.; Van Vaerenbergh, J.; Craeghs, T. Feedback control of selective laser melting. In Proceedings of the 3rd International Conference on Advanced Research in Virtual and Rapid Prototyping, Leiria, Portugal, 24–29 September 2007; CRC Press: Boca Raton, FL, USA, 2007; pp. 521–527.
39. Clijsters, S.; Craeghs, T.; Buls, S.; Kempen, K.; Kruth, J.-P. In situ quality control of the selective laser melting process using a high-speed, real-time melt pool monitoring system. *Int. J. Adv. Manuf. Technol.* **2014**, *75*, 1089–1101. [CrossRef]
40. Craeghs, T.; Bechmann, F.; Berumen, S.; Kruth, J.-P. Feedback control of Layerwise Laser Melting using optical sensors. *Phys. Procedia* **2010**, *5*, 505–514. [CrossRef]
41. Kanko, J.A.; Sibley, A.P.; Fraser, J.M. In situ morphology-based defect detection of selective laser melting through inline coherent imaging. *J. Mater. Process. Technol.* **2016**, *231*, 488–500. [CrossRef]
42. Krauss, H.; Eschey, C.; Zaeh, M. Thermography for monitoring the selective laser melting process. In *2012 Solid Freeform Fabrication Symposium*; University of Texas at Austin: Austin, TX, USA, 2012; pp. 999–1014.
43. Le, T.N.; Lee, M.H.; Lin, Z.H.; Tran, H.C.; Lo, Y.L. Vision-based in-situ monitoring system for melt-pool detection in laser powder bed fusion process. *J. Manuf. Process.* **2021**, *68*, 1735–1745. [CrossRef]
44. Price, S.; Cooper, K.; Chou, K. Evaluations of temperature measurements by near-infrared thermography in powder-based electron-beam additive manufacturing. In *2012 Solid Freeform Fabrication Symposium*; University of Texas at Austin: Austin, TX, USA, 2012; pp. 761–773.
45. Pavlov, M.; Doubenskaia, M.; Smurov, I. Pyrometric analysis of thermal processes in SLM technology. *Phys. Procedia* **2010**, *5*, 523–531. [CrossRef]
46. Cheng, B.; Price, S.; Lydon, J.; Cooper, K.; Chou, K. On process temperature in powder-bed electron beam additive manufacturing: Model development and validation. *J. Manuf. Sci. Eng.* **2014**, *136*, 061018. [CrossRef]
47. Price, S.; Cheng, B.; Lydon, J.; Cooper, K.; Chou, K. On process temperature in powder-bed electron beam additive manufacturing: Process parameter effects. *J. Manuf. Sci. Eng.* **2014**, *136*, 061019. [CrossRef]
48. Price, S.; Lydon, J.; Cooper, K.; Chou, K. Experimental temperature analysis of powder-based electron beam additive manufacturing. In Proceedings of the 24th Annual International Solid Freeform Fabrication Symposium, Austin, TX, USA, 12–14 August 2013; pp. 162–173.
49. Gong, X.; Cheng, B.; Price, S.; Chou, K. Powder-bed electron-beam-melting additive manufacturing: Powder characterization, process simulation and metrology. In Proceedings of the Early Career Technical Conference, Birmingham, AL, USA, 2–3 November 2013; pp. 55–66.
50. Krauss, H.; Zeugner, T.; Zaeh, M.F. Thermographic process monitoring in powder bed based additive manufacturing. In *AIP Conference Proceedings*; American Institute of Physics: College Park, MD, USA, 2015; pp. 177–183.
51. Krauss, H.; Zeugner, T.; Zaeh, M.F. Layerwise monitoring of the selective laser melting process by thermography. *Phys. Procedia* **2014**, *56*, 64–71. [CrossRef]
52. Raplee, J.; Plotkowski, A.; Kirka, M.M.; Dinwiddie, R.; Okello, A.; Dehoff, R.R.; Babu, S.S. Thermographic microstructure monitoring in electron beam additive manufacturing. *Sci. Rep.* **2017**, *7*, 43554. [CrossRef]
53. Williams, R.J.; Piglione, A.; Rønneberg, T.; Jones, C.; Pham, M.-S.; Davies, C.M.; Hooper, P.A. In situ thermography for laser powder bed fusion: Effects of layer temperature on porosity, microstructure and mechanical properties. *Addit. Manuf.* **2019**, *30*, 100880. [CrossRef]
54. Dinwiddie, R.B.; Dehoff, R.R.; Lloyd, P.D.; Lowe, L.E.; Ulrich, J.B. Thermographic in-situ process monitoring of the electron-beam melting technology used in additive manufacturing. In *Thermosense: Thermal Infrared Applications XXXV*; SPIE: Baltimore, MD, USA, 2013; p. 87050K.
55. Price, S.; Lydon, J.; Cooper, K.; Chou, K. Temperature measurements in powder-bed electron beam additive manufacturing. In Proceedings of the ASME 2014 International Mechanical Engineering Congress and Exposition, Montreal, QC, Canada, 14–20 November 2014.
56. Rodriguez, E.; Medina, F.; Espalin, D.; Terrazas, C.; Muse, D.; Henry, C.; MacDonald, E.; Wicker, R.B. Integration of a thermal imaging feedback control system in electron beam melting. In *WM Keck Center for 3D Innovation*; University of Texas at El Paso: Austin, TX, USA, 2012; pp. 945–961.
57. Rodriguez, E.; Mireles, J.; Terrazas, C.A.; Espalin, D.; Perez, M.A.; Wicker, R.B. Approximation of absolute surface temperature measurements of powder bed fusion additive manufacturing technology using in situ infrared thermography. *Addit. Manuf.* **2015**, *5*, 31–39. [CrossRef]
58. Schwerdtfeger, J.; Singer, R.F.; Körner, C. In situ flaw detection by IR-imaging during electron beam melting. *Rapid Prototyp. J.* **2012**, *18*, 259–263. [CrossRef]

59. Ridwan, S.; Mireles, J.; Gaytan, S.; Espalin, D.; Wicker, R. Automatic layerwise acquisition of thermal and geometric data of the electron beam melting process using infrared thermography. In *2014 International Solid Freeform Fabrication Symposium*; University of Texas at Austin: Austin, TX, USA, 2014; pp. 343–352.
60. Mireles, J.; Ridwan, S.; Morton, P.A.; Hinojos, A.; Wicker, R.B. Analysis and correction of defects within parts fabricated using powder bed fusion technology. *Surf. Topogr. Metrol. Prop.* **2015**, *3*, 034002. [CrossRef]
61. Erler, M.; Streek, A.; Schulze, C.; Exner, H. Novel machine and measurement concept for micro machining by selective laser sintering. In *2014 International Solid Freeform Fabrication Symposium*; University of Texas at Austin: Austin, TX, USA, 2014; pp. 4–6.
62. Imani, F.; Gaikwad, A.; Montazeri, M.; Rao, P.; Yang, H.; Reutzel, E. Layerwise in-process quality monitoring in laser powder bed fusion. In Proceedings of the ASME 2018 13th International Manufacturing Science and Engineering Conference, College Station, TX, USA, 18–22 June 2018.
63. Watt, I.M. *The Principles and Practice of Electron Microscopy*; Cambridge University Press: Cambridge, MA, USA; New York, NY, USA; Port Chester, NY, USA; Melbourne, Australia; Sydney, Australia, 1997.
64. Arnold, C.; Pobel, C.; Osmanlic, F.; Körner, C. Layerwise monitoring of electron beam melting via backscatter electron detection. *Rapid Prototyp. J.* **2018**, *24*, 1401–1406. [CrossRef]
65. Wong, H.; Neary, D.; Jones, E.; Fox, P.; Sutcliffe, C. Pilot capability evaluation of a feedback electronic imaging system prototype for in-process monitoring in electron beam additive manufacturing. *Int. J. Adv. Manuf. Technol.* **2019**, *100*, 707–720. [CrossRef]
66. Wong, H.; Neary, D.; Jones, E.; Fox, P.; Sutcliffe, C. Benchmarking spatial resolution in electronic imaging for potential in-situ Electron Beam Melting monitoring. *Addit. Manuf.* **2019**, *29*, 100829. [CrossRef]
67. Rieder, H.; Dillhöfer, A.; Spies, M.; Bamberg, J.; Hess, T. Online monitoring of additive manufacturing processes using ultrasound. In Proceedings of the 11th European Conference on Non-Destructive Testing, Prague, Czech Republic, 6–10 October 2014; pp. 6–10.
68. Ye, D.; Hong, G.S.; Zhang, Y.; Zhu, K.; Fuh, J.Y.H. Defect detection in selective laser melting technology by acoustic signals with deep belief networks. *Int. J. Adv. Manuf. Technol.* **2018**, *96*, 2791–2801. [CrossRef]
69. Smith, R.J.; Hirsch, M.; Patel, R.; Li, W.; Clare, A.T.; Sharples, S.D. Spatially resolved acoustic spectroscopy for selective laser melting. *J. Mater. Process. Technol.* **2016**, *236*, 93–102. [CrossRef]
70. Shevchik, S.A.; Kenel, C.; Leinenbach, C.; Wasmer, K. Acoustic emission for in situ quality monitoring in additive manufacturing using spectral convolutional neural networks. *Addit. Manuf.* **2018**, *21*, 598–604. [CrossRef]

Article

Selective Laser Melting and Mechanical Properties of Stainless Steels

Daniel Gatões *, Ricardo Alves, Bernardo Alves and Maria Teresa Vieira

CEMMPRE—Centre for Mechanical Engineering Materials and Processes, Department of Mechanical Engineering, University of Coimbra, Rua Luís Reis Santos, 3030-788 Coimbra, Portugal
* Correspondence: daniel.gatoes@uc.pt; Tel.: +351-239790765

Abstract: Metal additive manufacturing (AM) has been evolving in response to industrial and social challenges. However, new materials are hindered in these technologies due to the complexity of direct additive manufacturing technologies, particularly selective laser melting (SLM). Stainless steel (SS) 316L, due to its very low carbon content, has been used as a standard powder in SLM, highlighting the role of alloying elements present in steels. However, reliable research on the chemical impact of carbon content in steel alloys has been rarely conducted, despite being the most prevalent element in steel. Considering the temperatures involved in the SLM process, the laser–powder interaction can lead to a significant carbon decrease, whatever the processing atmosphere. In the present study, four stainless steels with increasing carbon content—AISI 316L, 630 (17-4PH), 420 and 440C—were processed under the same SLM parameters. In addition to roughness and surface topography, the relationship with the microstructure (including grain size and orientation), defects and mechanical properties (hardness and tensile strength) were established, highlighting the role of carbon. It was shown that the production by SLM of stainless steels with similar packing densities and different carbon contents does not oblige the changing of processing parameters. Moreover, alterations in material response in stainless steels produced under the same volumetric energy density mainly result from microstructural evolution during the process.

Keywords: additive manufacturing; selective laser melting; stainless steel; AISI 316L; AISI 630 (17-4PH); AISI 420; AISI 440C; carbon content

Citation: Gatões, D.; Alves, R.; Alves, B.; Vieira, M.T. Selective Laser Melting and Mechanical Properties of Stainless Steels. *Materials* **2022**, *15*, 7575. https://doi.org/10.3390/ma15217575

Academic Editors: Wai Yee Yeong and Swee Leong Sing

Received: 17 September 2022
Accepted: 17 October 2022
Published: 28 October 2022

1. Introduction

Metal additive manufacturing (MAM), and particularly selective laser melting (SLM), is an effective way of producing steel metal parts [1]. In SLM, a laser is used to melt and fuse metallic powder particles, layer by layer, to build the desired three-dimensional shape (3D object) [2]. SLM was shown to be able to produce low-carbon stainless steel with good quality and reliability [3]. However, some properties in SLM are primarily dependent on the processing step, since a large number of variables affect the quality of final parts, such as powder characteristics, atmosphere, the response of the material to the volumetric energy density (VED) and the rapid heating and cooling cycles [4,5]. As a result, the complex physical and chemical behavior in the melting pool leads to an anisotropic microstructure and the appearance of voids, affecting the final properties [6,7]. The carbon content must be carefully controlled during the manufacturing process, since its presence and quantity can significantly impact the properties of the resulting steel. While the carbon content of SLM-produced stainless steels can be controlled to some degree and generally leads to better mechanical properties, typically associated with less corrosion resistance [8], these steels have a tendency to form cracks and pores during the SLM process [9,10].

Besides porosity, which has a significant impact on mechanical properties, interstitial defects (chemical composition of powder, processing atmosphere, etc.) during SLM and microstructures can also contribute to substantial variations in the properties of the stainless steels. Thus, it is essential to carefully consider the properties of an SLM-produced alloy before selecting it for application [11–13].

Stainless steel 316L has been the subject of many studies since the dawn of SLM [14]. The ease of production associated with the low carbon content, no phase transformation (austenitic matrix) in the SLM process, and good mechanical properties make it a standard material in SLM [15,16], and it was used as a standard stainless steel in this study. SS 630 studies have been growing in the last few years due to its excellent mechanical properties [17]. Note that 420 martensitic stainless steel has also been of significant interest in SLM, since its properties are suitable for specific applications, due to its high strength and corrosion resistance [18]. Consequently, many studies have been published concerning the microstructure, mechanical properties, and roughness of these stainless steels, produced by SLM ([19–21] −316L, [22–24] −630, [25–27] −420). SS 440C is also a martensitic stainless steel, but has high carbon content; it is used in applications where high hardness and corrosion resistance are necessary, such as bearings, knives and automotive parts [28]. As far as the authors are aware, this steel has not been subject to any studies regarding SLM technology. Moreover, a detailed study comparing various stainless steels produced with the same parameters in SLM is not available.

In addition to 316L, the standard, in the present study, three stainless steels with similar chromium contents and various levels of carbon content were processed through SLM, using the same atmosphere and set of SLM parameters, as a way to highlight the role of carbon content in AM mechanical properties. A detailed study of the microstructures of these steels compared with the powder was performed. It is worth noting that powder production was also attained in the same atmosphere. The occurrence of defects and their consequences on the mechanical properties can be highlighted by microcomputed tomography (μCT). This non-destructive technique can be useful to study 3D objects pores, voids and impurities distribution, whatever the material [29–31].

2. Materials and Methods

Four stainless steel powders with increasing carbon content were selected for this study, attained by gas atomization. The 316L and 630 (17-4PH) stainless steels powder were from SLM Solutions GmbH (SLM Solutions Group AG, Lübeck, Germany), and 420 and 440C powders were from Sandvik Osprey Ltd. (Sandvik AB, Sandviken, Sweden). The powders' chemical compositions are summarized in Table 1. Particle size and particle size distribution (PSD) were evaluated by laser diffraction spectrometry LDS, Malvern Mastersizer 3000 (Malvern Panalytical, Egham, UK). SEM Quanta 400 FEG STEM (FEI Company, Hillsboro, OR, USA) was used for powder shape-factor evaluation. Powder density was measured (5 measurements per steel) by helium pycnometry with Accupyc 1330 (Micrometrics, Norcross, GA, USA).

Table 1. Chemical compositions of the different stainless steels (wt.%).

Element	C	Cr	Ni	Cu	Mo	Nb + Ta	Si	Mn	N	P	S	O
316L	0.030	16–18	10–14	-	2–3	-	1	2	0.100	0.045	0.030	0.100
630	0.070	15–17	3–5	3–5	-	0.150–0.450	1	1	0.100	-	0.030	0.100
420	min 0.150	12–14	-	-	-	-	1	1	-	0.040	0.030	-
440C	0.950–1.200	16–18	-	-	0.750	-	1	1	-	0.040	0.030	0.100

X-ray diffraction to evaluate types of phases involved a Philips X'Pert diffractometer (Philips, Egham, UK) at 40 kV, Bragg–Brentano geometry (θ–2θ), cobalt anticathode ($\lambda(k\alpha1)$ = 0.178897 nm and $\lambda(k\alpha2)$ = 0.179285 nm) and a current intensity of 35 mA. The X-ray diffraction scans were carried out from 40 to 100° in steps of 0.025°, with an acquisition time of 1 s per step.

The SLM equipment was an EOS M290 system (EOS GmbH, Krailling, Germany) equipped with a Yb-fiber laser (λ = 1064 nm) with a maximum power of 400 W and a spot size of 100 µm. The SLM processing was undertaken with an oxygen content below 0.1% in the working chamber using a continuous flow of nitrogen. The laser power was set to 260 W, scanning speed was 1060 mm/s, hatch space was 100 µm and layer thickness was 30 µm (VED = 82 $\frac{J}{mm^3}$). The scanning strategy was a zigzag pattern with a rotation angle of 67° between adjacent layers. Each batch included density cubes (10 × 10 × 10 mm^3) and tensile test specimens. Additionally, a 10° rotation relative to the substrate position (Figure 1) was added to avoid contamination by spattering. All 3D objects were studied as SLMed, without post-processing treatment.

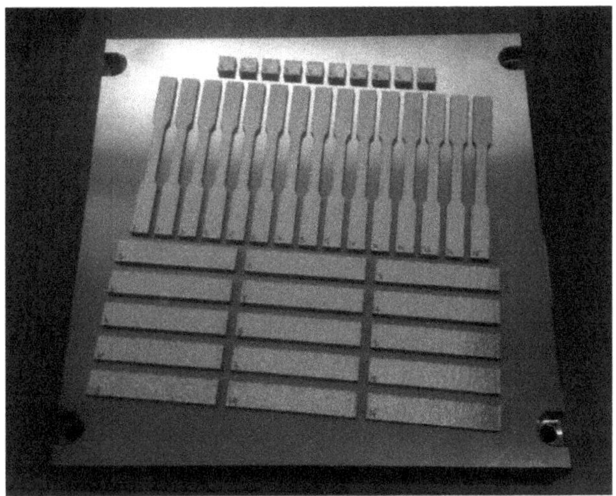

Figure 1. Positions of the specimens on the SLM bed.

The final density was evaluated through the Archimedes method and averaged for 10 specimens (density cubes).

Surface and inside defects on tensile specimens were evaluated by X-ray micro-computed tomography using a Bruker SkyScan 1275 (Bruker, Kontich, Belgium). Specimens were polished on both the top and bottom surfaces of tensile test specimens until a thickness of 2 mm was achieved. An acceleration voltage of 100 kV and a beam current of 100 µA were set using a 1 mm copper filter with step-and-shoot mode. Pixel size was set to 10 µm, and the random mode was used. The images were acquired at a 0.4° angular step with 10 frames on average per step using an exposure time of 245 ms. The µCT images were reconstructed with the dedicated manufacturer software.

Optical microscopy was done using a Leica DM 4000 M LED (Leica Microsystems AG, Wetzlar, Germany) with a Leica camera, model MC 120 HD.

For 316L, 630 and 420 steels, etching used a Vilella solution (2 g of picric acid, 5 mL of HCl and 100 mL of ethanol). For 440C, Kalling solution etching (5 g of $CuCl_2$, 100 mL of HCl and 100 mL of ethanol) was selected.

Surface roughness was evaluated using focus variation microscopy Alicona Infinite Focus (Bruker, Kontich, Belgium) following ISO 4287 and 4288.

Microhardness measurements were performed on a Fisherscope H100 (Fischer Instrumentation LTD, Pershore, UK), equipped with a Vickers indentor (10 measurements per sample, maximum load of 1000 mN, holding time of 30 s).

Tensile tests were performed on a SHIMADZU Autograph (Shimadzu, Kyoto, Japan), with a 100 kN load cell, according to ISO 6892, at room temperature, at a strain rate of 10 MPa per second. Tensile specimens' dimensions were in accordance with Figure 2.

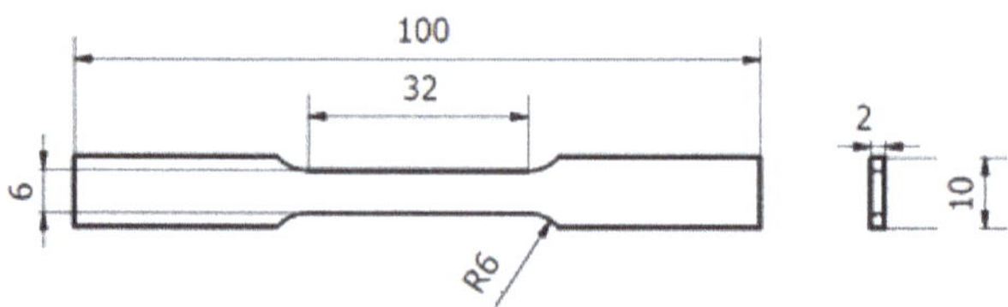

Figure 2. Tensile test specimen measurements, in mm (ISO 6892).

3. Results

3.1. Powder

All four studied stainless steel powders showed symmetric and narrow normal particle size distributions. Table 2 summarizes the d_{10}, d_{50} and d_{90} powder particle size and density.

Table 2. Powder size distribution and density of each stainless steel powder.

Powder	d_{10} (µm)	d_{50} (µm)	d_{90} (µm)	$\rho\left(\frac{Kg}{m^3}\right)$
316L	22.7	32.4	45.2	7880
630	17.8	26.2	37.6	7880
420	17.0	24.3	34.3	7820
440C	18.1	26.4	37.9	7940

Powder shape was almost spherical for all powders; there were some satellites, particularly in 316L and 420 steel powders (Figure 3).

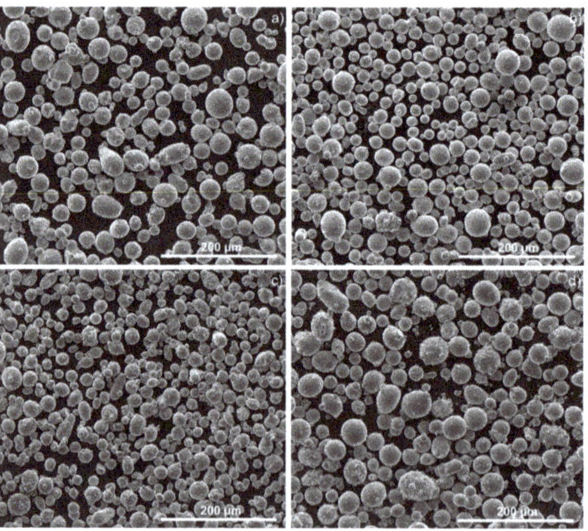

Figure 3. Particle shapes of the 316L (**a**), 630 (**b**), 420 (**c**) and 440C (**d**) powders (SEM).

The selected stainless steels presented martensitic and austenitic phases (Figure 4).

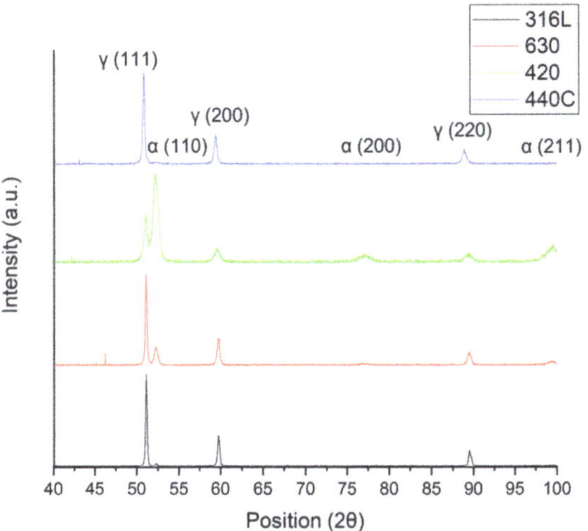

Figure 4. X-ray diffractograms of the 316L, 630, 420 and 440C powders (from bottom to top, respectively).

3.2. 3D Object

3.2.1. Porosity Evaluation

Porosity is directly related to density. However, the measured values for the four steels are not only related to the porosity, but mainly to the phase difference between the 3D object and starting powder.

The reduced section (cross-section) of the tensile specimen can be observed in Figure 5. Figure 6 shows a rendering of the pore distribution within the reduced section of the tensile specimen. It is noteworthy that only defects bigger than the pixel size (10 μm) were observable in this case.

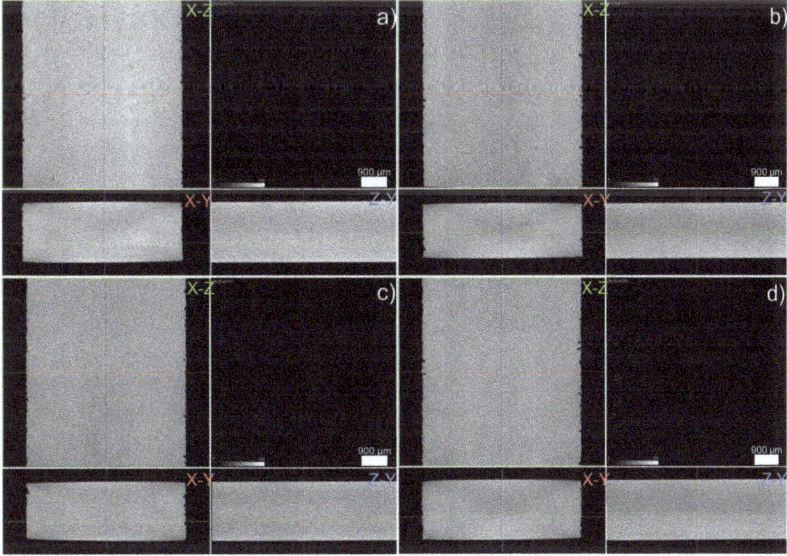

Figure 5. 2D visualization of the pores location in cross-section for 316L (**a**), 630 (**b**), 420 (**c**) and 440C (**d**).

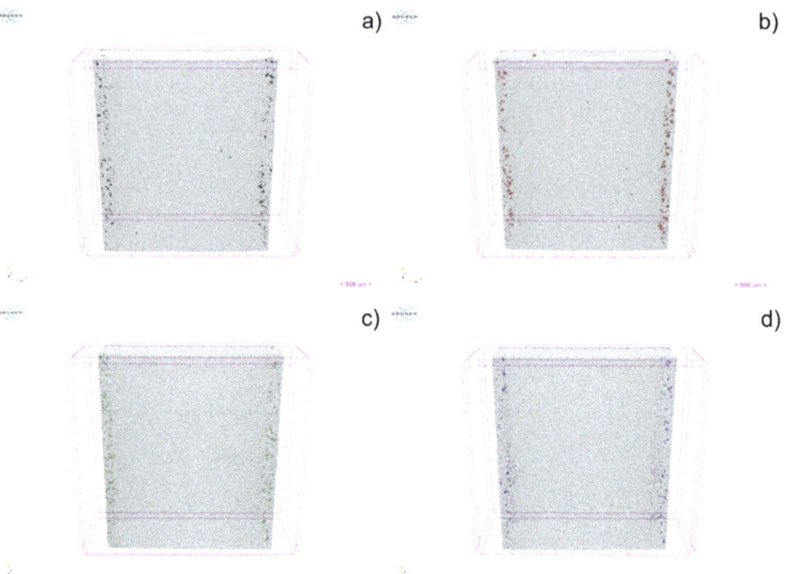

Figure 6. 3D visualization of the pores location within the 3D object volume for 316L (**a**), 630 (**b**), 420 (**c**) and 440C (**d**).

Table 3 summarizes the values of density of the 3D object and the relative porosity, compared to powder density.

Table 3. Density comparison between powder and 3D object.

Type	Powder Density $\left(\frac{Kg}{m^3}\right)$	3D Object Density $\left(\frac{Kg}{m^3}\right)$
316L	7880	7790
630	7880	7660
420	7820	7590
440C	7940	7490

3.2.2. Microstructure

X-ray diffractions of the 3D objects are shown in Figure 7. Stainless steel 316L showed only an austenitic phase but a strong orientation in [220]. The texture corresponds to <011> direction, which has a major influence on mechanical properties [32]. SS 630 showed an increase in martensite, and 420 showed a shift in the austenite phase when compared to the powder.

All observed stainless steels, besides SS 316L, tended to form elongated grains in the scanning direction, which is common in metals processed by SLM (Figure 8). It was possible to distinguish narrower elongated grains in 630, 420 and 440C, which allowed the distinction between the previous layer (rotated 90°) and the current layer.

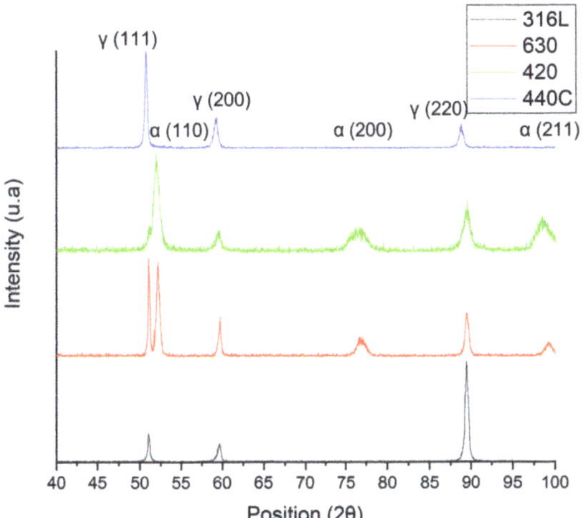

Figure 7. X-ray diffractograms of the 316L, 630, 420 and 440C 3D objects (from bottom to top, respectively).

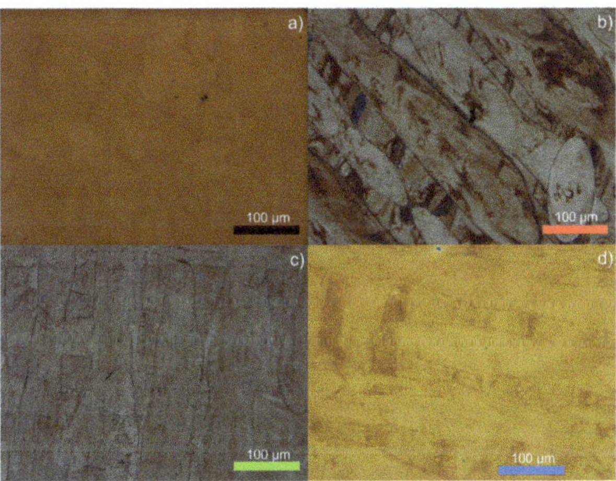

Figure 8. Micrographies of the etched surfaces of 316L (**a**), 630 (**b**), 420 (**c**) and 440C (**d**) stainless steels.

3.2.3. Roughness, Topography and Geometrical Evaluation

Table 4 shows the attained roughness values of each specimen.

In what concerns the surface topography, a group of beads corresponding to the scanning direction could be observed in Figure 9. Moreover, the not-fully melted powder could be observed scattered throughout the surface.

Produced tensile specimens were measured in order to ensure the adequacy of mechanical tests. Table 5 summarizes the size comparison between CAD and 3D objects.

Table 4. Roughness measurements of the tensile test specimens, for each stainless steel.

Type	Ra (µm)	Rq (µm)	Rz (µm)
316L	16.204	19.864	94.401
630	12.988	16.030	79.347
420	7.669	9.440	53.783
440C	6.369	8.093	46.391

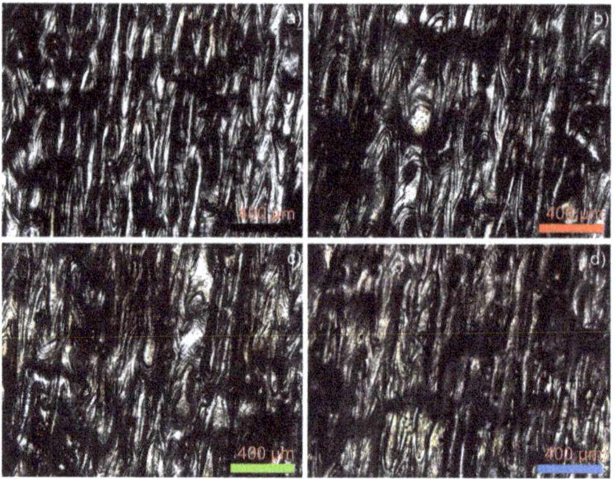

Figure 9. Micrographies of the 316L (**a**), 630 (**b**), 420 (**c**) and 440C (**d**) surfaces.

Table 5. Comparison between CAD and final 3D object sizes of tensile test specimens, for each material type.

Type	Width at Grip (%)	Width at Reduced Section (%)	Length (%)
316L	99.40 ± 0.24	98.50 ± 0.32	99.70 ± 0.03
630	99.60 ± 0.19	99.00 ± 0.27	99.81 ± 0.05
420	99.70 ± 0.25	98.67 ± 0.27	100.00 ± 0.06
440C	99.20 ± 0.03	98.83 ± 0.25	99.58 ± 0.05

3.2.4. Mechanical Properties

Microhardness

Table 6 summarizes the measured hardness ($HV_{0.1}$) for the different stainless steels and compares it to the bulk.

Table 6. Microhardness levels of the selected stainless steel 3D objects, compared to the bulk [33].

Type	3D Object $HV_{0.1}$	Bulk $HV_{0.1}$
316L	133 ± 17	170–220
630	306 ± 11	250–460
420	647 ± 27	260–641
440C	803 ± 26	510–760

Tensile Tests

Figure 10 shows the stress–strain results for five tensile test specimens per stainless steel type.

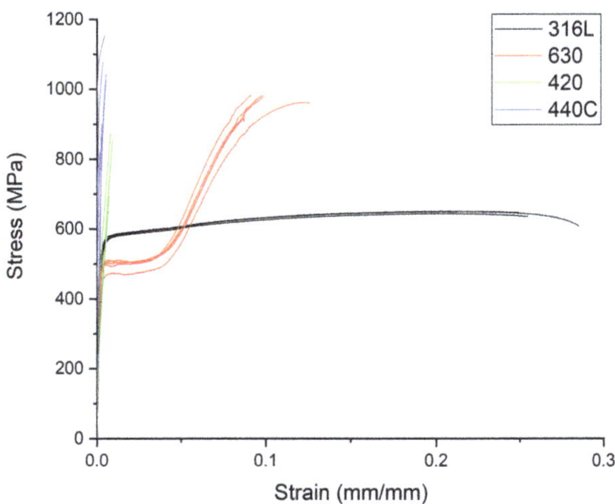

Figure 10. Strain–stress curves for the 316L (black), 630 (red), 420 (green) and 440C (blue) steels.

Table 7 summarizes the elongation and ultimate tensile strength (UTS) of the stainless steel tensile specimens.

Table 7. Elongation and ultimate tensile strength for each steel-type 3D object, compared to bulk [34].

Type	3D Object ϵ (%)	Bulk ϵ (%)	3D Object UTS (MPa)	Bulk UTS (MPa)
316L	25.4 ± 2.4	30–50	645 ± 10	550
630	10.0 ± 1.3	4–6	1020 ± 33	943
420	0.7 ± 0.2	5–11	814 ± 74	655
440C	0.4 ± 0.1	0.5–4	1164 ± 13	760

4. Discussion

4.1. Powder

All selected powder particles had similar particle sizes, particle size distributions and shape factors. However, some satellites were observed for 316L and 420 steel powders. These characteristics did not contribute to a significant discrepancy in the powder flowability, and consequently, packing behavior in the powder bed. The difference in measured powder density was mainly related to the phase composition of the steel and its atomization atmosphere. Thus, the austenite phase is always present, resulting from the atomization atmosphere—nitrogen. This is supported by the X-ray diffraction (Figure 4). The SS 440C powder presented mainly the austenite phase and the highest powder density. In opposition, the 420 powder showed the highest martensite content of all selected powder, and therefore, the lowest density.

4.2. 3D Object

4.2.1. Density and Microstructure

A representative area of defects (Figure 6) shows that the main pores in all stainless steel 3D objects were relatively similar, independently of carbon content. Surface finishing is mostly obligatory in SLM, since close-to-surface defects can be the main driving force for crack initiation, if not removed. Thus, the presence of these defects is not relevant in real-world applications. However, due to µCT resolution, it can be concluded that microporosity (<10 µm) could also have been present. Nevertheless, this is not enough to justify the measured 3D object's density.

Considering that the nitrogen atmosphere was similar for powder and 3D objects, and the fact that the carbon content's effects on the porosity distribution and percentage were negligible, the density must be justified by taking into account the microstructure of the steel (before and after SLM). This means that the density modification can be attributed mainly to the content of the prevalent austenitic phase and the presence of carbide phases, particularly in high-carbon content steels.

SLM processing of the 3D objects, despite 440C not sustaining a phase alteration, affected the phase compositions of 316L, 630 and 420 steels. Particularly, 316L and 420 suffered a shift in the crystallographic orientation, and 630 had an increase in the martensitic phase. This deviation in crystallographic texture resulted from short, deep melt pools, resulting from higher laser power, as shown by [32].

As is common in metals processed by SLM, the microstructure is oriented toward the scanning strategy. This is due to the heating and cooling rates involved in the SLM process, which are the main drivers of microstructure growth. However, high carbon presence in the melt pool was shown to have a particular effect on microstructure due to the carbon movement in the direction of the melt pool frontier, resulting from Marangoni convection [35,36]. Moreover, carbon is a significant element of austenite stabilization. Thus, the carbon depletion of 630 may have been the driver for the increase in martensitic content when compared to the same material powder.

Figure 8 displays apparent grain anisotropy in the 3D objects, whatever the steel. SS 316L's clear etched surface was consistent with the fact that little-to-no martensite was present in the steel, but grain orientation along the scanning direction was still visible. SS 630 presented a microstructure close to that of dual-phase steel, with austenite and martensite distributed almost equally within the steel and strong processed-caused anisotropy. SS 420 had a dominant martensitic phase with residual austenite. Martensite grew alongside austenite in these steels, as a virtue of the complex melt pool temperature dynamics that occur in the SLM production.

4.2.2. Roughness, Geometry and Topography

Melt pool beads could be seen all over the surfaces. However, it is possible to relate the reduced z-growth of the melt pool with increased carbon content, resulting in lower roughness for higher carbon steels. All steels presented fused powders on their surface, which is detrimental to roughness but can usually be mitigated by shot-peening [37].

The geometrical deviation between the steels, after SLM, was low and consistent, which can mean that the selected VED is indicated for all the steels. VED has a significant effect on melt pool width. If too low, a clear frontier within adjacent melt pools should be observed, along with oriented porosity along the scanning direction. If too high, an uncontrolled melt pool with varying lengths is achieved, and growth in XY direction is expected. Consequently, VED must be considered to be within the optimal range for all the studied stainless steels, corroborating the previous statements.

4.2.3. Mechanical Properties

The microhardness values of the SLM 3D objects produced from the powders of 630, 420 and 440C stainless steels were higher than the maximum hardness of the bulk steels processed by conventional approaches, including heat treatment. Nevertheless, SS 316L showed a slight decrease in hardness compared to the bulk due to the texture assumed during SLM processing. In concomitance with hardness, UTS values for all stainless steels studied, processed by SLM, show a tendency to be higher than the values of conventional processing. It must be highlighted that SS 630 had hardness within the range of the values assumed, considering the various heat treatments that can be used. However, the tensile test showed a particular behavior consistent with the strain-induced martensite formation (TRIP) effect observed by [1,23,24], resulting from a high retained austenite content. This effect led to higher ductility than conventionally heat-treated SS 630. SS 420 and 440C have higher hardness, and the UTS values are outstanding.

Nevertheless, SS 420 processed by SLM is a brittle material with negligible elongation due to an unconventional microstructure.

5. Conclusions

Whatever the carbon composition, the SLM parameters generally used for 316L induce in other stainless steels with low nickel and higher carbon content low porosity and high densification. The microstructures resulting from SLM without post-processing treatments are mandatory for assessing the mechanical behavior of 3D objects. The hardness and UTS of SLM 3D objects are higher than those of bulk stainless steels with the same compositions, after heat treatment. This is due to a direct relationship between carbon and elements with high affinity to it (e.g., chromium, iron, and molybdenum); however, the content of other elements of selected stainless steels is insufficient for carbide formation. Moreover, it must be highlighted that the stabilization of residual austenite present in the steels with higher carbon content can result from the processing atmosphere; nitrogen is more effective than other elements in the matrix. Furthermore, the strong anisotropy observed in all stainless steels resulted from the selected scanning strategy and VED values. In the studied stainless steels, the microstructural difference, when compared to conventionally processed bulk materials, is mainly due to the processing atmosphere, so a constant VED can be used to process different stainless steels with varying carbon compositions. Future studies regarding the influence of carbon content in indirect additive manufacturing will be compared to the results reported here.

Author Contributions: Conceptualization, D.G. and M.T.V.; methodology, D.G.; validation, M.T.V.; formal analysis, D.G. and M.T.V.; investigation, D.G., B.A. and R.A.; resources, M.T.V.; data curation, D.G. and B.A.; writing—original draft preparation, D.G. and R.A.; writing—review and editing, D.G., B.A. and M.T.V.; visualization, D.G.; supervision, M.T.V.; project administration, M.T.V.; funding acquisition, M.T.V. All authors have read and agreed to the published version of the manuscript.

Funding: This work was supported by the European Regional Development Fund (ERDF) under the Portuguese program Programa Operacional Factores de Competitividade (COMPETE) (grant agreement number POCI-01-0247-FEDER-039910). This research is sponsored by national funds through FCT—Fundação para a Ciência e a Tecnologia—under the project UIDB/00285/2020 and LA/P/0112/2020.

Informed Consent Statement: Not applicable.

Data Availability Statement: Data sharing not applicable to this article.

Conflicts of Interest: The authors declare no conflict of interest.

References

1. Bajaj, P.; Hariharan, A.; Kini, A.; Kürnsteiner, P.; Raabe, D.; Jägle, E. Steels in additive manufacturing: A review of their microstructure and properties. *Mater. Sci. Eng. A* **2020**, *772*, 138633. [CrossRef]
2. *ISO/ASTM 52900:2021(en)*; Additive Manufacturing—General Principles—Fundamentals and Vocabulary. ISO/ASTM: Geneva, Switzerland, 2021.
3. Sun, H.; Chu, X.; Liu, Z.; Gisele, A.; Zou, Y. Selective Laser Melting of Maraging Steels Using Recycled Powders: A Comprehensive Microstructural and Mechanical Investigation. *Metall. Mater. Trans. A* **2021**, *52*, 1714–1722. [CrossRef]
4. Wang, Y.; Luo, L.; Liu, T.; Wang, B.; Luo, L.; Zhao, J.; Wang, L.; Su, Y.; Guo, J.; Fu, H. Tuning process parameters to optimize microstructure and mechanical properties of novel maraging steel fabricated by selective laser melting. *Mater. Sci. Eng. A* **2021**, *823*, 141740. [CrossRef]
5. Waqar, S.; Guo, K.; Sun, J. FEM analysis of thermal and residual stress profile in selective laser melting of 316L stainless steel. *J. Manuf. Process.* **2021**, *66*, 81–100. [CrossRef]
6. Ansari, P.; Rehman, A.U.; Pitir, F.; Veziroglu, S.; Mishra, Y.K.; Aktas, O.C.; Salamci, M.U. Selective Laser Melting of 316L Austenitic Stainless Steel: Detailed Process Understanding Using Multiphysics Simulation and Experimentation. *Metals* **2021**, *11*, 1076. [CrossRef]
7. Trejos-Taborda, J.; Reyes-Osorio, L.; Garza, C.; del Carmen Zambrano-Robledo, P.; Lopez-Botello, O. Finite element modeling of melt pool dynamics in laser powder bed fusion of 316L stainless steel. *Int. J. Adv. Manuf. Technol.* **2022**, *120*, 3947–3961. [CrossRef]
8. Adeyemi, A.; Akinlabi, E.T.; Mahamood, R.M. Powder Bed Based Laser Additive Manufacturing Process of Stainless Steel A Review. *Mater. Today: Proc.* **2018**, *5*, 18510–18517. [CrossRef]

9. Chang, C.; Yan, X.; Deng, Z.; Lu, B.; Bolot, R.; Gardan, J.; Deng, S.; Chemkhi, M.; Liu, M.; Liao, H. Heat treatment induced microstructural evolution, oxidation behavior and tribological properties of Fe-12Cr-9Ni-2Al steel (CX steel) prepared using selective laser melting. *Surf. Coatings Technol.* **2022**, *429*, 127982. [CrossRef]
10. Saewe, J.; Gayer, C.; Vogelpoth, A.; Schleifenbaum, J.H. Feasability Investigation for Laser Powder Bed Fusion of High-Speed Steel AISI M50 with Base Preheating System. *BHM Berg-Und Hüttenmännische Monatshefte* **2019**, *164*, 101–107. [CrossRef]
11. Ronneberg, T.; Davies, C.M.; Hooper, P.A. Revealing relationships between porosity, microstructure and mechanical properties of laser powder bed fusion 316L stainless steel through heat treatment. *Mater. Des.* **2020**, *189*, 108481. [CrossRef]
12. Pasebani, S.; Ghayoor, M.; Badwe, S.; Irrinki, H.; Atre, S.V. Effects of atomizing media and post processing on mechanical properties of 17-4 PH stainless steel manufactured via selective laser melting. *Addit. Manuf.* **2018**, *22*, 127–137. [CrossRef]
13. Maharjan, N.; Zhou, W.; Zhou, Y.; Wu, N. Decarburization during laser surface processing of steel. *Appl. Phys. A* **2018**, *124*, 682. [CrossRef]
14. Ahmed, N.; Barsoum, I.; Haidemenopoulos, G.; Al-Rub, R.K.A. Process parameter selection and optimization of laser powder bed fusion for 316L stainless steel: A review. *J. Manuf. Process.* **2022**, *75*, 415–434. [CrossRef]
15. Yang, X.; Ma, W.J.; Ren, Y.J.; Liu, S.F.; Wang, Y.; Wang, W.L.; Tang, H.P. Subgrain microstructures and tensile properties of 316L stainless steel manufactured by selective laser melting. *J. Iron Steel Res. Int.* **2021**, *28*, 1159–1167. [CrossRef]
16. Zhai, W.; Zhou, W.; Zhu, Z.; Nai, S.M.L. Selective Laser Melting of 304L and 316L Stainless Steels: A Comparative Study of Microstructures and Mechanical Properties. *Steel Res. Int.* **2022**, *93*, 2100664. [CrossRef]
17. Zhang, Y.; Roch, A. Fused filament fabrication and sintering of 17-4PH stainless steel. *Manuf. Lett.* **2022**, *33*, 29–32. [CrossRef]
18. Zhao, X.; Wei, Q.S.; Gao, N.; Zheng, E.L.; Shi, Y.S.; Yang, S.F. Rapid fabrication of TiN/AISI 420 stainless steel composite by selective laser melting additive manufacturing. *J. Mater. Process. Technol.* **2019**, *270*, 8–19. [CrossRef]
19. Huang, Y.; Yang, S.; Gu, J.; Xiong, Q.; Duan, C.; Meng, X.; Fang, Y. Microstructure and wear properties of selective laser melting 316L. *Mater. Chem. Phys.* **2020**, *254*, 123487. [CrossRef]
20. Jiang, H.Z.; Li, Z.Y.; Feng, T.; Wu, P.Y.; Chen, Q.S.; Feng, Y.L.; Chen, L.F.; Hou, J.Y.; Xu, H.J. Effect of Process Parameters on Defects, Melt Pool Shape, Microstructure, and Tensile Behavior of 316L Stainless Steel Produced by Selective Laser Melting. *Acta Metall. Sin. (Engl. Lett.)* **2021**, *34*, 495–510. [CrossRef]
21. Greco, S.; Gutzeit, K.; Hotz, H.; Kirsch, B.; Aurich, J.C. Selective laser melting (SLM) of AISI 316L—impact of laser power, layer thickness, and hatch spacing on roughness, density, and microhardness at constant input energy density. *Int. J. Adv. Manuf. Technol.* **2020**, *108*, 1551–1562. [CrossRef]
22. Vunnam, S.; Saboo, A.; Sudbrack, C.; Starr, T.L. Effect of powder chemical composition on the as-built microstructure of 17-4 PH stainless steel processed by selective laser melting. *Addit. Manuf.* **2019**, *30*, 100876. [CrossRef]
23. Ponnusamy, P.; Sharma, B.; Masood, S.H.; Rahman Rashid, R.A.; Rashid, R.; Palanisamy, S.; Ruan, D. A study of tensile behavior of SLM processed 17-4 PH stainless steel. *Mater. Today Proc.* **2021**, *45*, 4531–4534. [CrossRef]
24. Ali, U.; Esmaeilizadeh, R.; Ahmed, F.; Sarker, D.; Muhammad, W.; Keshavarzkermani, A.; Mahmoodkhani, Y.; Marzbanrad, E.; Toyserkani, E. Identification and characterization of spatter particles and their effect on surface roughness, density and mechanical response of 17-4 PH stainless steel laser powder-bed fusion parts. *Mater. Sci. Eng. A* **2019**, *756*, 98–107. [CrossRef]
25. Nath, S.D.; Irrinki, H.; Gupta, G.; Kearns, M.; Gulsoy, O.; Atre, S. Microstructure-property relationships of 420 stainless steel fabricated by laser-powder bed fusion. *Powder Technol.* **2019**, *343*, 738–746. [CrossRef]
26. Yang, X.H.; Jiang, C.M.; Ho, J.R.; Tung, P.C.; Lin, C.K. Effects of Laser Spot Size on the Mechanical Properties of AISI 420 Stainless Steel Fabricated by Selective Laser Melting. *Materials* **2021**, *14*, 4593. [CrossRef]
27. Jiang, C.M.; Ho, J.R.; Tung, P.C.; Lin, C.K. Tempering Effect on the Anisotropic Mechanical Properties of Selective Laser Melted 420 Stainless Steel. *J. Mater. Eng. Perform.* **2022**. [CrossRef]
28. Kultamaa, M.; Mönkkönen, K.; Saarinen, J.J.; Suvanto, M. Corrosion Protection of Injection Molded Porous 440C Stainless Steel by Electroplated Zinc Coating. *Coatings* **2021**, *11*, 949. [CrossRef]
29. Wang, X.; Zhao, L.; Fuh, J.Y.H.; Lee, H.P. Effects of statistical pore characteristics on mechanical performance of selective laser melted parts: X-ray computed tomography and micromechanical modeling. *Mater. Sci. Eng. A* **2022**, *834*, 142515. [CrossRef]
30. Zheng, Z.; Peng, L.; Wang, D. Defect Analysis of 316 L Stainless Steel Prepared by LPBF Additive Manufacturing Processes. *Coatings* **2021**, *11*, 1562. [CrossRef]
31. Liverani, E.; Fortunato, A. Additive manufacturing of AISI 420 stainless steel: Process validation, defect analysis and mechanical characterization in different process and post-process conditions. *Int. J. Adv. Manuf. Technol.* **2021**, *117*, 809–821. [CrossRef]
32. Sun, Z.; Tan, X.; Tor, S.B.; Chua, C.K. Simultaneously enhanced strength and ductility for 3D-printed stainless steel 316L by selective laser melting. *NPG Asia Mater.* **2018**, *10*, 127–136. [CrossRef]
33. *The CES Edupack 2013*; Granta's Design Ltd.: Cambridge, UK, 2013.
34. Online Materials Information Resource—MatWeb. Available online: https://www.matweb.com/ (accessed on 8 September 2022).
35. Zhao, X.; Song, B.; Zhang, Y.; Zhu, X.; Wei, Q.; Shi, Y. Decarburization of stainless steel during selective laser melting and its influence on Young's modulus, hardness and tensile strength. *Mater. Sci. Eng. A* **2015**, *647*, 58–61. [CrossRef]
36. Karthik, G.M.; Kim, H.S. Heterogeneous Aspects of Additive Manufactured Metallic Parts: A Review. *Met. Mater. Int.* **2021**, *27*, 1–39. [CrossRef]
37. Aqilah, D.N.; Sayuti, A.K.M.; Farazila, Y.; Suleiman, D.Y.; Amirah, M.A.N.; Izzati, W.B.W.N. Effects of Process Parameters on the Surface Roughness of Stainless Steel 316L Parts Produced by Selective Laser Melting. *J. Test. Eval.* **2018**, *46*, 20170140. [CrossRef]

Article

Feasibility Study on Additive Manufacturing of Ferritic Steels to Meet Mechanical Properties of Safety Relevant Forged Parts

Linda Mally [1,2,*], Martin Werz [2] and Stefan Weihe [1,2]

[1] Institute for Materials Testing, Materials Science and Strength of Materials (IMWF), University of Stuttgart, Pfaffenwaldring 32, 70569 Stuttgart, Germany; stefan.weihe@mpa.uni-stuttgart.de

[2] Materials Testing Institute, University of Stuttgart, Pfaffenwaldring 32, 70569 Stuttgart, Germany; martin.werz@mpa.uni-stuttgart.de

* Correspondence: linda.mally@mpa.uni-stuttgart.de

Abstract: Additive manufacturing processes such as selective laser melting are rapidly gaining a foothold in safety-relevant areas of application such as powerplants or nuclear facilities. Special requirements apply to these applications. A certain material behavior must be guaranteed and the material must be approved for these applications. One of the biggest challenges here is the transfer of these already approved materials from conventional manufacturing processes to additive manufacturing. Ferritic steels that have been processed conventionally by forging, welding, casting, and bending are widely used in safety-relevant applications such as reactor pressure vessels, steam generators, valves, and piping. However, the use of ferritic steels for AM has been relatively little explored. In search of new materials for the SLM process, it is assumed that materials with good weldability are also additively processible. Therefore, the processability with SLM, the process behavior, and the achievable material properties of the weldable ferritic material 22NiMoCr3-7, which is currently used in nuclear facilities, are investigated. The material properties achieved in the SLM are compared with the conventionally forged material as it is used in state-of-the-art pressure water reactors. This study shows that the ferritic-bainitic steel 22NiMoCr3-7 is suitable for processing with SLM. Suitable process parameters were found with which density values > 99% were achieved. For the comparison of the two materials in this study, the microstructure, hardness values, and tensile strength were compared. By means of a specially adapted heat treatment method, the material properties of the printed material could be approximated to those of the original block material. In particular, the cooling medium/cooling method was adapted and the cooling rate reduced. The targeted ferritic-bainitic microstructure was achieved by this heat treatment. The main difference found between the two materials relates to the grain sizes present. For the forged material, the grain size distribution varies between very fine and slightly coarse grains. The grain size distribution in the printed material is more uniform and the grains are smaller overall. In general, it was difficult and only minimal possible to induce grain growth. As a result, the hardness values of the printed material are also slightly higher. The tensile strength could be approximated to that of the reference material up to 60 MPa. The approximation of the mechanical-technological properties is therefore deemed to be adequate.

Keywords: ferritic steel; heat treatment; SLM; additive manufacturing; nuclear application

Citation: Mally, L.; Werz, M.; Weihe, S. Feasibility Study on Additive Manufacturing of Ferritic Steels to Meet Mechanical Properties of Safety Relevant Forged Parts. *Materials* **2022**, 15, 383. https://doi.org/10.3390/ma15010383

Academic Editor: Swee Leong Sing

Received: 30 November 2021
Accepted: 25 December 2021
Published: 5 January 2022

1. Introduction

Additive manufacturing (AM) processes have evolved in a short time from rapid prototyping—the production of prototypes and sample components, to industrially applicable components and structures [1]. The national and international effort to bring these additively manufactured components into use in safety-relevant areas such as power plants or nuclear facilities is great [2]. The production processes developed for the second generation of nuclear power plants during the first peak of nuclear technology (Generation II Peak) have become obsolete due to dismantling of existing plants, restructuring of production facilities (scrapping of machinery, takeover, etc.), and loss of knowledge (retirement of employees), and can no longer be applied without considerable effort and expense [3]. Currently, we are seeing a second peak (Generation III Peak) in the nuclear power sector, with some countries like Germany dealing with the dismantling of nuclear facilities, while in other areas of the world new plant systems are being built (Gen. III), developed (Gen. IV) and commissioned. Therefore, new advanced production processes and manufacturing techniques such as additive manufacturing are being relied upon for future spare parts, the development, and new construction of power plants worldwide [3]. According to the German Gesellschaft für Reaktorsicherheit (GRS) [4], the trend in the development of new reactors and reactor concepts is moving towards Small Modular Reactors (SMR). As a result, component sizes are likely to decrease in the future and new manufacturing methods such as additive manufacturing could be applied. Among the most important advanced manufacturing technologies in this regard, based on industry interest, is selective laser beam melting (SLM/L-PBF), according to the United States Nuclear Regulatory Commission (USNRC) [2]. Currently, the SLM process is mainly used for the production of small to medium-sized components. In the field of nuclear technology, additively manufactured valves and other non-safety-critical small to medium-sized components are already being used. The industry's interest in also being able to manufacture large components with this process is significant and is leading to the development of ever-larger machines. For example, the SLM®800 developed by SLM Solutions already has a build envelope with dimensions of 500 × 290 × 850 mm [5]. Another example of the rapid growth of the available built envelope dimensions is the SLM laboratory system developed by the Aachen University of Applied Sciences and the Fraunhofer Institute for Laser Technology ILT. Their system has an effectively usable built envelope of 1000 × 800 × 500 mm, which is considerably larger than other previous commercial SLM systems [6]. This trend makes the use of additively manufactured components in the reactor sector feasible and could lead to the manufacturing of large components.

The principle of the laser and powder bed-based manufacturing process (SLM/L-PBF) is a layer-by-layer build-up, in which a component is created using selective melting of metal powder with a laser beam [1]. The resulting component properties are determined by a variety of influencing factors, such as machine size, powder properties, process parameters, resulting imperfections, etc. In total, there are more than 130 quality-deciding influencing factors [7]. In addition, each material exhibits different behavior in the additive manufacturing process and must therefore be investigated separately. The most commonly used materials for L-PBF are AlSi10Mg [8,9], Ti6Al4V [10,11], Inconel [12] or 316L [13,14]. These have been studied in more detail in the past, so optimized process parameter sets for these materials are available in the literature. However, the parameter sets presented in the published studies are often incomplete or are not transferable to other manufacturing equipment because of the selected performance settings. In addition, it has been found that even the same parameter sets on identical production equipment in different sites can lead to different results, for example in terms of porosity and mechanical strength, so that the parameters must be adapted separately for each machine [15].

This study focuses on the ferritic reactor steel 22NiMoCr3-7 as this material was used almost exclusively in reactor pressure vessel construction in the Federal Republic of Germany [16]. The optimized reactor steel 22NiMoCr3-7 was derived from the American material ASTM A 508 Cl.2 (20NiMoCr3-6). The ASTM A 508 Cl.2 low-alloy (higher-strength) steel was selected for use in German nuclear power plants on account of its mechanical properties (especially toughness), which are adequate even for large wall thicknesses, its good through-temperability, its low tendency to radiation embrittlement, and it is at this time considered good weldability. The optimized reactor steel 22NiMoCr3-7 had also been considered well suited for welded joints. Later, however, this steel proved to be sensitive to embrittlement and cracking in weld-affected zones under certain circumstances [16]. Experiments were conducted at the Materials Testing Institute (MPA) when this welding issue was discovered analyzing the reason and influencing factors for relaxation embrittlement and crack formation in 22NiMoCr3-7. These experiments confirmed the only cause for the development of relaxation embrittlement and crack formations to be overheating of the material with coarse grain formation and if the initially dissolved metallic carbides are precipitated at the grain boundaries because of rapid cooling followed by subsequent heat treatment. These precipitates are responsible for additional hardening of the material and reduce both its creep tendency and its ductility in the grain boundary zones. The residual stresses generated during welding are relieved during conventional two-stage stress relief annealing by plastic creep deformation. In damaged grain boundary regions, however, the creep and strain capacity is quickly exhausted, and inter-granular cracking may occur. It was found out that the main decisive influencing factors are, in addition to the acting stresses, the chemical steel composition, as well as the kinetics of the precipitation processes in the heat-affected zone (HAZ), which depend on the temperature-time processes during welding and the subsequent heat treatment [16].

The problems with the welding safety of the steel 22NiMoCr3-7 and its tendency to undercladding and side cracks led to a discussion of whether 20MnMoNi5-5 might not be more suitable. However, the discussion about the competing reactor steels eased after a strong narrowing of the chemical analysis limits, preferably for molybdenum and trace elements accompanying the steel, which was recommended after the investigations of the MPA Stuttgart, so that the steel 22NiMoCr3-7 could be substantially improved in its properties. The limit values that should not be exceeded are shown in Table 1. According to the Reactor Safety Committee, the modified 22NiMoCr3-7 is a desirable alternative to the steel 20MnMoNi5-5 [16]. Because of its still important role as one of the main materials for reactor pressure vessels erected in the Federal Republic of Germany, it is therefore of interest for future research endeavors with regard to spare parts production using additive manufacturing processes. Possible problems arising with regard to weldability and thus processability using additive manufacturing were investigated in this study. The material behavior of the conventionally manufactured 22NiMoCr3-7 has already been extensively investigated in past studies and research projects at the MPA of the University of Stuttgart. The material properties, as well as the material and failure behavior of the conventionally manufactured material, with which the obtained results of this study are compared, were for example elaborated by Seebich [17].

Table 1. Crack initiation after exceeding two or more of these values [18].

Limit Contents in Weight Percent					
Mo	**P**	**S**	**Cu**	**Sn**	**N**
0.62	0.008	0.008	0.12	0.011	0.013

In literature, there are only a few published studies on the subject of processing ferritic materials using L-PBF [19–24]. To date, there are no empirical values or published parameter windows for the processability of 22NiMoCr3-7 using the L-PBF method. The parameter sets given in the literature cannot be applied unchanged for the reasons already mentioned, but serve as a data basis for future process parameter variations. Moreover, the use of ferritic powder materials such as 22NiMoCr3-7 in the L-PBF process makes the manufacturing process and the creation of the desired microstructure and the associated material properties even more complex. Due to the higher cooling rates > 10^6 °C/s [25] during the manufacturing process, subsequent heat treatment is required to obtain the desired ferritic-bainitic microstructure instead of the expected martensitic structure present after printing. Initial studies on microstructure adjustment with subsequent heat treatment for ferritic materials exist [19] and are used for comparison purposes.

However, to establish additive manufactured components (AM components) and structures in safety-critical and certification-relevant application areas, the process-related material properties and the resulting material behavior must be fundamentally understood and mastered. The challenges that have to be overcome in the future include anisotropic microstructures, non-uniform stress distributions and pore formations as well as many other effects. Only when additive manufacturing processes have been sufficiently validated can the economic and technical advantages of this technology be used without hesitation in safety-critical areas such as nuclear power plants.

2. Materials and Methods

2.1. Material

The material block of 22NiMoCr3-7 used in this study was taken from the upper forged core ring of the reactor pressure vessel originally planned for the nuclear Biblis-C plant. The material is a ferritic-bainitic fine-grained structural steel. The microstructure of the block material displays the lines and textures typical for a forged material. The mentioned lines can be seen in the vertical color differences visible in the left image in Figure 1. For this material, the heat treatment takes place after the forging process, which makes the texture slightly less obvious. The grain size varies from very fine to slightly coarse [17], as can be seen in the right image in Figure 1.

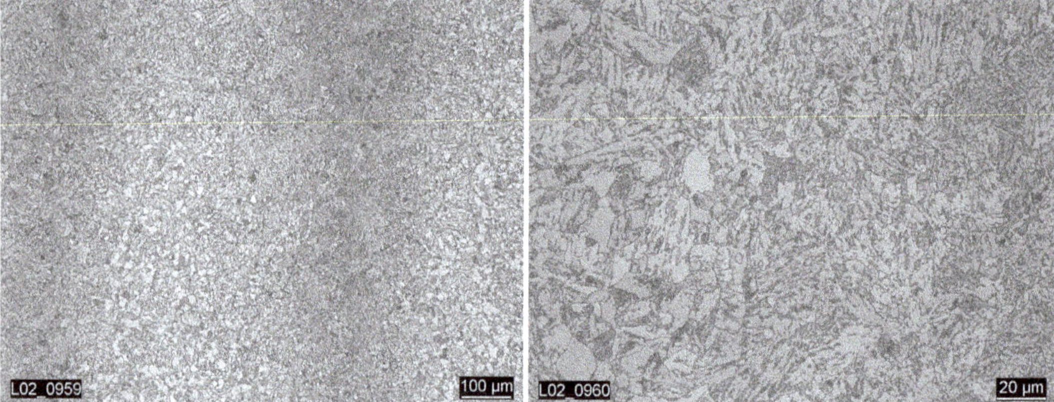

Figure 1. Textures and lines of the forged block material [17].

The hardness resulted in an average value of 189 HV 30. Further investigations of the material in previous projects show that the material has very homogeneous and isotropic material properties [17]. The results of the performed tensile test are shown in Table 2.

Table 2. Strength and deformation characteristics from tensile test [17].

L_0 [mm]	UTS [MPa]	UYP [MPa]	A [%]	A_g [%]	Z [%]
50	584	445	23.0	10.7	68

The steel powder was produced by atomizing the 22NiMoCr3-7 block material provided by the Materials Testing Institute (MPA) Stuttgart at Höganäs. The influence of the particle size on the flowability of metal powder material was taken into account when selecting the particle size range. In this respect, it was considered that fine powder particles tend to form agglomerates and thus impair flowability. However, fine particles can fill the voids of the coarser particles and help to increase the density of the powder layer, which leads to a better process result [26]. Based on experience and typical industry specifications for steel powder for L-PBF production, 15 µm was chosen as the lower limit. The upper limit was set at 45 µm.

The specification for the chemical composition of 22NiMoCr3-7 including limit values and an overview of the chemical compositions of the block and powder material is shown in Table 3.

Table 3. Specification [27] and overview of chemical composition for 22NiMoCr3-7.

	C	Si	Mn	Cr	Mo	Ni	P	S	Cu	Sn	Al	V	Ta	Co	As
Spec min.	0.17	0.10	0.50	0.25	0.50	0.60	-	-	-	-	-	-	-	-	-
Spec max.	0.25	0.35	1.00	0.50	0.75	1.00	0.012	0.015	0.1	-	0.05	0.05	0.03	0.03	-
block	0.21	0.20	0.88	0.4	0.53	0.83	0.006	0.002	0.039	0.007	0.016	0.007	<0.003	0.011	0.005
powder	0.19	0.22	0.93	0.28	0.51	0.67	0.006	0.004	0.007	0.02	0.02	<0.01	<0.01	0.01	-
as-printed	0.197	0.21	0.79	0.43	0.547	0.93	0.007	0.004	0.042	0.007	0.014	0.007	0.004	0.013	0.006

The chemical composition of the original block material is compared with the composition of the resulting target grain to guarantee the correct alloy composition even after pulverization. In addition, information on welding behavior, microstructure development and the effect of a subsequent heat treatment strategy can be derived from the listed composition. The chemical composition of the block material was determined using a Quantovac analysis at MPA Stuttgart [17], while the powder was analyzed at the manufacturer's site. Since losses of certain alloying elements can occur during the production process, for example due to burn-off, the chemical composition was also checked after the production of the specimens by means of Spark Optical Emission Spectrometry (F-OES). The results for the chemical composition of the as-printed material are discussed in Section 3.1.

2.2. Manufacturing

At present, there is almost no information to be found on the additive processing of reactor steel, especially on 22NiMoCr3-7. Therefore, no parameter sets for the processing of 22NiMoCr3-7 by means of SLM are available. There is also little information on the general processing of ferrites or high-temperature steels. One study dealing with a material similar in area of application, namely a strong and ductile Reduced Activation Ferritic/Martensitic (RAFM) steel employed for fusion reactors [23], is used as a starting point for the parameter variation of this study. Jiang et al. provides a complete defined process parameter window with parameters adaptable to the equipment available for this study. The laser power varied between 160 W and 320 W at scanning speeds of 400 to 1200 mm/s, while keeping the layer thickness constant at 30 µm and a hatch spacing of 85 µm. The parameter combinations of 200 W and scanning speed of 800 mm/s, as well as laser power of 320 W and scanning speed of 600 mm/s proved to be the most suitable for the RAFM material [23]. To verify whether this parameter window deviates strongly from other materials frequently used in additive manufacturing, the process parameters of the austenitic steel 1.4404 were also considered. 1.4404 is one of the most common steels processed by SLM because the material is very inexpensive and has very good processing properties. For this reason, there are already process parameter sets available. Laser powers between 150 W and 250 W and scanning speeds of 600–950 mm/s have proven suitable for this material on the Aconity Mini System used in this study. It was found that even for different materials with different properties, the parameter windows overlap to a large extent. Therefore, parameters from this range or slightly above or below were selected for the parameter determination of this study. The process parameter variations carried out are listed in Table 4.

The processability of 22NiMoCr3-7 was investigated on cube-shaped specimens with different process parameter variations using an Aconity Mini system (Aconity3D). The printed cubes were designed with wedge-shaped solid support structures, as seen in Figure 2, to ensure a secure bond to the build plate. An edge length of 10 mm was chosen for the cubes. The layer thickness was set constant at 30 µm for all cubes. The hatch spacing and laser spot diameter were not changed during the parameter variation. A simple hatching scan strategy was chosen for the entire cross-section, which was rotated 90°

for each subsequent layer. The contour was exposed after filling. The manufacturing was performed under an argon atmosphere at an oxygen content below 100 ppm. The build plate was initially not preheated to investigate the influence of the high cooling rate during the printing process on the resulting microstructure.

Table 4. Parameter variation.

Parameter Set	Laser Power [W]	Scan Speed [mm/s]
PV1	250	700
PV2	200	800
PV3	150	800
PV4	300	1000
PV5	250	1525

Figure 2. Specimen cube with a wedge-shaped support structure.

2.3. Density Measurement

The relative density of the additively manufactured cube specimens achieved with the selected process parameters was used as a determining criterion for assessing the suitability of the parameters used. The density was determined according to Archimedes' principle (ASTM B962-17) and additionally by microscopic image analysis of a cut surface of the cube specimens. For this purpose, one cube per parameter set was cut in the center parallel to the direction of build-up. Subsequently, the cut surfaces of each specimen were polished to visualize the pore size and distribution.

2.4. Heat Treatment and Hardness Measurement

To approximate the mechanical properties of the additively manufactured material to those of the original forging block material, various heat treatment strategies were investigated. The performed heat treatment variations are listed in Table 5. The temperature profile for the specimens cooled in the oven in a controlled environment is shown in Figure 3.

Following the heat treatment, sections were again prepared to analyze the resulting microstructure. In addition, Vickers hardness tests were carried out in accordance with DIN EN ISO 6507-1.

Table 5. Heat treatment strategies.

Strategy	Austenitization Temperature [°C]	Austenitization Time [min]	Cooling Medium	Tempering Temperature [°C]	Tempering Time [min]
Z	900		water	650	450
A0	-	-	-	-	-
B1/B2	900	20	oil/water	-	-
C1/C2	900	20	oil/water	650	60
D1	900	20	air	650	60
D2	900	20	air	650	240
D3	900	240	air	650	60
E1	900	60	oven (100 °C)	-	-
E2	900	60	oven (100 °C)	700	240
E3	900	60	oven (100 °C)	700	480

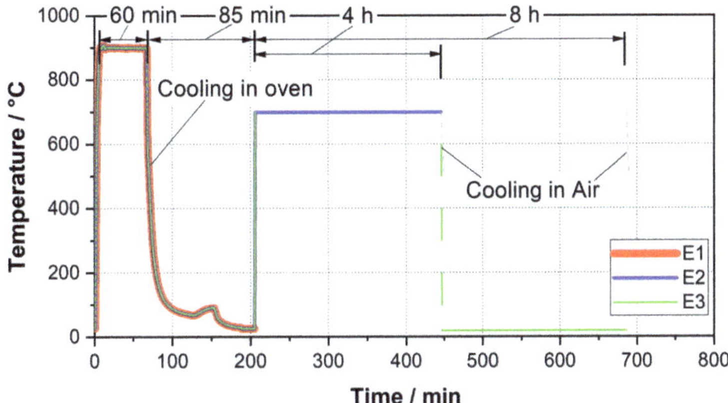

Figure 3. Controlled heat treatment for E1–E3.

2.5. Tensile Testing

For further material characterization, tensile specimen blanks (TSB) with dimensions 12 mm × 12 mm × 86 mm were manufactured in horizontal orientation (0°). The parameter set PV1 (laser power = 250 W, scan speed = 700 mm/s) was used for the production. The layer thickness was left at 30 µm. Fabrication also took place in an argon atmosphere. For the fabrication of these specimens, to minimize residual stresses, the build platform was heated to 150 °C. In addition to the tensile test specimens printed with the simple hatching scan strategy (designation according to pre-process software Netfabb Autodesk) three additional scanning strategies (see Figure 4) were tested to estimate their effect on the resulting material properties.

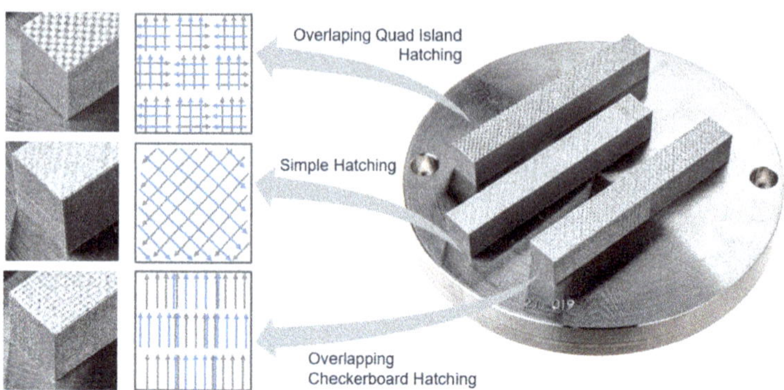

Figure 4. Specimens with different scan strategies.

The tensile specimen blanks were heat-treated before machining the target contour by lathing according to B 8 X 40 DIN 50125 (see Figure 5) to prevent decarbonization of the layers near the surface. The tensile tests were performed in accordance with DIN EN ISO 6892-1:2020-06 on a Zwick Roell 100 kN tensile testing machine Typ BZ1-MMZ100.ZW02.

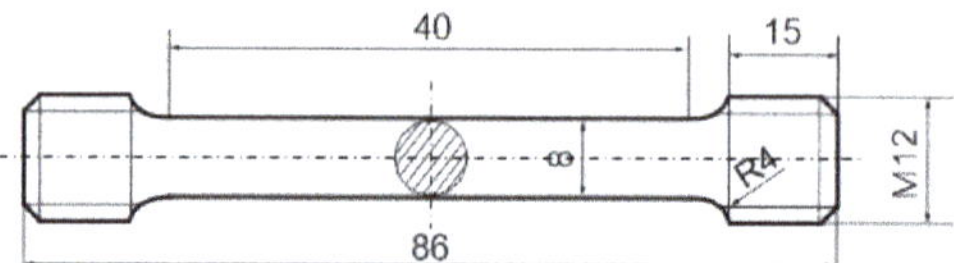

Figure 5. Tensile test specimen B 8 X 40 DIN 50125.

3. Experimental Results

3.1. Material

The results of the chemical analysis of the printed material show that the amount of alloying elements present is still within specification, see Table 2. However, significant deviations from the powder analysis are noticeable in some cases. The manganese content has decreased because of vaporization during by the manufacturing process. An increase in chromium, nickel and copper was detected compared to the results of the atomized powder material. As the measured values of these elements in the printed material are very similar to those of the block material, it is assumed that the chemical analysis of the powder material is inaccurate due to the different material state (powder versus solid).

The fact that material burns off during the process is confirmed by the fact that the filters in the area of the process gas circulation are completely black and clogged after the process. The results of the chemical analysis of the material deposited on the filters are still pending and will provide further information on the loss of alloying elements.

3.2. Density Measurement and Parameter Evaluation

The results of the density test are shown below in Table 6. There were significant differences in the relative density for the different process parameter sets used. Parameter set PV1 showed the highest density values. Moreover, the specimens showed a good surface quality. PV2 and PV3 showed a considerably worse surface quality compared to PV1.

Table 6. Density measurement according to the Archimdedean principle.

Specimen	Laser Power [W]	Scan Speed [mm/s]	Density [%]
PV1-1	250	700	99.10
PV1-2	250	700	98.07
PV1-3	250	700	99.60
PV2-1	200	800	90.74
PV2-2	200	800	95.17
PV2-3	200	800	91.57
PV3-1	150	800	90.74
PV3-2	150	800	99.58
PV3-3	150	800	95.93

The measurements according to the Archimedean principle could be quantitatively confirmed by light microscopic images of the micrographs. The micrographs of parameter set 2 (PV2) show a relatively large number of flaws and attachment defects (see Figure 6a), whereas PV1 (see Figure 6b) and PV3 hardly show any discontinuities.

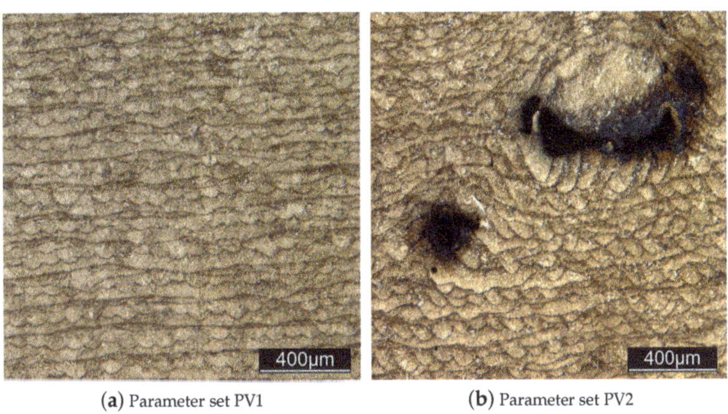

(**a**) Parameter set PV1 (**b**) Parameter set PV2

Figure 6. Micrograph without defects (**a**) and with defects (**b**).

Except for the area of influence of the flaws, all samples show a uniform layer thickness. The tested parameter variations PV4 and PV5 are not suitable for the processing of the investigated material. In the case of PV4 unwanted material accumulations on the surface occurred during the manufacturing process and in the case of PV5, only a poor bond to the building board was achieved. Parameter set 1 (PV1) was therefore used for all further tests.

3.3. Heat Treatment and Hardness Measurement

The aim of the heat treatment was to approximate the microstructure of the as-printed material, as well as the mechanical properties to that of the forged material. The target microstructure of the forged block material (Z) is displayed in Figure 7a and shows a ferritic-bainitic microstructure with fine to slightly coarse grain size. In Figure 7b the as-printed state (A0), the individual melting traces of the manufacturing process are clearly visible. The results of the different heat treatment strategies are shown in Figures 8–12.

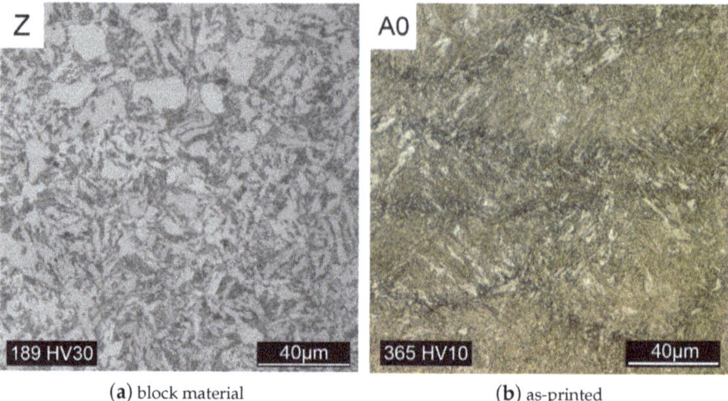

(a) block material (b) as-printed

Figure 7. Microstructure forged block material (**a**) and as-printed material (**b**).

In contrast to the as-printed material, the melt traces are no longer visible in the cross-section of all heat-treated specimens. Homogenization of the microstructure takes place during the heat treatment. Initially, heat treatments were carried out without subsequent tempering (Figure 8). Quenching after austenitizing took place in oil Figure 8a or water Figure 8b. With these heat treatment strategies, only a martensitic microstructure could be produced (see Figure 8).

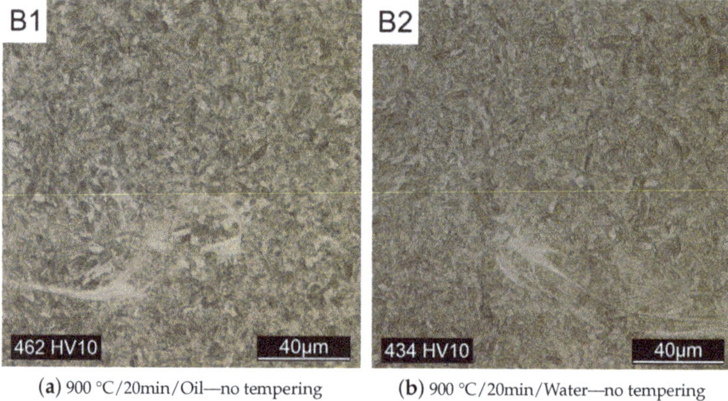

(a) 900 °C/20min/Oil—no tempering (b) 900 °C/20min/Water—no tempering

Figure 8. Heat treated but not tempered specimens, quenched with oil (**a**) or water (**b**).

The bright streaks in the microstructure image probably originate from particle splashes that were ejected from the melt pool and landed back on the specimen surface, resulting in a different cooling rate than the rest of the sample. These streaks are no longer visible in the micrographs after other heat treatment strategies with subsequent tempering.

An additional tempering process at 650 °C for 60 min followed by cooling in air lowers the hardness value, but only a or fine-grained martensitic microstructure could be achieved, see Figure 9a,b. The additional tempering process at 650 °C for 60 min followed by cooling in air lowers the hardness value, but the microstructure is still martensitic. The results show that quenching in oil and water is too fast so that the bainite region cannot be reached. Therefore, for the subsequent heat treatment tests, cooling in the air was carried out after austenitizing at 900 °C and then tempering at 650 °C for 60 min.

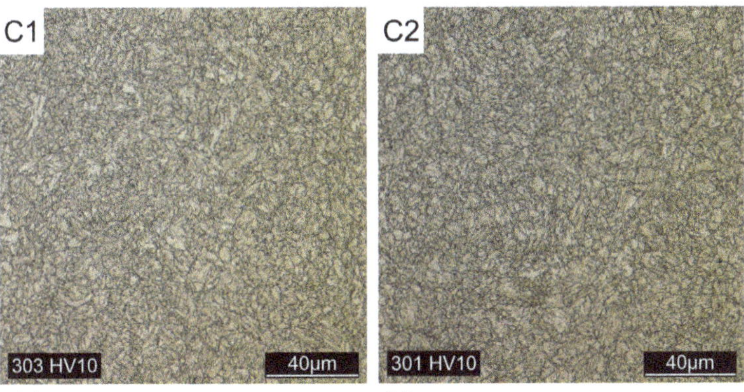

(**a**) 900 °C/20 min/Oil/650 °C/60 min/Air (**b**) 900 °C/20 min/Water/650 °C/60 min/Air

Figure 9. Heat treated and tempered specimens, quenched with oil (**a**) or water (**b**).

The slower cooling process in air allowed to reach the bainitic region, see Figure 10a. However, a very fine structure with small grains is still present here.

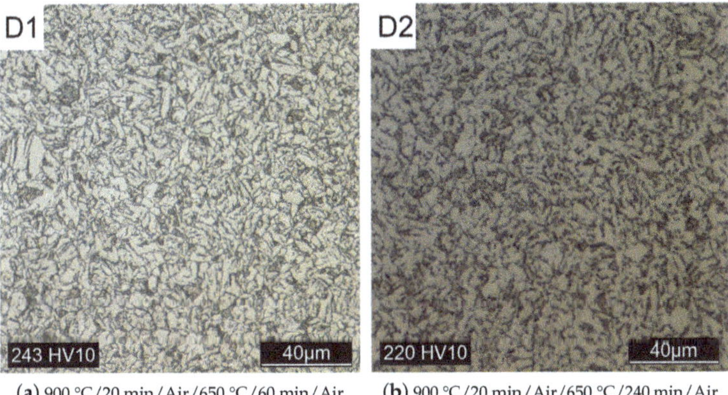

(**a**) 900 °C/20 min/Air/650 °C/60 min/Air (**b**) 900 °C/20 min/Air/650 °C/240 min/Air

Figure 10. Heat treated with different tempering times and cooling in air.

At this point, a significant increase in grain size is estimated to be only possible through overheating. This heat treatment, however, deviates from the specifications of the material manufacturer. Therefore, the grain size was supposed to be increased by a longer tempering time (specimen D2, see Figure 10b) or austenitizing time (specimen D3, see Figure 11a). This goal could not be achieved, only the hardness could be lowered further.

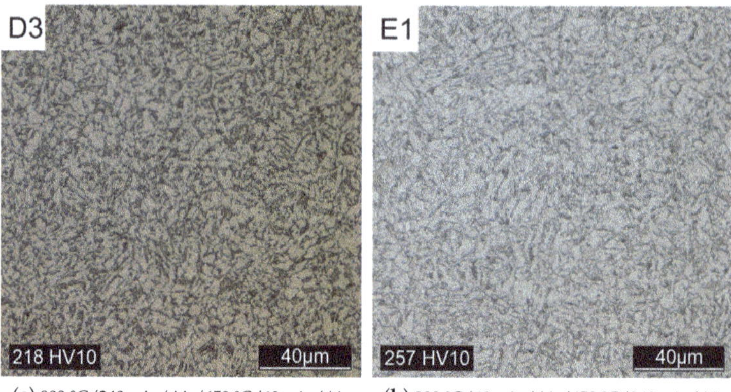

(**a**) 900 °C/240 min/Air/650 °C/60 min/Air (**b**) 900 °C/60 min/Air/650 °C/240 min/Air

Figure 11. Optimized cooling in air (**a**) and first attempt at controlled cooling in the oven (**b**).

Heat treatment with a controllable cooling process for both austenitizing and tempering allows further reduction of hardness and slight grain growth (specimen E2 and E3, see Figure 12).

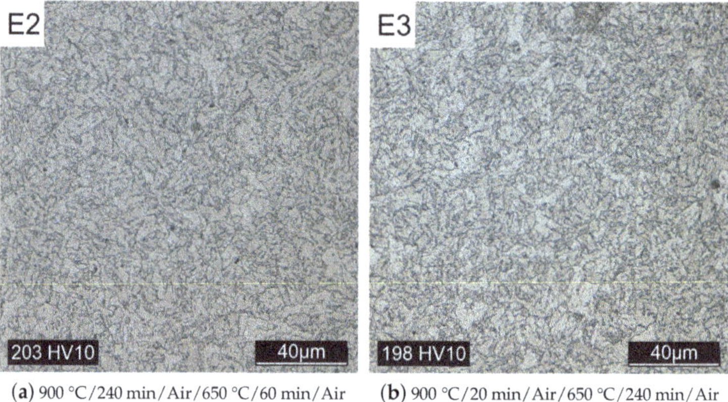

(**a**) 900 °C/240 min/Air/650 °C/60 min/Air (**b**) 900 °C/20 min/Air/650 °C/240 min/Air

Figure 12. Different controlled cooling strategies in the oven.

For subsequent component tests, the duration of austenitizing and tempering must be determined as a function of the dimensions. The heat treatment of the component tests can therefore deviate from the heat treatment strategy determined here.

3.4. *Tensile Performance*

The results of the tensile tests showed a substantial difference between the block material and the as-printed material (see Figure 13), which was to be expected because of the difference in the presented microstructure and the corresponding mechanical properties. The dashed black curve (block material) corresponds to the conventionally produced forged material and the black curve to the as-printed condition. The as-printed specimen was produced using the simple hatching scanning strategy, parameter set PV1, and was not heat treated. The forged block material possesses a tensile strength of UTS = 563 MPa and an elongation at fracture of A = 26%. In comparison the as-printed material has a significantly higher tensile strength with a value of UTS = 1230 MPa and an elongation at fracture of A = 15%. All other specimens shown in Figure 13 were manufactured using the same parameter set (PV1) as the as-printed material and in addition, subject to the D3 heat treatment strategy (900 °C/240 min/Air/650 °C/60 min/Air) available at the time of testing. Further on the D3 heat treatment is also referred to as HT.1 because it was considered the first successful heat treatment strategy. Applying heat treatment D3 significantly reduced the tensile strength. The tensile strength of the specimens could thereby be approximated to a difference of 140 MPa to the forged block material. Moreover, the plots of the stress-strain curves of the specimens with different scan strategies confirm a correlation of the applied scanning strategy with the ductility of the additively manufactured material.

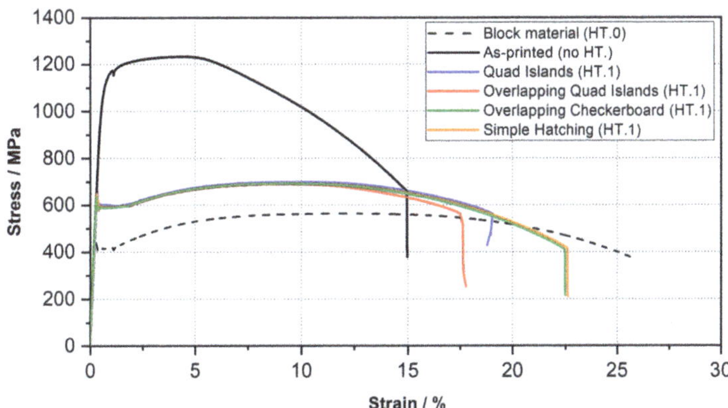

Figure 13. Stress-Strain curves - 22NiMoCr3-7 at RT for specimens with different scan strategies.

The first three tensile test specimens were manufactured using the quad island strategy. The results of the tensile tests showed strong scattering with regard to the achieved elongation at fracture, as can be seen in Table 7. Analysis of the fracture surface of specimen RTS-1 under the electron microscope reveals lack of fusion between the quad islands (see Figures 14 and 15) resulting from the selected scan strategy. The result from the specimen RTS 1 is plotted in dark blue in Figure 13.

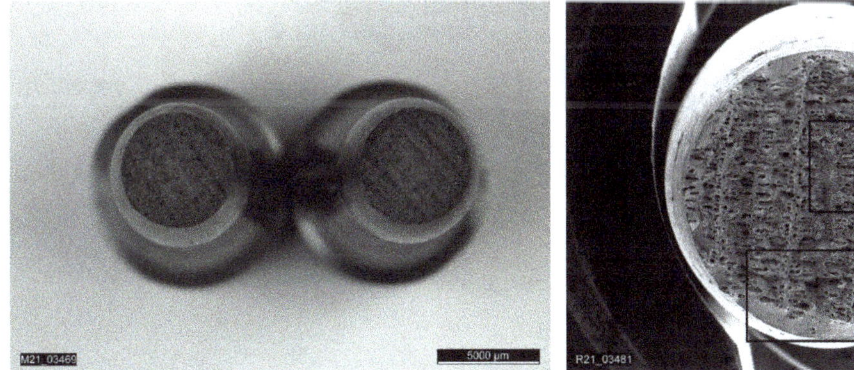

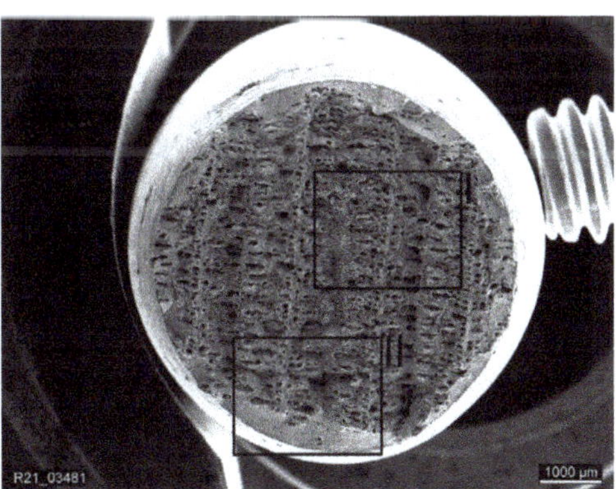

Figure 14. Overview of the fracture surface of a specimen printed with the quad island scan strategy.

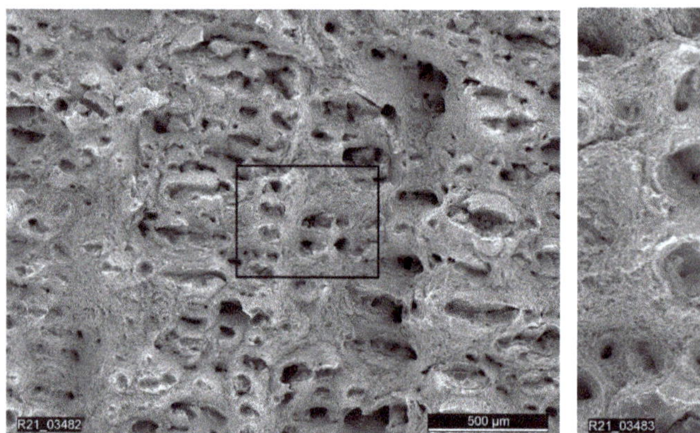

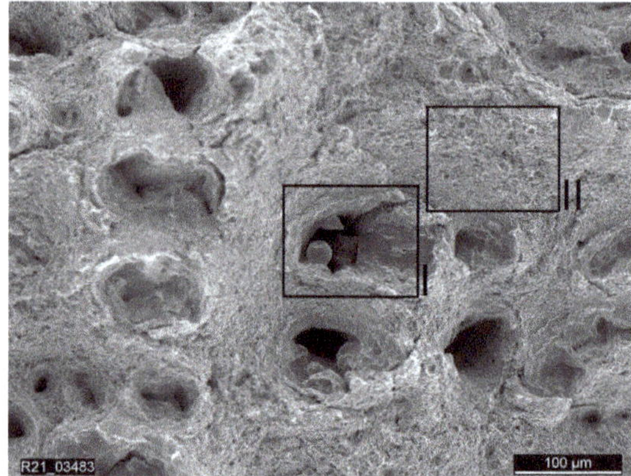

Figure 15. Detailed image showing lack of fusion and uniform material sections on fracture surface using quad island scan strategy.

To further increase the elongation at fracture an overlap of individual islands within the quad island scan strategy was tested. For this purpose, the hatch spacing was reduced, while the laser spot diameter remained the same, to achieve a larger overlap of the individual fusion tracks. The stress-strain curve in Figure 14 shows, however, that this adjustment, plotted in red, does not lead to an improvement for the elongation at break. In contrast, the scanning strategies of the specimens RTS-5 (simple-hatching, orange) and RTS-6 (overlapping checkerboard, dark green) visibly improve the toughness.

Continued optimization of the D3 heat treatment (HT.1) lead to the E3 strategy (HT.2) with even better results concerning the tensile performance (see the green curve in Figure 16).

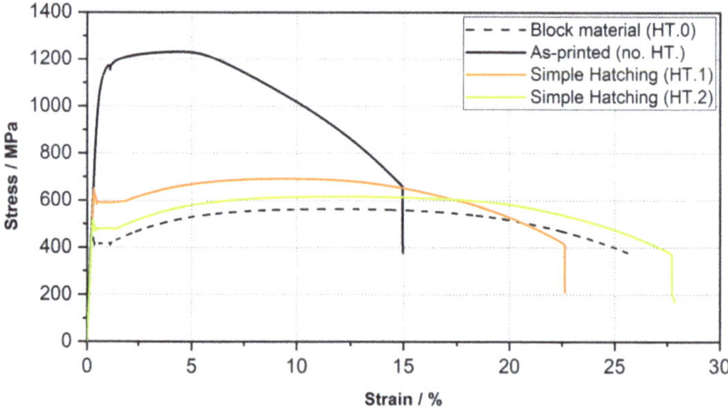

Figure 16. Stress-Strain curves—22NiMoCr3-7 at RT for specimens with different heat treatment strategies.

An evaluation of the results of the new tensile tests shows that the elongation at break could be approximated to within 4% of that of the forged block material using the HT.2 method. The goal of achieving the same tensile strength of the bulk material could not be fully achieved due to the grain refinement caused by L-PBF. There is a remaining deviation of 60 MPa.

Table 7. Strength and deformation characteristics.

Specimen	Diameter [mm]	L_0 [mm]	UTS [MPa]	UYP [MPa]	A [%]	A_g [%]	Z [%]
Block material	no information	50	584	445	23.0	10.7	68
As-printed	7.96	40	1230	-	15.0	-	68
RTS-2	8.00	40	700	665	19.0	-	51
RTS-3	8.01	40	703	636	18.0	-	48
RTS-4	7.98	40	690	646	17.5	9.6	55
RTS-5	7.98	40	693	651	22.0	9.7	71
RTS-6	7.96	40	693	635	22.0	9.8	71
RTS-7	8.00	40	623	492	24	12.0	57
RTS-8	8.00	40	618	510	23	11.8	57
RTS-9	8.00	40	618	519	28	12.2	72
RTS-10	8.00	40	618	517	27	12.0	70
RTS-11	8.00	40	616	510	26	11.8	73
RTS-12	8.00	40	622	469	27	12.2	72

4. Conclusions

In this study, the ferritic reactor steel 22NiMoCr3-7 was successfully processed additively using the L-PBF process. The focus of the investigation was on defining a suitable process parameter window for selective laser melting of the selected material. The laser power, scan speed, and scan strategy were varied to develop a suitable process window. A set of process parameters (PV1) was developed that achieved a relative density of over 99%. In addition, different scanning strategies were tested and their effect on the resulting tensile strength and toughness was investigated. The scan strategy named simple hatching showed the best results. The parameter set PV1 and the named scan strategy were used for subsequent sample fabrication and investigation of various post-treatment heat strategies. The additively manufactured samples with no heat treatment showed insufficient ductile material behavior in the as-printed state, as well as a martensitic microstructure. The goal to approximate the microstructure and the mechanical-technological properties of the forged block material was first successful with the D1 heat treatment. The desired ferritic-bainitic microstructure was obtained. Even the hardness values (218 HV10) were in a similar range to those of the forged material (189 HV30). Further optimization of the heat treatment strategy adding a controlled cooling process in an oven achieved even better results concerning the grain size of the microstructure as well as the hardness values (198 HV10). The development of the E3 heat treatment concludes the search for a suitable heat-treatment process. With this heat treatment, it was possible to achieve the original microstructure of the forged block material as well as its mechanical-technological properties. Based on these investigations, tensile specimens were manufactured and tested. The results showed very good agreement between the additively manufactured and forged specimens. In addition, further experiments on notch tensile, shear tensile, and flat tensile specimens with cracks are in preparation and should show that the material behavior with varying stress multiaxiality is similar for additively manufactured and forged material.

Author Contributions: Conceptualization, L.M. and M.W.; methodology, L.M.; software, L.M.; validation, L.M.; formal analysis, L.M.; investigation, L.M.; resources, S.W.; data curation, L.M.; writing—original draft preparation, L.M.; writing—review and editing, L.M. and M.W.; visualization, L.M.; supervision, M.W.; project administration, S.W.; funding acquisition, S.W. All authors have read and agreed to the published version of the manuscript.

Funding: This research work was funded in the framework of the German reactor safety research program by the Federal Ministry for Economic Affairs and Energy (BMWi) under contract No. 1501596A.

Data Availability Statement: Not applicable.

Acknowledgments: This publication was supported by the Open Access Publishing Fund of the University of Stuttgart.

Conflicts of Interest: The authors declare no conflict of interest.

Abbreviations

The following abbreviations are used in this manuscript:

AM	Additive manufacturing
SLM	Selective Laser Melting
L-PBF	Laser Powder Bed Fusion
GRS	Gesellschaft für Reaktorsicherheit
SMR	Small Modular Reactors
USNRC	United States Nuclear Regulatory Commission
MPA	Materials Testing Institute
HAZ	heat-affected zone
L0	Measuring length
UTS	Ultimate tensile strength
UYP	Upper yield point
A	Elongation at fracture
Ag	Uniform elongation
UTS	Ultimate tensile strength
Z	Necking
F-OES	Spark Optical Emission Spectrometry
RAFM	Reduced Activation Ferritic/Martensitic
TSB	tensile specimen blank

References

1. Gebhardt, A.; Kessler, J.; Thurn, L. *3D-Drucken: Grundlagen und Anwendungen des Additive Manufacturing (AM)*; Carl Hanser Verlag GmbH Co KG: München, Germany, 2016. [CrossRef]
2. Hull, A.; Moyer, C.; Harris, B.; Christensen, J. Proceedings of the Public Meeting on Additive Manufacturing for Reactor Materials and Components, Rockville, MA, USA, 2019. Available online: https://www.nrc.gov/docs/ML1921/ML19214A205.pdf (accessed on 18 December 2021).
3. Idaho National Laboratory. 3D Printing Could Transform Nuclear Industry. 2019. Available online: https://inl.gov/article/3d-printing-could-transform-nuclear-industry-inl-results-indicate/ (accessed on 18 December 2021).
4. Bundesministerium für Wirtschaft und Klimaschutz, Konzept der Bundesregierung zur Kompetenz- und Nachwuchsentwicklung für die Nukleare Sicherheit. 2020. Available online: https://www.bmwi.de/Redaktion/DE/Publikationen/Energie/konzept-zur-kompetenz-und-nachwuchsentwicklung-fuer-die-nukleare-sicherheit.pdf?__blob=publicationFile&v=8 (accessed on 18 December 2021).
5. SLM Solutions. SLM®800. 2021. Available online: https://www.slm-solutions.com/products-and-solutions/machines/slm-800/ (accessed on 18 December 2021).
6. Aachen University of Applied Sciences and the Fraunhofer Institute for Laser Technology ILT. Aachen Center for 3D Printing: Official Launch of the World's Largest SLM Facility. 2016. Available online: https://www.ilt.fraunhofer.de/content/dam/ilt/en/documents/Publication-and-Press/press_release/pr2017/PR_Official-launch-of-XLine-2000R.pdf (accessed on 18 December 2021).
7. Yadroitsev, I. Selective Laser Melting: Direct Manufacturing of 3D-Objects by Selective Laser Melting of Metal Powders. Ph.D. Thesis, Saint-Etienne, London, UK, 2009.
8. Brandl, E.; Heckenberger, U.; Holzinger, V.; Buchbinder, D. Additive manufactured AlSi10Mg samples using Selective Laser Melting (SLM): Microstructure, high cycle fatigue, and fracture behavior. *Mater. Des.* **2012**, *34*, 159–169. [CrossRef]
9. Buchbinder, D. Selective Laser Melting von Aluminiumgusslegierungen. Ph.D. Thesis, Shaker, Aachen, Germany, 2013.
10. Greitemeier, D. *Untersuchung der Einflussparameter auf die Mechanischen Eigenschaften von Additiv Gefertigtem TiAl6V4*; Springer Fachmedien Wiesbaden: Wiesbaden, Germany, 2016. [CrossRef]
11. Günther, J.; Krewerth, D.; Lippmann, T.; Leuders, S.; Tröster, T.; Weidner, A.; Biermann, H.; Niendorf, T. Fatigue life of additively manufactured Ti–6Al–4V in the very high cycle fatigue regime. *Int. J. Fatigue* **2017**, *94*, 236–245. [CrossRef]
12. Kunze, K.; Etter, T.; Grässlin, J.; Shklover, V. Texture, anisotropy in microstructure and mechanical properties of IN738LC alloy processed by selective laser melting (SLM). *Mater. Sci. Eng. A* **2015**, *620*, 213–222. [CrossRef]
13. Röttger, A.; Geenen, K.; Windmann, M.; Binner, F.; Theisen, W. Comparison of microstructure and mechanical properties of 316 L austenitic steel processed by selective laser melting with hot-isostatic pressed and cast material. *Mater. Sci. Eng. A* **2016**, *678*, 365–376. [CrossRef]

14. Liverani, E.; Toschi, S.; Ceschini, L.; Fortunato, A. Effect of selective laser melting (SLM) process parameters on microstructure and mechanical properties of 316L austenitic stainless steel. *J. Mater. Process. Technol.* **2017**, *249*, 255–263. [CrossRef]
15. Röttger, A.; Boes, J.; Theisen, W.; Thiele, M.; Esen, C.; Edelmann, A.; Hellmann, R. Microstructure and mechanical properties of 316L austenitic stainless steel processed by different SLM devices. *Int. J. Adv. Manuf. Technol.* **2020**, *108*, 769–783. [CrossRef]
16. Laufs, P. *Reaktorsicherheit für Leistungskernkraftwerke*; Springer: Berlin/Heidelberg, Germany, 2013. [CrossRef]
17. Seebich, H. Mikromechanisch Basierte Schädigungsmodelle zur Beschreibung des Versagensablaufs Ferritischer Bauteile. Ph.D. Thesis, Universität Stuttgart, Stuttgart, Germany, 2007.
18. Ewald, J.E.A. *Forschungsprogramm "Reaktordruckbehälter" Dringlichkeitsprogramm 22NiMoCr3-7*; Technical Report; MPA Stuttgart: Stuttgart, Germany, 1975.
19. Raffeis, I.; Vrommen, U; Westhoff, E.; Bremen, S.W.; Schaberger-Zimmermann, E. *Konfiguration der Mikrostruktur Eines Ferritisch-Austenitischen Duplexstahls in der SLM-Fertigung*; Fraunhofer-Institut für Lasertechnik-ILT: Potsdam, Germany, 2018. Available online: https://publications.rwth-aachen.de/record/741740/files/741740.pdf (accessed on 18 December 2021).
20. Feng, S.; Chen, X.; Zhang, P.; Ji , Z.; Ming, F.; Ren, S.; Qu, X. Novel Ferritic Stainless Steel with Advanced Mechanical Properties and Significant Magnetic Responses Processed by Selective Laser Melting. *Mater. Trans.* **2019**, *60*, 1096–1102. [CrossRef]
21. Boegelein, T.; Dryepondt, S.N.; Pandey, A.; Dawson, K.; Tatlock, G.J. Mechanical response and deformation mechanisms of ferritic oxide dispersion strengthened steel structures produced by selective laser melting. *Acta Mater.* **2015**, *87*, 201–215. [CrossRef]
22. Garibaldi, M.; Ashcroft, I.; Simonelli, M.; Hague, R. Metallurgy of high-silicon steel parts produced using Selective Laser Melting. *Acta Mater.* **2016**, *110*, 207–216. [CrossRef]
23. Jiang, M.G.; Chen, Z.W.; Tong, J.D.; Liu, C.Y.; Xu, G.; Liao, H.B.; Wang, P.; Wang, X.Y.; Xu, M.; Lao, C.S. Strong and ductile reduced activation ferritic/martensitic steel additively manufactured by selective laser melting. *Mater. Res. Lett.* **2019**, *7*, 426–432. [CrossRef]
24. Karlsson, D.; Chou, C.Y.; Pettersson, N.H.; Helander, T.; Harlin, P.; Sahlberg, M.; Lindwall, G.; Odqvist, J.; Jansson, U. Additive manufacturing of the ferritic stainless steel SS441. *Addit. Manuf.* **2020**, *36*, 101580. [CrossRef]
25. Qiu, C.; Kindi, M.A.; Aladawi, A.S.; Hatmi, I.A. A comprehensive study on microstructure and tensile behaviour of a selectively laser melted stainless steel. *Sci. Rep.* **2018**, *8*, 7785. [CrossRef] [PubMed]
26. Karapatis, N.P.; Egger, G.; Gygax, P.E.; Glardon, R. *Optimization of Powder Layer Density in Selective Laser Sintering*; The University of Texas at Austin: Austin, TX, USA, 1999. [CrossRef]
27. TÜV Bayern. *Anlage 2 zum Abnahmeprüfzeugnis*; Prüfungs-Nr. 1415113/7; TÜV Bayern: Füssen, Germany, 1976.

Article

Use of Fumed Silica Nanostructured Additives in Selective Laser Melting and Fabrication of Steel Matrix Nanocomposites

Hwee Kang Koh [1], James Guo Sheng Moo [2], Swee Leong Sing [3,*] and Wai Yee Yeong [1,*]

1 Singapore Centre for 3D Printing, School of Mechanical & Aerospace Engineering, Nanyang Technological University, 50 Nanyang Avenue, Singapore 639798, Singapore; m170010@e.ntu.edu.sg
2 Evonik (SEA) Pte. Ltd., Asia Research Hub, 21 Biopolis Road, Singapore 138567, Singapore; james.moo@evonik.com
3 Department of Mechanical Engineering, National University of Singapore, 9 Engineering Drive 1, Singapore 117575, Singapore
* Correspondence: sweeleong.sing@nus.edu.sg (S.L.S.); wyyeong@ntu.edu.sg (W.Y.Y.); Tel.: +65-651-6255 (S.L.S.); +65-6790-4343 (W.Y.Y.)

Abstract: The advancement of additive manufacturing (AM) for metal matrix nanocomposites (MM-NCs) is gaining enormous attention due to their potential improvement of physical and mechanical performance. When using nanostructured additives as reinforcements in 3D printed metal composites and with the aid of selective laser melting (SLM), the mechanical properties of the composites can be tailored. The nanostructured additive AEROSIL® fumed silica is both cost-effective and beneficial in the production of MMNCs using SLM. In this study, both hydrophobic and hydrophilic fumed silicas were shown to successfully achieve homogenous blends with commercial 316L stainless steel powder. The powder blends, which exhibited better flow, were then used to fabricate samples using SLM. The samples' microstructure demonstrated that smaller grains were present in the composites, resulting in improvements in mechanical properties by grain refinement compared to those of 316L stainless steel samples.

Keywords: additive manufacturing; 3D printing; selective laser melting; powder bed fusion; metal matrix composites

Citation: Koh, H.K.; Moo, J.G.S.; Sing, S.L.; Yeong, W.Y. Use of Fumed Silica Nanostructured Additives in Selective Laser Melting and Fabrication of Steel Matrix Nanocomposites. *Materials* **2022**, *15*, 1869. https://doi.org/10.3390/ma15051869

Academic Editor: Irina Hussainova

Received: 28 January 2022
Accepted: 28 February 2022
Published: 2 March 2022

Publisher's Note: MDPI stays neutral with regard to jurisdictional claims in published maps and institutional affiliations.

1. Introduction

Additive manufacturing (AM), also commonly known as 3D printing or formally known as rapid prototyping (RP), is a family of processes that produces parts by joining materials together, usually layer-by-layer, based on information retrieved from computer aided design (CAD) software without the use of physical moulds. They can be directly used to fabricate functional parts without the need for post-processing. AM also circumvents some of the limitations of conventional methods, such as the need for multiple exhaustive processes that must be used to produce sophisticated and complex parts [1].

AM has been increasingly used over the years and provides key benefits that product manufacturers over conventional manufacturing. These benefits include, but are not limited to, cost effectiveness from the short downtime in terms of prototyping, reduced material wastage, vast potential for product design innovation, and the ability to build complex workpieces which cannot be achieved through conventional processes [1]. AM is capable of fabricating parts using metals, ceramic, and polymers [2,3]. Among the many AM methods currently available in the industry, several processes such as selective laser melting (SLM) and direct metal laser sintering (DMLS) are used for the fabrication of metals. SLM and DMLS are also commonly called laser powder bed fusion (LPBF). Due to the inherent benefits of AM over conventional manufacturing, it has been adopted widely for manufacturing by many industries, including aerospace, medical, and automotive [4].

Stainless steel has been used widely across many industries including construction, automotive, and pharmaceutical. It has been one of the most essential and important materials to keep costs low compared to other available materials [5–8]. Among all grades of stainless steel, 316L stainless steel is commonly used due to its mechanical properties' advantages, such as being highly resistant to corrosion and oxidation. Many papers in the literature reported on SLM 316L stainless steel samples with regard to the effect of processing parameters, such as laser power, scanning speed, hatch distance, and scanning pattern [5,6]. Many engineers have leveraged the advantages of this technology and have successfully experimented with adding in different kinds of additives to improve the characteristics of stainless steel by forming metal matrix composites (MMCs). MMCs have been effectively applied to aerospace, automotive, and medical industries, which require the latest material improvements [5]. MMCs are normally fabricated with a metallic matrix phase and with one or more different kinds of phases for reinforcement purposes. The reinforcements come in the form of particulates, whiskers, or fibres from materials such as ceramics, carbon or metallic materials. Studies have been conducted to improve the properties by introducing solid solution [9] and grain refinement [10]. Past research has involved adding in reinforcement such as silicon carbide (SiC), titanium diboride (TiB_2), and titanium carbide (TiC) in 316L stainless steel to increase wear resistance [3,11,12]. Furthermore, investigations have also been conducted on the use of nanostructured materials such as silica, titania, carbon nanotubes, and layered silicates as reinforcements to modify the properties of different matrices [13,14]. There are reports regarding the use of small amounts of nanostructured materials to polymers, which was shown to improve the mechanical properties [15–17] and dimensional stability under creep conditions [18]. Among all the additives, fumed silica is a prime candidate that can be produced with different surface areas (50–400 m^2/g) using a variety of surface treatments from hydrophilic to hydrophobic. These silica structures have a fractal structure, and can form a network of interconnecting particles [19].

The ability to be customised to suit the industry's needs in comparison to conventional metals has increased the use of MMCs in the industry [20]. In the past twenty years, the use of MMCs has grown significantly [21]. Numerous researchers have conducted intensive tests and proven, with the aid of suitable reinforcement, that the shortcomings of properties in materials such as 316L stainless steel can be improved [11,22,23]. This led to an increase in demand for customised material, specifically with lightweight and high-performance parts, which has steered the advancement of metal matrix nanocomposites (MMNCs).

MMNCs, a version of MMCs, use nanostructured materials as reinforcements. Due to their nanostructure morphology, MMNCs are capable of performing better than MMCs for wear resistance, damping properties, and mechanical strength, which are key for various applications [24,25]. Experiments have been conducted and the results have shown that for aluminium matrix composites with only 3 vol % of Al_2O_3 nanostructured materials, the parts performed better in comparison to the same parts made of 10 vol % Al_2O_3 and 10 vol.% SiC reinforcement in the microscale [26,27].

Despite the exceptional gains in interest from these industries, the complex processes and lack of economic efficiency have restricted the use of MMNCs. However, SLM has become a viable processing technique due to its ability to fabricate near-net-shaped parts and recycle unused powder which lowered the cost of producing MMNCs. Thus, SLM shows rising popularity in the production of MMNCs with complex structures and properties at a fair cost [28–31]. This has led to various studies concentrating on the exploration and development of this process [32,33].

In this study, the properties of MMNCs fabricated under the same processing conditions with 316L stainless steel combined with reinforcement materials, which consisted of hydrophilic and hydrophobic fumed silicas from the same manufacturer, are analysed. The investigation also evaluate the compatibility of different reinforced materials in terms of distribution and agglomeration. Finally, conclusions were drawn on the effect of different properties of additives (hydrophilic or hydrophobic) on the MMNCs properties, fabricated using SLM by benchmarking them with 316L stainless steel.

2. Materials & Methods

2.1. Preparation of Powder

The 316L stainless steel powder used in this experiment was produced by TLS Technik Spezialpulver (Bitterfeld-Wolfen, Germany) and was spherical in shape with a particle size distribution between 20 and 63 μm. The two kinds of fumed silica powder used were AEROSIL® R 812 S hydrophobic fumed silica and AEROSIL® 200 hydrophilic fumed silica. AEROSIL® 200 is a hydrophilic fumed silica with surface silanol groups; AEROSIL® R 812 S is a hydrophobic fumed silica surface modified with hexamethyldisilazane. The same amount (0.05 wt %) of silica was added to the 316L stainless steel powder in this experiment to determine if one produced more advantages over the other by evaluating the characteristics of the specimens. The powders were mixed using an Inversina Tumbler Mixer (Bioengineering, Inc., Massachusetts, MA, United States). The mixing speed was set at 140 rpm and the duration was set at 4 h for the powder to flow and mix freely while using three-dimensional inversion kinematics, otherwise known as the Paul Schatz principle, to homogenize the powders. Powder flowability was measured using the ISO4490 standard with a Hall flowmeter (MZ-102, Mayzun, Shenzhen, China), which measures the time needed for 50 g of a sample to pass through a funnel with an aperture of 2.5 mm.

2.2. Selective Laser Melting

The SLM machine used to fabricate the specimens was an SLM 250HL (SLM Solutions AG, Lubeck, Germany). Its build volume can reach 250 × 250 × 250 mm^3 and produce homogeneous metal components using fine metal powders. During the process of fabricating the specimens, argon gas was introduced into the system to prevent specimens from oxidising by keeping the oxygen level below 2%. During the process, a recoater was used to lay a coat of powder across the building platform. As it moved backwards, the recoater blade smoothed out the surface, and excess powder was collected in a container, which would be recycled. The laser beam scanned each layer of powder directed by the data of the stereolithography (STL) file. The building platform was lowered by the layer thickness defined after each scan before a new coat of powder was laid by the recoater. This cycle repeated until the final parts were fabricated. The processing parameters used for fabricating the specimens were laser power of 400 W, scanning speed of 760 mm/s, hatch spacing of 0.120 mm, and layer thickness of 0.05 mm. The stripes laser scanning pattern was used.

A total of three sets of five cubic specimens, with each set for 316L stainless steel, 316L stainless steel with AEROSIL® R 812 S, and 316L stainless steel with AEROSIL® 200, were fabricated for testing and evaluation. Each cube had dimensions of 10 × 10 × 10 mm^3 and were fabricated to conduct density, surface roughness, microhardness tests and microstructure analysis.

2.3. Density

The Archimedes method was used to measure the density of the specimens. The measurements were taken using a XS204 density test kit (Mettler Toledo, Columbus, OH, USA), with 99.9% ethanol at 23.8 °C. Five replicates for each material were used in the density measurement.

2.4. Surface Roughness

The surface roughness of the samples was measured using a confocal microscope (VK-X130K, Keyence Corporation, Osaka, Japan) to find the roughness average (Ra). Five replicates were used in the measurement with five Ra values obtained for each surface. Each specimen's measurement was taken straight after removal from the substrate to evaluate the surface produced by SLM and to identify any characteristic changes due to the different additives.

2.5. Grinding and Polishing

After the samples were mounted, grinding and polishing were performed prior to performing microhardness and microstructure tests. The samples were first processed on a grinding and polishing machine (Tegramin-25, Struers, Copenhagen, Denmark) with 320 grit SiC paper for 5 min, followed by MD-Largo and DiaPro Largo 9 μm suspension for 5 min. The specimens then underwent polishing steps with MD-Dac and DiaPro Dac 3 μm suspension for 5 min and another polish with MD-Nap and DiaPro Nap 1 μm suspension for 3 min. Lastly, in order to achieve a mirror finish for the samples, they were polished with Chem and OP-S 0.25 μm suspension for 1 min. All the polishing cloths and suspensions are from Struers (Copenhagen, Denmark). All grinding and polishing steps were completed with 150 rpm co-rotation and 25 N force pressing on the samples.

2.6. Microstructure Analysis

Microstructure analysis was carried out on the mirror polished and etched surfaces (using etchant containing HCl and HNO_3 in a ratio of 1:1). Microstructural analysis was also performed on the powders of 316L stainless steel, 316L stainless steel with AEROSIL® R 812 S and 316L stainless steel with AEROSIL® 200. All (5 replicates) top and side surfaces of the cubes and fractured surfaces of the tensile test coupons were also evaluated.

2.7. Microhardness Testing

The Vickers microhardness testing was performed using an FM-300E microhardness tester (Future-Tech Corp., Kanagawa, Japan). Five replicates were used, with 10 measurements obtained from each top and side surface.

2.8. Tensile Test

Tensile test coupons were fabricated to evaluate the performance of 316L stainless steel, 316L stainless steel with AEROSIL® R 812 S, and 316L stainless steel with AEROSIL® 200. Five test coupons following the ASTM E8/E8M standard for the tensile testing of each material were fabricated. The dimensions of the samples followed the subsize specimen stated in the standard, with a gauge length of 40 mm. The tensile tests were conducted using an Instron Static Tester Series 5569 (Instron, Norwood, MA, USA) with a 50 kN load cell and a strain rate of 1 mm/min.

3. Results and Discussion

Fumed silica products of AEROSIL® were mixed homogenously with 316L stainless steel powders. The two additives were homogenously coated onto the 316L stainless steel powder particles, as shown in Figure 1. The nanostructure fumed silica is shown as a white spot that spread evenly on 316L stainless steel particles, indicated by the white arrows.

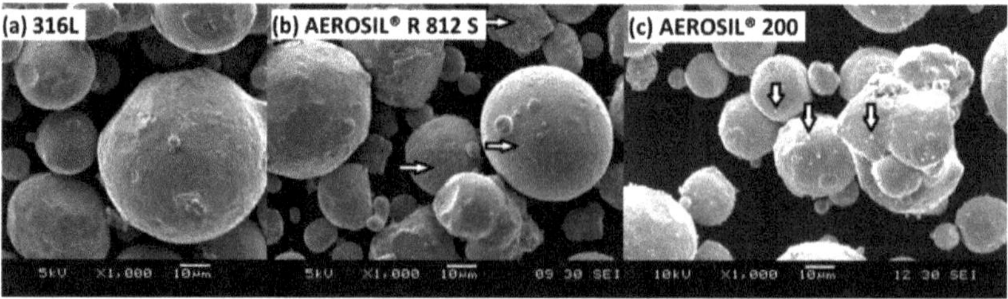

Figure 1. Scanning electron microscopy of (**a**) 316L stainless steel, (**b**) 316L stainless steel with 0.05 wt % AEROSIL® R 812 S, and (**c**) 316L stainless steel with 0.05 wt % AEROSIL® 200 powder at 1000× magnification.

In Figure 2, the addition of 0.05 wt % AEROSIL® 200 or AEROSIL® R 812 S had a slight effect on the flowability of the powders. This is attributed to the fact that the 316L stainless steel powder particles were already spherical, hence possessed adequate flowability. The total flow time taken for a 50 g sample was used as a gauge for flowability using a Hall flowmeter. There was little improvement in the metal powder's flow, with a total flow time of 16.8 s for 316L stainless steel powder compared to 17.2 s with the addition of 0.05 wt % AEROSIL® 200. However, a reduction to a flow time of 14.8 s with the addition of 0.05 wt % AEROSIL® R 812 S was observed, corresponding to a 12% improvement. The hydrophobic modification of the silica surface in AEROSIL® R 812 S fumed silica particles reduced cohesive forces between the powder particles, enabling the silica particles to serve as spacers to improve powder flow.

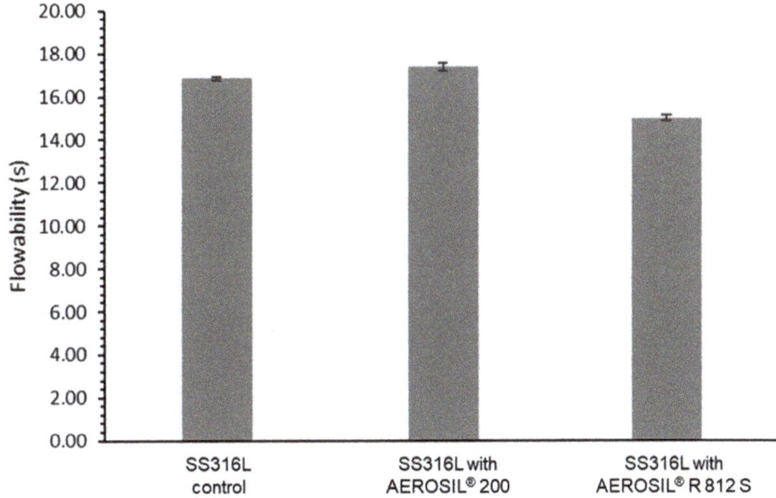

Figure 2. Flowability of 316L stainless steel, 316L stainless steel with 0.05 wt % AEROSIL® 200, and 316L stainless steel with 0.05 wt % AEROSIL® R 812 S powders.

The density of the SLM fabricated samples is shown in Table 1. With a small amount of additives added, there was an insignificant change in the densities in terms of average density, which was expected. The small amount of additives was not effective in modifying the thermal properties of the powder. However, AEROSIL® 200 significantly lowered the density's standard deviation, which may be attributed to more stable melt flows during SLM. A higher stability is desirable in terms of fabrication. Smaller deviations in density refer to the ability to produce consistent parts, which can eventually reduce quality control issued and downtime.

Table 1. Density of SLM specimens.

Sample	SS316L	AEROSIL® R 812 S	AEROSIL® 200
Density (g/cm^3)	7.722 ± 0.012	7.718 ± 0.011	7.718 ± 0.005

In this experiment, marginal improvement in the surface roughness was observed in the specimens mixed with AEROSIL® R 812 S in terms of average value and smaller standard deviation compared to that of 316L stainless steel. The average value and standard deviation taken from the samples mixed with AEROSIL® 200 showed a slightly rough texture, as shown in Table 2. The surface roughness of the samples could have been due to the partially melted particles adhering to the specimens.

Table 2. Surface roughness of SLM specimens.

Sample	SS316L		SS316L with AEROSIL® R 812 S		SS316L with AEROSIL® 200	
Surface	Top	Side	Top	Side	Top	Side
Ra (μm)	15.6 ± 1.5	16.6 ± 1.4	14.0 ± 1.7	13.8 ± 0.7	17.6 ± 7.0	15 ± 1.5

Microhardness tests were performed to investigate the surface hardness of the produced samples. The results shown in Figure 3 show improvement in the microhardness of specimens with the addition of 0.05 wt % AEROSIL® 200 and 0.05 wt % AEROSIL® R 812 S. This shows that there was a significant improvement in the microhardness of the parts manufactured with silica additives. The microhardness of the 316L stainless steel, 316L stainless steel with AEROSIL® 200, and 316L stainless steel with AEROSIL® R 812 S are 177, 247, and 242 HV, respectively. As such, 40% and 37% improvements due to the silica additives AEROSIL® 200 and AEROSIL® R 812 S were achieved, respectively. Such improvements present an opportunity for enhancing the wear and durability of functional parts. The low standard deviation showed that the hardness of the specimen was relatively consistent throughout the surface, which is desirable. These consistent hardness values lead to accurate material selection, which might have otherwise resulted in premature mechanical failure.

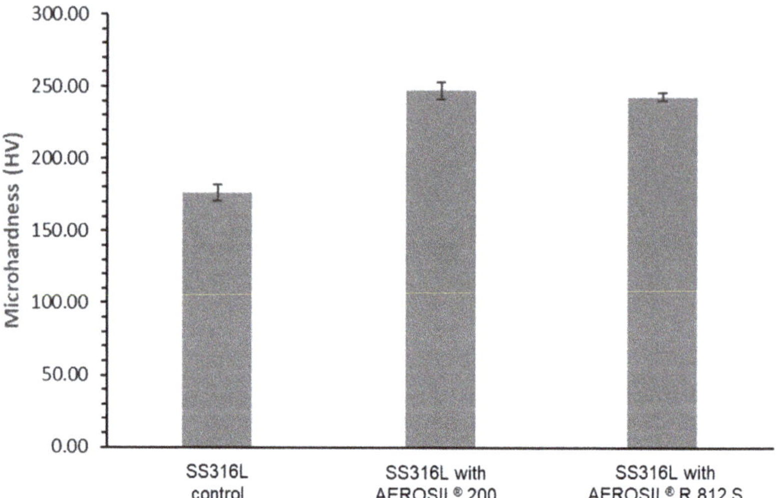

Figure 3. Microhardness of SS316L, SS316L with 0.05 wt % AEROSIL® 200, and SS316L with 0.05 wt % AEROSIL® R 812 S.

A comparison of the tensile properties of 316L stainless steel and 316L stainless steel with AEROSIL® R 812 S and AEROSIL® 200 is shown in Figure 4 and Table 3. The specimens with AEROSIL® R 812 S additives had properties that were comparable to those of 316L stainless steel, while specimens with AEROSIL® 200 showed poorer performance compared to those with AEROSIL® R 812 S additives, but they still performed better than the 316L stainless steel samples. The test results demonstrated that the MMNCs produced have better tensile properties than 316L stainless steel.

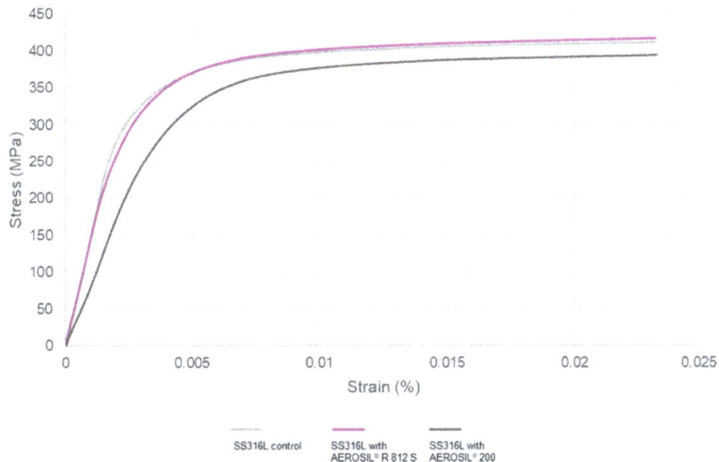

Figure 4. Tensile properties of 316L stainless steel, 316L stainless steel with 0.05 wt % hydrophobic AEROSIL® R 812 S, and 316L stainless steel with 0.05 wt % hydrophilic AEROSIL® 200.

Table 3. Young's modulus, ultimate tensile strength, yield strength, and elongation of SLM-built samples.

Specimen	SS316L	SS316L with AEROSIL® R 812 S	SS316L with AEROSIL® 200
Young's modulus (GPa)	147.91 ± 5.67	143.14 ± 8.71	81.84 ± 6.25
Ultimate tensile strength (MPa)	491.76 ± 11.05	494.78 ±16.54	485.05 ± 14.56
Yield strength (MPa)	363.4 ± 8.22	358.9 ± 12.63	283.7 ± 10.29
Elongation (%)	38.14 ± 3.86	35.18 ± 5.23	37.47 ± 2.71

As stated in Table 3, 316L stainless steel with AEROSIL® R 812 S had better ultimate tensile strength at 495 MPa than 316L stainless steel with AEROSIL® 200 at 485 MPa and 316L stainless steel at 492 MPa, while maintaining a satisfactory Young's modulus, yield strength, and elongation. However, SS316L with AEROSIL® 200 suffered from a reduction in Young's modulus to 82 GPa and a yield strength of 284 MPa compared to the control of SS316L at 148 GPa and 363 MPa, which are reductions of 44% and 22%, respectively. This phenomenon is explained later using Figure 5.

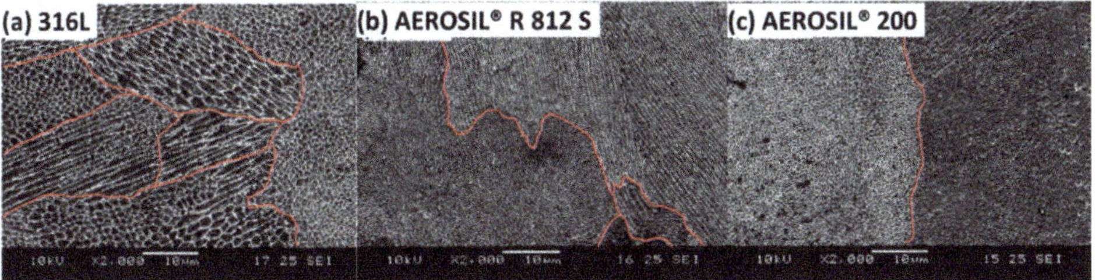

Figure 5. Scanning electron micrograph of (**a**) 316L stainless steel, (**b**) 316L stainless steel with 0.05 wt % AEROSIL® R 812 S, and (**c**) 316L stainless steel with 0.05 wt % AEROSIL® 200 after etching in 2000× magnification.

The differences in tensile properties between the samples can be explained by the differences in their microstructure. Samples with columnar grains have lower strength but higher plasticity, while those with equiaxed grains have higher strength and lower plasticity. A mix of equiaxed grains and columnar grains and, to an extent, an even distribution of the two grain types, would provide the optimal balance of strength and plasticity [34]. Specimens of 316L stainless steel with AEROSIL® 200 had a more distinct segregation of equiaxed grains and columnar grains compared to pure 316L stainless steel specimens, as shown in Figure 5. The distinct segregation could be a reason for the reduction in the mechanical properties of the specimens of 316L stainless steel with AEROSIL® 200.

Another significant difference worth mentioning is the grain size observed in Figure 6. Based on image measurements, the 316L stainless steel specimens contained five grains within 5 μm, while both 316L stainless steel with AEROSIL® R 812 S and 316L stainless steel with AEROSIL® 200 specimens showed eight grains in 5 μm. This implies that the additives resulted in smaller grains, which likely led to the improvements in microhardness observed in Figure 2. The increase in their mechanical properties is due to the reduction in grain size, which potentially leads to improved microhardness of the specimens when accompanied by ideal grain segregations [34].

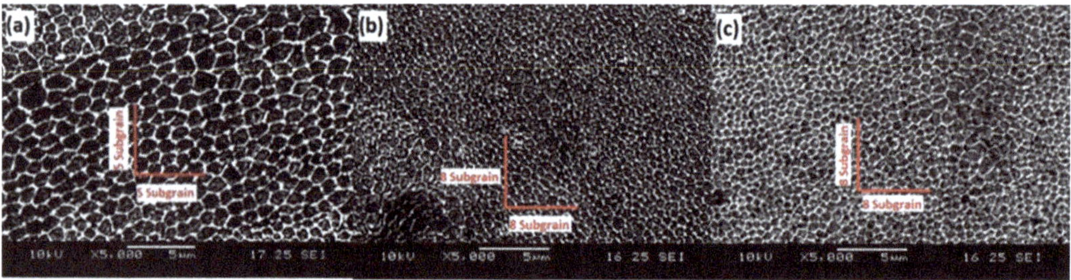

Figure 6. Scanning electron micrograph of (**a**) 316L stainless steel, (**b**) 316L stainless steel with 0.05 wt % AEROSIL® R 812 S, and (**c**) 316L stainless steel with 0.05 wt % AEROSIL® 200 after etching in 5000× magnification.

4. Conclusions

In this investigation, adding fumed silica to 316L stainless steel showed the potential to improve the mechanical properties of parts fabricated by SLM. This was achieved by using additives to manipulate the microstructure of the MMNCs, which resulted in grain refinement and distinct segregation between the equiaxed and columnar grain structures. The 316L stainless steel with hydrophobic AEROSIL® R 812 S fumed silica specimens had lower surface roughness and improved microhardness. Furthermore, the 316L stainless steel with AEROSIL® R 812 S specimens achieved tensile properties comparable to those of 316L stainless steel. This shows that the mechanical properties of the material can be improved with the introduction of fumed silica as nanostructured reinforcements in MMNCs. However, the 316L stainless steel with hydrophilic AEROSIL® 200 fumed silica specimens performed poorly compared to 316L stainless steel in several aspects and, thus, may not be a suitable additive for MMNCs produced by SLM. Future work can consider adding different weightages of fumed silica to investigate and understand the relationship with regard to the changes in material characteristics.

Author Contributions: Conceptualization, S.L.S., W.Y.Y. and J.G.S.M.; methodology, S.L.S., W.Y.Y. and J.G.S.M.; formal analysis, H.K.K., S.L.S. and J.G.S.M.; investigation, H.K.K. and S.L.S.; resources, W.Y.Y. and J.G.S.M.; data curation, H.K.K.; writing—original draft preparation, H.K.K. and J.G.S.M.; writing—review and editing, S.L.S., W.Y.Y. and J.G.S.M.; supervision, S.L.S. and W.Y.Y.; project administration, W.Y.Y.; funding acquisition, S.L.S., W.Y.Y. and J.G.S.M. All authors have read and agreed to the published version of the manuscript.

Funding: This research received no external funding.

Institutional Review Board Statement: Not applicable.

Informed Consent Statement: Not applicable.

Data Availability Statement: Data are made available upon request, subjected to confidentiality clauses.

Conflicts of Interest: The authors declare no conflict of interest.

References

1. Takezawa, A.; Koizumi, Y.; Kobashi, M. High-stiffness and strength porous maraging steel via topology optimization and selective laser melting. *Addit. Manuf.* **2017**, *18*, 194–202. [CrossRef]
2. Frazier, W.E. Metal additive manufacturing: A review. *J. Mater. Eng. Perform.* **2014**, *23*, 1917–1928. [CrossRef]
3. Wu, C.L.; Zhang, S.; Zhang, C.H.; Zhang, J.B.; Liu, Y.; Chen, J. Effects of sic content on phase evolution and corrosion behavior of sic-reinforced 316l stainless steel matrix composites by laser melting deposition. *Opt. Laser Technol.* **2019**, *115*, 134–139. [CrossRef]
4. Tomus, D.; Tian, Y.; Rometsch, P.A.; Heilmaier, M.; Wu, X. Influence of post heat treatments on anisotropy of mechanical behaviour and microstructure of hastelloy-x parts produced by selective laser melting. *Mater. Sci. Eng. A* **2016**, *667*, 42–53. [CrossRef]
5. Nickels, L. Am and aerospace: An ideal combination. *Met. Powder Rep.* **2015**, *70*, 300–303. [CrossRef]
6. Wang, H. Stainless steel as bipolar plate material for polymer electrolyte membrane fuel cells. *J. Power Sources* **2003**, *115*, 243–251. [CrossRef]
7. Hanzl, P.; Zetek, M.; Bakša, T.; Kroupa, T. The influence of processing parameters on the mechanical properties of slm parts. *Procedia Eng.* **2015**, *100*, 1405–1413. [CrossRef]
8. Boldridge, D. Morphological characterization of fumed silica aggregates. *Aerosol Sci. Technol.* **2010**, *44*, 182–186. [CrossRef]
9. Rawers, J.; Croydon, F.; Krabbe, R.; Duttlinger, N. Tensile characteristics of nitrogen enhanced powder injection moulded 316l stainless steel. *Powder Metall.* **2013**, *39*, 125–129. [CrossRef]
10. Lei, Y.B.; Wang, Z.B.; Xu, J.L.; Lu, K. Simultaneous enhancement of stress- and strain-controlled fatigue properties in 316l stainless steel with gradient nanostructure. *Acta Mater.* **2019**, *168*, 133–142. [CrossRef]
11. AlMangour, B.; Grzesiak, D.; Yang, J.-M. In-situ formation of novel tic-particle-reinforced 316l stainless steel bulk-form composites by selective laser melting. *J. Alloys Compd.* **2017**, *706*, 409–418. [CrossRef]
12. Nahme, H.; Lach, E.; Tarrant, A. Mechanical property under high dynamic loading and microstructure evaluation of a tib2 particle-reinforced stainless steel. *J. Mater. Sci.* **2009**, *44*, 463–468. [CrossRef]
13. Bondioli, F.; Dorigato, A.; Fabbri, P.; Messori, M.; Pegoretti, A. Improving the creep stability of high-density polyethylene with acicular titania nanoparticles. *J. Appl. Polym. Sci.* **2009**, *112*, 1045–1055. [CrossRef]
14. Bondioli, F.; Dorigato, A.; Fabbri, P.; Messori, M.; Pegoretti, A. High-density polyethylene reinforced with submicron titania particles. *Polym. Eng. Sci.* **2008**, *48*, 448–457. [CrossRef]
15. Cai, L.F.; Lin, Z.Y.; Qian, H. Dispersion of nano silica in monomer casting nylon6 and its effect on the structure and properties of composites. *Express Polym. Lett.* **2010**, *4*, 397–403. [CrossRef]
16. Mandalia, T.; Bergaya, F. Organo clay mineral–melted polyolefin nanocomposites effect of surfactant/cec ratio. *J. Phys. Chem. Solids* **2006**, *67*, 836–845. [CrossRef]
17. Pegoretti, A.; Dorigato, A.; Penati, A. Tensile mechanical response of polyethylene–clay nanocomposites. *Express Polym. Lett.* **2007**, *1*, 123–131. [CrossRef]
18. Starkova, O.; Yang, J.; Zhang, Z. Application of time–stress superposition to nonlinear creep of polyamide 66 filled with nanoparticles of various sizes. *Compos. Sci. Technol.* **2007**, *67*, 2691–2698. [CrossRef]
19. Dorigato, A.; Pegoretti, A.; Penati, A. Linear low-density polyethylene/silica micro- and nanocomposites: Dynamic rheological measurements and modelling. *Express Polym. Lett.* **2010**, *4*, 115–129. [CrossRef]
20. Shi, J.; Wang, Y. Development of metal matrix composites by laser-assisted additive manufacturing technologies: A review. *J. Mater. Sci.* **2020**, *55*, 9883–9917. [CrossRef]
21. Dadbakhsh, S.; Mertens, R.; Hao, L.; Van Humbeeck, J.; Kruth, J.P. Selective laser melting to manufacture "in situ" metal matrix composites: A review. *Adv. Eng. Mater.* **2019**, *21*, 1801244. [CrossRef]
22. Li, B.; Qian, B.; Xu, Y.; Liu, Z.; Zhang, J.; Xuan, F. Additive manufacturing of ultrafine-grained austenitic stainless steel matrix composite via vanadium carbide reinforcement addition and selective laser melting: Formation mechanism and strengthening effect. *Mater. Sci. Eng. A* **2019**, *745*, 495–508. [CrossRef]
23. Song, B.; Wang, Z.; Yan, Q.; Zhang, Y.; Zhang, J.; Cai, C.; Wei, Q.; Shi, Y. Integral method of preparation and fabrication of metal matrix composite: Selective laser melting of in-situ nano/submicro-sized carbides reinforced iron matrix composites. *Mater. Sci. Eng. A* **2017**, *707*, 478–487. [CrossRef]
24. Mortensen, A.; Llorca, J. Metal matrix composites. *Annu. Rev. Mater. Res.* **2010**, *40*, 243–270. [CrossRef]
25. Moya, J.; Lopezesteban, S.; Pecharroman, C. The challenge of ceramic/metal microcomposites and nanocomposites. *Prog. Mater. Sci.* **2007**, *52*, 1017–1090. [CrossRef]
26. Tjong, S.C. Novel nanoparticle-reinforced metal matrix composites with enhanced mechanical properties. *Adv. Eng. Mater.* **2007**, *9*, 639–652. [CrossRef]

27. Kang, Y.-C.; Chan, S.L.-I. Tensile properties of nanometric al2o3 particulate-reinforced aluminum matrix composites. *Mater. Chem. Phys.* **2004**, *85*, 438–443. [CrossRef]
28. Kühn, U.; Mattern, N.; Gebert, A.; Kusy, M.; Boström, M.; Siegel, U.; Schultz, L. Nanostructured zr- and ti-based composite materials with high strength and enhanced plasticity. *J. Appl. Phys.* **2005**, *98*, 54307. [CrossRef]
29. Liu, L.H.; Yang, C.; Wang, F.; Qu, S.G.; Li, X.Q.; Zhang, W.W.; Li, Y.Y.; Zhang, L.C. Ultrafine grained ti-based composites with ultrahigh strength and ductility achieved by equiaxing microstructure. *Mater. Des.* **2015**, *79*, 1–5. [CrossRef]
30. Zhang, L.-C.; Xu, J.; Eckert, J. Thermal stability and crystallization kinetics of mechanically alloyed tic/ti-based metallic glass matrix composite. *J. Appl. Phys.* **2006**, *100*, 33514. [CrossRef]
31. Attar, H.; Bönisch, M.; Calin, M.; Zhang, L.C.; Zhuravleva, K.; Funk, A.; Scudino, S.; Yang, C.; Eckert, J. Comparative study of microstructures and mechanical properties of in situ ti–tib composites produced by selective laser melting, powder metallurgy, and casting technologies. *J. Mater. Res.* **2014**, *29*, 1941–1950. [CrossRef]
32. Gu, D.; Shen, Y. Direct laser sintered wc-10co/cu nanocomposites. *Appl. Surf. Sci.* **2008**, *254*, 3971–3978. [CrossRef]
33. AlMangour, B.; Baek, M.-S.; Grzesiak, D.; Lee, K.-A. Strengthening of stainless steel by titanium carbide addition and grain refinement during selective laser melting. *Mater. Sci. Eng. A* **2018**, *712*, 812–818. [CrossRef]
34. Wang, K.; Wang, D.; Han, F. Effect of crystalline grain structures on the mechanical properties of twinning-induced plasticity steel. *Acta Mech. Sin.* **2015**, *32*, 181–187. [CrossRef]

Article

Bone Conduction Capacity of Highly Porous 3D-Printed Titanium Scaffolds Based on Different Pore Designs

Ho-Kyung Lim [1], Miyoung Ryu [2], Su-Heon Woo [3], In-Seok Song [4], Young-Jun Choi [5] and Ui-Lyong Lee [5,*]

1 Department of Oral and Maxillofacial Surgery, Korea University Guro Hospital, Seoul 08308, Korea; ungassi@naver.com
2 CUSMEDI Inc., Sungnam 13217, Korea; miyung0829@gmail.com
3 R&D Center, Medyssey Co., Ltd., Jechon 27159, Korea; su-heon.woo@medyssey.com
4 Department of Oral and Maxillofacial Surgery, Korea University Anam Hospital, Seoul 02841, Korea; densis@korea.ac.kr
5 Department of Oral & Maxillofacial Surgery, Chungang University Hospital, Seoul 03080, Korea; oms@hanmail.net
* Correspondence: davidjoy76@gmail.com; Tel.: +82-2-6299-2875

Abstract: In porous titanium scaffolds manufactured via 3D printing, the differences in bone formation according to pore design and implantation period were studied. Titanium scaffolds with three types of different pore structures (Octadense, Gyroid, and Dode) were fabricated via 3D printing using the selective laser melting method. Mechanical properties of scaffolds were investigated. Prepared specimens were inserted into both femurs of nine rabbits and their clinical characteristics were observed. Three animals were sacrificed at the 2nd, 4th, and 6th weeks, and the differences in bone formation were radiologically and histologically analyzed. The percentage of new bone and surface density in the pore structure were observed to be approximately 25% and 8 mm^2/mm^3, respectively. There was no difference in the amount of newly formed bone according to the pore design at 2, 4, and 6 weeks. In addition, no differences in the amount of newly formed bone were observed with increasing time within the same pore design for all three designs. During the 6-week observation period, the proportion of new bones in the 3D-printed titanium scaffold was approximately 25%. Differences in bone formation according to the pore design or implantation period were not observed.

Keywords: titanium scaffolds; 3D printing; pore design; selective laser melting

Citation: Lim, H.-K.; Ryu, M.; Woo, S.-H.; Song, I.-S.; Choi, Y.-J.; Lee, U.-L. Bone Conduction Capacity of Highly Porous 3D-Printed Titanium Scaffolds Based on Different Pore Designs. *Materials* **2021**, *14*, 3892. https://doi.org/10.3390/ma14143892

Academic Editors: Swee Leong Sing and Wai Yee Yeong

Received: 9 June 2021
Accepted: 9 July 2021
Published: 12 July 2021

1. Introduction

Titanium has been successfully applied to humans in various fields currently. Titanium is widely used in various implantable medical devices such as dental implants, bone fixators, artificial joints, vascular stents, and shielding membranes [1]. Because titanium is not resorbed like polymers or some ceramics, the application of titanium as a scaffold for bone growth is still not extensive compared to that of ceramic materials [2]. However, titanium, which has higher tensile and compressive strength than ceramic and similar elastic modulus to human bone, is recognized for its high clinical value as bone grafting material [3].

The porous structures of titanium are considered to be effective for eliminating elastic modulus mismatches with human bone [4]. Additionally, the porous structure provides space for bone formation and regrowth [5]. However, accurately controlling the pore size, porosity, shape, and connectivity of the porous structures of titanium scaffolds manufactured by conventional casting or milling methods is difficult [6]. To overcome this problem, various manufacturing methods such as sintering of metal fibers, plasma spraying, anodic dissolution, or powder metallurgy have been introduced; however, constructing a homogenous porous structure is still difficult [7]. In addition, fabricating a small-sized scaffold with a microstructure using a conventional manufacturing method was impossible [8].

However, 3D printing, which has been actively researched and used in recent years, has made it possible to overcome such manufacturing imperfections. Based on a fast, uniform, and simple process through 3D printing, producing a metallic scaffold with a homogeneous porous structure has become possible [9].

It has been reported that the porosity of the ideal scaffold for bone regeneration is more than 66% [10]. Moreover, the pore shape of a scaffold affects bone formation through cell growth. Bidan et al. demonstrated that the optimization of the pore shape can increase the growth rate of the bone tissue in a porous scaffold, and the cells grow faster in square pores [11]. The results of Urda et al. indicate that the straight and low convexities in the pore structure are the least favorable for cell growth [12]. Van Bael et al. studied the pore shape and curvature, and the results showed that an obtuse angle between the pores was more prone to cause blockage of cell growth than an acute angle [13]. However, these studies were conducted at an in vitro level, and the results of in vivo experiments on the effect of pore structure and bone formation were inconsistent [14,15].

In this study, under the premise that the specimens have the same porosity, titanium alloy scaffolds with microstructures of three different shaped pore designs were fabricated using 3D printing. In addition, we analyzed whether there was a difference in bone formation according to pore design or implantation period in titanium scaffolds in vivo using rabbit animal model.

2. Materials and Methods

2.1. Selection of the Pore Design

Among the various mesh designs provided by the Magics 21.0 program (Materialise NV, Leuven, Belgium), 5 designs with a pore size of 1 mm were selected—Diamond, Cubic, Dode, Octadens, Gyroid. Before specimen production for the experiment, non-destructive analysis was performed using micro-computed tomography (μCT) (Quantum FX micro-CT, Perkin Elmer Inc., Hopkinton, MA, USA) to analyze the structure of the lattice shape. After obtaining μCT images of each pore design sample, 3D images were rendered with NRecon program (Bruker-CT, Kontich, Belgium). BoneJ tap of ImageJ software (NIH and LOCI, University of Wisconsin, Madison, WI, USA) for bone micro-architecture analysis was used for stereology analysis of the output design. After cropping a certain area (TV, 125 mm^3) from the entire μCT image, scaffold volume ratio was calculated on the program. It was observed that the Dode (0.708%), Octadens (0.828%), and Gyroid (0.775%) designs had relatively higher densities than the Diamond (0.455%) and Cubic (0.447%) designs. Therefore, it was decided to manufacture Dode, Octadens, and Gyroid designs for experimental specimens.

2.2. Manufacture of 3D-Printed Titanium Specimens

A 3D printing method was adopted with the selective laser melting (SLM) (Cusmedi©, Sungnam, Korea) fabrication method. All manufacturing processes were managed in accordance with the Good Manufacturing Practice standards of the Korean Ministry of Food and Drug Safety. The raw material for printing was Ti-6Al-4 V-ELI powder (Arcam A2, Arcam, Moindal, Sweden) suitable for the ASTM F3001-14 (Standard specification for additive manufacturing Ti-6Al-4 V-ELI with powder bed fusion). The composition of the powder was mainly titanium, which contained trace elements such as 5.94% aluminum, 4.14% vanadium, 0.008% carbon, 0.049% iron, 0.010% yttrium, 0.10% oxygen, 0.010% nitrogen, and less than 0.002% zinc and hydrogen (in wt.%). The manufacturing process based on the SLM system was as follows: Thin layers of atomized fine titanium powder were evenly distributed onto a substrate plate via selective laser melting using a coating mechanism. This process occurred inside a chamber with a strictly controlled atmosphere of inert nitrogen and argon. Once each layer had been stacked, each 2D slice of the part geometry was fused by melting the powder selectively. The laser energy was sufficiently intense to permit complete welding of the particles to form a solid and hard metal. This process was accomplished by a high-powered laser beam, which is usually composed an ytterbium fiber. The process was repeated layer-by-layer until the manufacturing process was complete (Figure 1). The standard of the manufactured product was the ASTM F136 (Grade 23, Standard specification for Ti-6Al-4 V-ELI alloy). Through the manufacturing process, cylinder-shaped specimens with four different pore designs were prepared: Solid (sham, no pore), Octadense, Gyroid, and Dode. All specimens have isotropy in which the same shape is repeated in the X, Y, and Z planes, and these structures also have similar mechanical strength along the X, Y, and Z axes. The specimens were 7 mm in height and 3 mm in diameter (Figure 2). Except for the solid, the specimens of the other designs were manufactured with the same porosity of 75%. The pore size of Octedens, Gyroid, and Dode design were 1.07, 0.30, and 0.76 mm, respectively. The strut thickness of Octedens, Gyroid, and Dode design were 0.12, 0.14, and 0.30 mm, respectively. The strut spacing of all specimens except solid was 0.8 mm, and the printing error was set within 0.05 mm. Surface roughness was less than 15um, and Vickers hardness was more than 310 VHN.

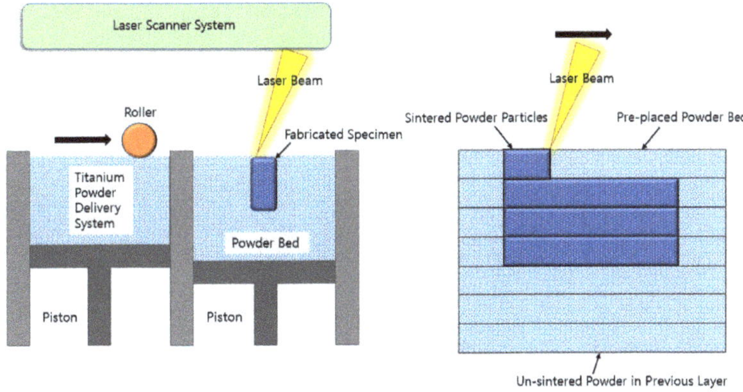

Figure 1. Selective laser melting type 3D-printing process of highly porous titanium specimens.

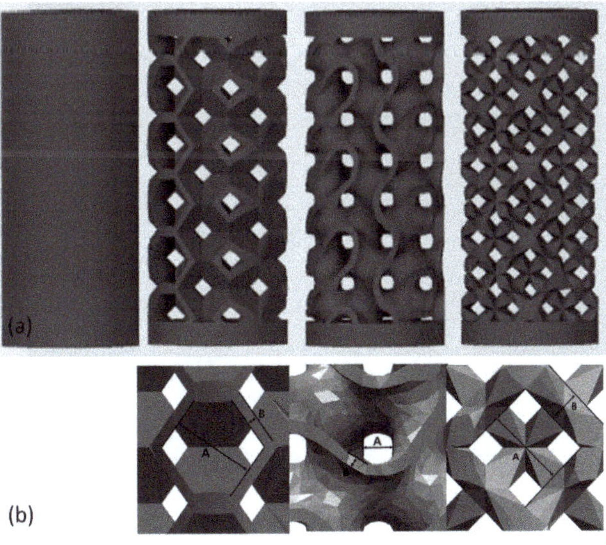

Figure 2. Stereolithographic image of four different pore designs. (**a**) Entire shape (Solid (sham), Octadens, Gyroid, and Dode). (**b**) Image of the unit cell (Octadens, Gyroid, and Dode) (A: pore size, B: strut thickness).

2.3. Animal Experiments

These animal experiments were approved by the Animal Laboratory Ethics Committee based on the law on laboratory animals (Approval No: CRONEX-IACUC 201908-001).

Nine twenty-week-old New Zealand white rabbits (weight: 3.6–3.8 kg) were used herein. After one week of acclimatization, general anesthesia using a combination of zolazepam and tiletamine (Zoletil®, 50 mg/kg, Virbac Korea, Seoul, Korea) and xylazine HCl (Rompun®, 10 mg/kg, Bayer Korea, Seoul, Korea) was administered intramuscularly. On one side of the shaved femur, a disinfection agent with betadine was applied. Subsequently, lidocaine with 1:100,000 epinephrine was administered subcutaneously for hemostatic purpose, and an incision of approximately 4 cm was made longitudinally on the inner thigh. Complete bony exposure of the medial aspect of the femur was performed through careful dissection between the rectus femoris and vastus medialis muscles. Four cuts at a width of 3 mm were made using drills for the installation of specimens. The gap between the cuts was 7 mm. The prepared specimens were individually inserted into each cut (Figure 3). To minimize the effect of the installation location, each different pore-designed specimen was inserted into a random location. After inserting the specimens, layer-to-layer suturing was performed using 4-0 Vicryl® and 5-0 Ethylon® (Johnson & Johnson, New Brunswick, NJ, USA). The same surgical procedure was performed on the opposite leg.

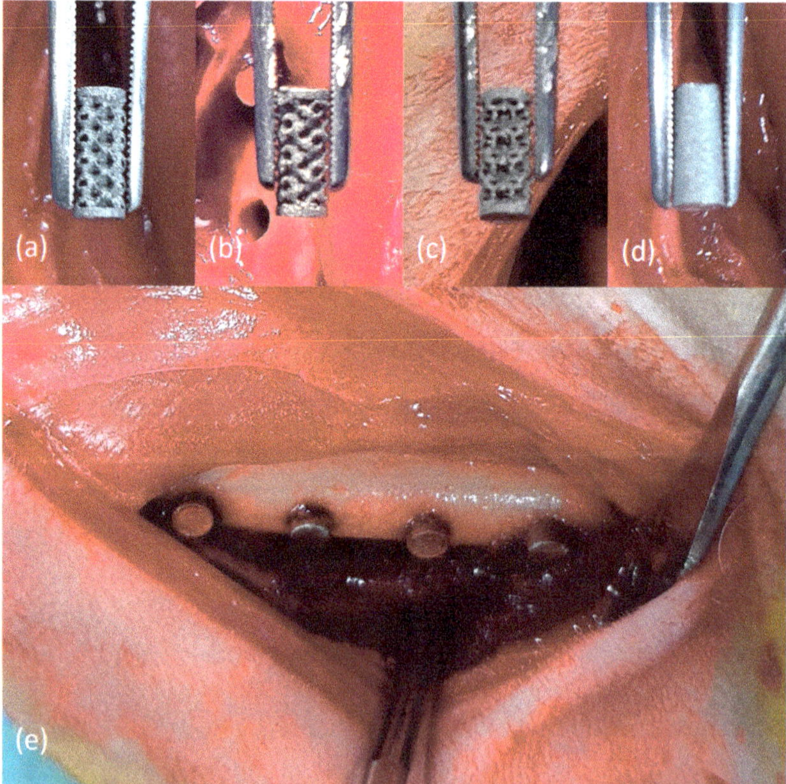

Figure 3. (**a**–**d**) Cylinder-shaped highly porous 3D-printed titanium specimens with four different pore designs (Octadens, Gyroid, Dode, and Solid (sham)). (**e**) Application and insertion of 3D-printed titanium specimens onto the rabbit medial femur.

For pain relief and prevention of infection, prednisolone (0.5 IM, IM, Yuhan, Seoul, Korea) and cephalexin (5 mg/kg, IM, Yuhan, Seoul, Korea) were postoperatively injected for two days. The animals were monitored weekly for infection, inflammation, wound dehiscence, delayed healing, and general health until they were euthanized. Out of a total of nine rabbits, three were euthanized at the 2nd, 4th, and 6th weeks. After being anesthetized using a zolazepam, tiletamine, and xylazine HCl, the animals were euthanized by administering potassium chloride to the marginal ear vein. After euthanasia process, the surgical site of the femur was dissected supra-periosteally, and the long bone and adjacent soft tissue of the femur were harvested.

2.4. Micro-CT Imaging and Volumetric Analysis

The extracted femurs were moved and fixed in a μCT machine (SkyScan1174®, Ver. 1.7; Bruker CT, Kontich, Belgium). Subsequently, μCT imaging was taken. The imaging settings were as follows—tube voltage of 130 kVp, tube current of 60 μA, aluminum filter of 1 mm, exposure time of 500 ms, and rotation angle of 0.3°. The pixel size of image was 13.86 μm. The number of pixels of each image cut was 2240 × 2240. Under these circumstances, a total of 800 high-resolution image cuts were obtained. NRecon (Ver 1.7.0.5, Bruker-CT, Kontich, Belgium) was used for reconfiguration of cross-sectional image cuts, and Dataviewer (Ver. 1.5.1.3, Bruker-CT, Kontich, Belgium) were used for the construction of 3D images. The volume of newly formed bone inside the pore of the specimen was calculated using the difference at the grayscale level. The formula for calculating the new bone volume is as follows:

(Percentage of) New bone volume (%) = (Percentage of) Total bone volume − (Percentage of) Grafted specimen volume

The surface density of newly formed bone inside the pores of the specimen was calculated in the same manner. The formula for calculating the surface density of a new bone is as follows:

Surface density of new bone (mm^2/mm^3) = segmented bone surface/total new bone volume of the region of interest.

2.5. Histological Findings

Tissue slides were prepared using a non-decalcified sample preparation method. The harvested long bone and adjacent soft tissue of the femur were fixed with 10% formalin solution. After soaking of the specimens in formalin solution for one week, they were rinsed meticulously with massive water for 12 h. The specimens were dehydrated using increasing concentrations of 70%, 90%, and 100% ethanol. Then, the tissue samples were infiltrated with resin solution (Technovit 7200 resin, Heraeus KULZER, Hanau, Germany) with increasing resin ratios. After penetration by vacuuming the resin solution for two days, curing of the resin was performed using a UV Embedding system (KULZER EXAKT 520®, Heraeus KULZER, Hanau, Germany), and cut of the center of the resin block was done using a diamond cutter to produce a cross-section of the embedded tissue until 40 μm thickness. Sectioned thin tissue was stained using hematoxylin and eosin (H&E), and tissue slides were produced. The images of tissue slides were scanned using an optical microscope (OLYMPUS BX®, Olympus Optical CO. Tokyo, Japan) equipped with a complementary metal–oxide–semiconductor sensored camera. Subsequently, the histological images were observed.

2.6. Statistical Interpretation

The Kruskal–Wallis test was used to analyze the differences in the percent volume and the surface density of new bone among the three different pored groups simultaneously. Moreover, the Kruskal–Wallis test was used to compare the percent volume and the surface density of new bone at 2, 4, and 6 weeks within the group. Statistical analysis was performed using SPSS 22 program (IBM SPSS Inc., Chicago, IL, USA). Statistical significance was designated at $p < 0.05$.

3. Results

3.1. Clinical Findings in the Animal Experiment

During the observational period, the animals did not show any life-threatening clinical signs. In addition, abnormal symptoms such as wound dehiscence, infection, or inflammation were not observed at the operated sites. Significant gait impairment was not observed during the observational period.

3.2. Radiological Findings

In all specimens of all groups, an uneven increase in radio-opacity in the pores was observed. This increase in opacity was lower than that of the cortical bone and higher than that of the marrow portion. There were no visual radiographic differences according to the time of sacrifice. No bony resorption pattern was observed around the specimens (Figure 4).

Figure 4. Micro-computed tomography findings of the titanium specimens with four different pore designs. (**a**) Longitudinal directional images (Dode, Gyroid, Octadens, and Solid (sham)). (**b**) Axial images (Dode, Gyroid, Octadens, and solid (sham)).

3.3. Histological Findings

Histological findings obtained using H&E staining showed that inflammatory reactions associated with titanium specimens did not occur in any of the groups. Necrosis, infiltration of immune cells, and granuloma formation were not observed. Over time, more bony growth and maturation appeared in the pores. An interesting observation is that the compact bony tissue was penetrated into the pores adjacent to the cortical layer and spongious bony tissue was penetrated into the pores adjacent to the marrow layer (Figure 5).

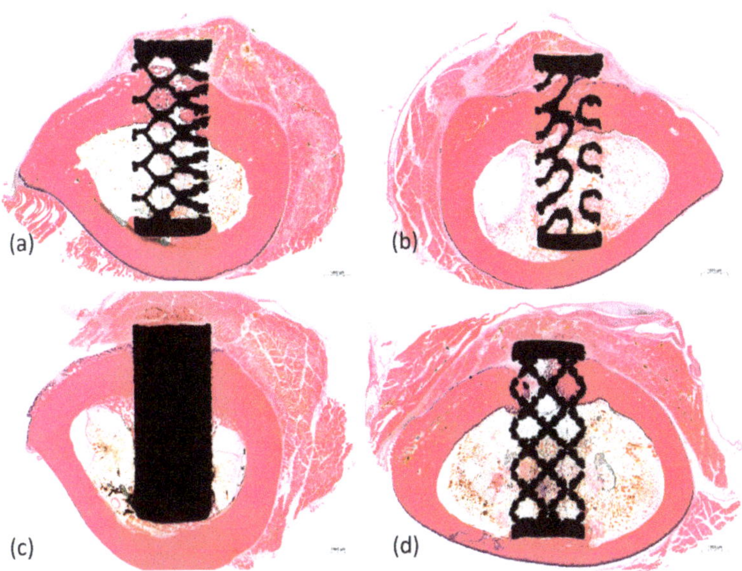

Figure 5. Histological images of the titanium specimens with four different pore designs. (**a**) Octadens pore design. (**b**) Gyroid pore design. (**c**) Solid. (**d**) Dode pore design (H&E staining, X1.0).

3.4. Statistical Interpretation of Bone Conduction Capacity

Based on the difference in grayscale level, the percentage of new bone in the pore structure was observed to be approximately 25% (Table 1). There was no difference in the amount of newly formed bone according to pore design at 2, 4, and 6 weeks ($p > 0.05$). In addition, no difference in the amount of new bone formation was observed with increasing time within the same pore design for all three designs ($p > 0.05$).

Table 1. New bone volume (%) of 3D-printed porous titanium with variable pore design.

	Octadense	Gyroid	Dode	*p*-Value
2 weeks	24.888 ± 0.872	25.069 ± 1.259	24.990 ± 2.715	0.957
4 weeks	27.874 ± 2.184	25.171 ± 1.656	26.527 ± 2.311	0.491
6 weeks	26.835 ± 2.078	27.591 ± 1.719	27.433 ± 4.143	0.733
p-value	0.252	0.193	0.670	

The surface density of the new bone in the pore structure was observed to be approximately 8 mm^2/mm^3 (Table 2). There was no difference in the surface density of new bone according to pore design at 2, 4, and 6 weeks ($p > 0.05$). In addition, no difference in the surface density of new bone was observed with increasing time within the same pore design for all three designs ($p > 0.05$).

Table 2. Surface density of new bone (mm^2/mm^3) of 3D-printed porous titanium with variable pore design.

	Octadense	Gyroid	Dode	*p*-Value
2 weeks	8.073 ± 0.170	8.100 ± 0.055	8.278 ± 0.205	0.393
4 weeks	8.117 ± 0.399	7.727 ± 0.081	8.263 ± 0.127	0.099
6 weeks	7.765 ± 0.181	7.746 ± 0.341	7.989 ± 0.084	0.587
p-value	0.193	0.301	0.113	

4. Discussion

As mentioned in the introduction, when titanium is applied as a porous structure, a space for bone formation and regrowth can be provided [5]. In addition, the porous structure lowers the weight of the substitute, lowers the stiffness to avoid stress shielding, has an elastic modulus similar to bone, and allows nutrient and oxygen exchange to occur along the connected pores [3,4,16].

However, until the introduction and development of 3D printers, imparting porosity to titanium has undergone many trials and errors. Casting and milling methods, which were the simplest manufacturing processes in the early days, are unsuitable for making homogeneous and fine pores in the internal structure of titanium [6]. To overcome such process limitations, various methods for manufacturing porous titanium have been studied, including hot isostatic pressing [17], metal injection molding [18], spark plasma sintering [19], space holder technique [20], plasma spraying [21], anodic dissolution [22] and powder metallurgy [23]. However, some techniques can produce only cavities and not interconnected pores. In addition, constructing a small and homogeneous pore structure using the above-mentioned methods was difficult.

3D printing, which is a revolutionary manufacturing method, has made it possible to overcome existing manufacturing imperfections. Unlike early printing methods, where only plastic materials were used, metal materials can now be processed through electron beam melting or SLM methods. In this study, the specimens were processed using SLM. The best advantages of constructing a bone substitute with titanium material having a porous structure using 3D printing are homogeneity and microstructure. An open and interconnected porous structure is typically needed for bone conduction with pore sizes ranging from 200 to 500 μm [24,25] and porosity ranging from 60% to 90% [26]. In addition, pore sizes larger than 300 μm could promote the formation of new bone and capillaries [27,28]. SLM allows for the production of bone implants made of metals such as, titanium and titanium-based alloys with a strut size as small as 200 μm [27]. Until now, residual powder, the so-called balling phenomenon, is a problem in the manufacturing of metal additives that causes adhering of particles in the 3D printing process. In other study, it has been reported that the use of hydrofluoric acid could overcome this problem [29].

The titanium specimens used herein were not subjected to special surface treatments. In fact, the current, widely used sandblasted or acid-etched techniques are known to increase the biocompatibility of titanium by maximizing the surface roughness of commercially pure titanium or titanium alloys [30]. Despite the advantages of such surface treatments, we did not perform it in consideration of the simplicity of the manufacturing process, and as excessive surface treatment may cause deformation of the set pore design. We envisioned that all used surfaces were treated the same to avoid additional effects of the surface treatment. Actually, according to one study, SLM-fabricated structures with native, sandblasted, acid-etched, polished, or vibratory grinded surfaces show identical cyto-compatibility with osteoblast compared with conventional titanium surfaces [27].

In this study, statistical differences in bone formation according to pore design were not identified. Regardless of the pore design, the percentage of new bone and surface density in the pore structure were observed in approximately 25% and 8 mm^2/mm^3, respectively, over 6 weeks. Research on titanium pore design has been actively conducted, even before the introduction of 3D printing. Under in vitro examination, there was a study that demonstrated better bone formation in square-shaped pores [11], and another study showed that tissue amplification was superior in hexagonal-shaped pores [31]. One study indicated that a larger pore throat size, which affects the interconnection, is more advantageous for bone ingrowth [32], and another study indicated that the straight edges and convexities in the pore design are the least favorable for cell growth [12].

The results were also inconsistent in a comparative study on the pore design of porous titanium made with 3D printing that can be searched in the current medline database. Deng et al. studied porous titanium manufactured by the SLM method with four pore designs (diamond, triangular, circular, cubic) having a size of 650 μm and a porosity of 65%. In a specimen of diamond pore shape, more bone formation was observed in animal experiments, and slower flow rates were observed with computational fluid dynamics, which would help blood vessel growth [33]. In addition, Zhao et al. compared the octahedral and tetrahedron structures of porous titanium fabricated by the SLM method and demonstrated that cell proliferation was better in the octahedral unit with immunofluorescence [34]. Van Bael et al. studied the pore shape and curvature, and their results showed that an obtuse angle between the pores was more prone to cause cell blockage than an acute angle [13].

Although statistical significance was not observed herein, several peculiarities were discovered. Histologically, ingrowth of the compact bone was observed in the specimen area close to the cortical layer, and ingrowth of the spongeous bone was observed in the specimen area close to the marrow layer. In addition, no difference in bone formation according to the implantation period was observed radiographically, suggesting that the maturation and calcification of bone tissue proceeded slowly before 6 weeks.

In this study, experiments related to cell affinity such as cell adhesion were not performed. Because except the pore design, the material and surface treatment were the same between the specimens, it was thought that there would be no significant difference in cell affinity. Even considering the existing literature, there is no disagreement about the high cell affinity of 3D-printed porous titanium. At 1 to 2 weeks after cell culture, more cell binding, filopodia, and lamellipodia were observed comparing to initial time [35]. In addition, it is reported that there was no significant difference in this tendency even when the specimen pore size was changed [25].

In addition to cell affinity, trabecular anisotropy may be considered in order for future porous specimens to adapt to loading conditions. As there are studies on bone mechanical integrity according to the open site of the pore [36,37], trabecular anisotropy according to the pore design should be investigated in the future. Besides that, the lack of sample size, short-term observation period, and non-loading environment are limitations of this study. Moreover, due to the absence of a device capable of superimposing the STL file used in the scaffold design and the CT data of the actually manufactured specimen, the printing accuracy could not be measured in this study. Studies addressing these aspects in the future would lead to more reliable results with regard to the effect of pore design on bone formation, which could not be observed herein.

5. Conclusions

High-porosity 3D printed titanium scaffolds that were constructed using the SLM method have high biocompatibility and bone conductivity in vivo. During the 6-week observation period, the proportion of new bones in the 3D-printed titanium scaffold was approximately 25%. However, radiologically and histologically, differences in bone formation according to the three types of pore design or implantation period were not observed.

Author Contributions: Conceptualization, U.-L.L.; methodology, M.R.; validation, I.-S.S.; formal analysis, Y.-J.C.; investigation, U.-L.L.; data curation, H.-K.L.; writing—original draft preparation, H.-K.L.; writing—review and editing, U.-L.L.; visualization, I.-S.S.; supervision, Y.-J.C. and U.-L.L.; project administration, S.-H.W.; funding acquisition, S.-H.W. All authors have read and agreed to the published version of the manuscript.

Funding: This study was funded by a grant from the Korea Health Technology R&D Project, Korea Health Industry Development Institute (KHIDI), Ministry of Health & Welfare, Republic of Korea (Grant number: HI18C1224), and 2019 Bio-Healthcare Industry Support from SeongNam Industry Promotion Agency.

Institutional Review Board Statement: This experiment was approved by the CRONEX Laboratory Animal Ethics Committee based on the law on laboratory animals (Animal Laboratory Approval No: CRONEX-IACUC 201908-001).

Informed Consent Statement: Not applicable.

Data Availability Statement: All results generated or analyzed during the present study are included in this published article. Data and materials will be made available upon request via email to the first author (davidjoy76@gmail.com).

Acknowledgments: Special thanks to Dong-Ryul Song for the design of the specimens.

Conflicts of Interest: The authors declare no conflict of interest.

References

1. Rotaru, H.; Schumacher, R.; Kim, S.G.; Dinu, C. Selective laser melted titanium implants: A new technique for the reconstruction of extensive zygomatic complex defects. *Maxillofac. Plast. Reconstr. Surg.* **2015**, *37*, 1. [CrossRef] [PubMed]
2. Khan, Y.; El-Amin, S.F.; Laurencin, C. In vitro and in vivo evaluation of a novel polymer-ceramic composite scaffold for bone tissue engineering. *Conf. Proc. IEEE Eng. Med. Biol. Soc.* **2006**, *2006*, 529–530. [CrossRef]
3. Li, G.; Wang, L.; Pan, W.; Yang, F.; Jiang, W.; Wu, X.; Kong, X.; Dai, K.; Hao, Y. In vitro and in vivo study of additive manufactured porous Ti6Al4V scaffolds for repairing bone defects. *Sci. Rep.* **2016**, *6*, 34072. [CrossRef] [PubMed]
4. Murr, L.E. Open-cellular metal implant design and fabrication for biomechanical compatibility with bone using electron beam melting. *J. Mech. Behav. Biomed. Mater.* **2017**, *76*, 164–177. [CrossRef] [PubMed]
5. Nam, J.W.; Kim, M.Y.; Han, S.J. Cranial bone regeneration according to different particle sizes and densities of demineralized dentin matrix in the rabbit model. *Maxillofac. Plast. Reconstr. Surg.* **2016**, *38*, 27. [CrossRef] [PubMed]
6. Jung, H.D.; Yook, S.W.; Jang, T.S.; Li, Y.; Kim, H.E.; Koh, Y.H. Dynamic freeze casting for the production of porous titanium (Ti) scaffolds. *Mater. Sci. Eng. C Mater. Biol. Appl.* **2013**, *33*, 59–63. [CrossRef] [PubMed]
7. Torres, Y.; Lascano, S.; Bris, J.; Pavon, J.; Rodriguez, J.A. Development of porous titanium for biomedical applications: A comparison between loose sintering and space-holder techniques. *Mater. Sci. Eng. C Mater. Biol. Appl.* **2014**, *37*, 148–155. [CrossRef]
8. Yan, L.; Wu, J.; Zhang, L.; Liu, X.; Zhou, K.; Su, B. Pore structures and mechanical properties of porous titanium scaffolds by bidirectional freeze casting. *Mater. Sci. Eng. C Mater. Biol. Appl.* **2017**, *75*, 335–340. [CrossRef]
9. Salah, M.; Tayebi, L.; Moharamzadeh, K.; Naini, F.B. Three-dimensional bio-printing and bone tissue engineering: Technical innovations and potential applications in maxillofacial reconstructive surgery. *Maxillofac. Plast. Reconstr. Surg.* **2020**, *42*, 18. [CrossRef]
10. Kujala, S.; Ryhanen, J.; Danilov, A.; Tuukkanen, J. Effect of porosity on the osteointegration and bone ingrowth of a weight-bearing nickel-titanium bone graft substitute. *Biomaterials* **2003**, *24*, 4691–4697. [CrossRef]
11. Bidan, C.M.; Kommareddy, K.P.; Rumpler, M.; Kollmannsberger, P.; Fratzl, P.; Dunlop, J.W. Geometry as a factor for tissue growth: Towards shape optimization of tissue engineering scaffolds. *Adv. Healthc. Mater.* **2013**, *2*, 186–194. [CrossRef] [PubMed]
12. Rudrich, U.; Lasgorceix, M.; Champion, E.; Pascaud-Mathieu, P.; Damia, C.; Chartier, T.; Brie, J.; Magnaudeix, A. Pre-osteoblast cell colonization of porous silicon substituted hydroxyapatite bioceramics: Influence of microporosity and macropore design. *Mater. Sci. Eng. C Mater. Biol. Appl.* **2019**, *97*, 510–528. [CrossRef] [PubMed]
13. Van Bael, S.; Chai, Y.C.; Truscello, S.; Moesen, M.; Kerckhofs, G.; Van Oosterwyck, H.; Kruth, J.P.; Schrooten, J. The effect of pore geometry on the in vitro biological behavior of human periosteum-derived cells seeded on selective laser-melted Ti6Al4V bone scaffolds. *Acta Biomater.* **2012**, *8*, 2824–2834. [CrossRef] [PubMed]
14. de Vasconcellos, L.M.; Leite, D.D.; Nascimento, F.O.; de Vasconcellos, L.G.; Graca, M.L.; Carvalho, Y.R.; Cairo, C.A. Porous titanium for biomedical applications: An experimental study on rabbits. *Med. Oral Patol. Oral Cir. Bucal* **2010**, *15*, e407–e412. [CrossRef] [PubMed]
15. do Prado, R.F.; Esteves, G.C.; Santos, E.L.S.; Bueno, D.A.G.; Cairo, C.A.A.; Vasconcellos, L.G.O.; Sagnori, R.S.; Tessarin, F.B.P.; Oliveira, F.E.; Oliveira, L.D.; et al. In vitro and in vivo biological performance of porous Ti alloys prepared by powder metallurgy. *PLoS ONE* **2018**, *13*, e0196169. [CrossRef] [PubMed]
16. Heinl, P.; Muller, L.; Korner, C.; Singer, R.F.; Muller, F.A. Cellular Ti-6Al-4V structures with interconnected macro porosity for bone implants fabricated by selective electron beam melting. *Acta Biomater.* **2008**, *4*, 1536–1544. [CrossRef] [PubMed]
17. Benzing, J.; Hrabe, N.; Quinn, T.; White, R.; Rentz, R.; Ahlfors, M. Hot isostatic pressing (HIP) to achieve isotropic microstructure and retain as-built strength in an additive manufacturing titanium alloy (Ti-6Al-4V). *Mater. Lett.* **2019**, *257*. [CrossRef]
18. Melli, V.; Juszczyk, M.; Sandrini, E.; Bolelli, G.; Bonferroni, B.; Lusvarghi, L.; Cigada, A.; Manfredini, T.; De Nardo, L. Tribological and mechanical performance evaluation of metal prosthesis components manufactured via metal injection molding. *J. Mater. Sci. Mater. Med.* **2015**, *26*, 5332. [CrossRef]
19. Sakamoto, Y.; Asaoka, K.; Kon, M.; Matsubara, T.; Yoshida, K. Chemical surface modification of high-strength porous Ti compacts by spark plasma sintering. *Bio-Med. Mater. Eng.* **2006**, *16*, 83–91.
20. Prananingrum, W.; Naito, Y.; Galli, S.; Bae, J.; Sekine, K.; Hamada, K.; Tomotake, Y.; Wennerberg, A.; Jimbo, R.; Ichikawa, T. Bone ingrowth of various porous titanium scaffolds produced by a moldless and space holder technique: An in vivo study in rabbits. *Biomed. Mater.* **2016**, *11*, 015012. [CrossRef]
21. Likibi, F.; Assad, M.; Coillard, C.; Chabot, G.; Rivard, C.H. [Bone integration and apposition of porous and non porous metallic orthopaedic biomaterials]. *Ann. Chir.* **2005**, *130*, 235–241. [CrossRef]

22. Wen, C.E.; Yamada, Y.; Shimojima, K.; Chino, Y.; Asahina, T.; Mabuchi, M. Processing and mechanical properties of autogenous titanium implant materials. *J. Mater. Sci. Mater. Med.* **2002**, *13*, 397–401. [CrossRef] [PubMed]
23. Ryan, G.E.; Pandit, A.S.; Apatsidis, D.P. Porous titanium scaffolds fabricated using a rapid prototyping and powder metallurgy technique. *Biomaterials* **2008**, *29*, 3625–3635. [CrossRef] [PubMed]
24. Taniguchi, N.; Fujibayashi, S.; Takemoto, M.; Sasaki, K.; Otsuki, B.; Nakamura, T.; Matsushita, T.; Kokubo, T.; Matsuda, S. Effect of pore size on bone ingrowth into porous titanium implants fabricated by additive manufacturing: An in vivo experiment. *Mater. Sci. Eng. C Mater. Biol. Appl.* **2016**, *59*, 690–701. [CrossRef] [PubMed]
25. Ran, Q.; Yang, W.; Hu, Y.; Shen, X.; Yu, Y.; Xiang, Y.; Cai, K. Osteogenesis of 3D printed porous Ti6Al4V implants with different pore sizes. *J. Mech. Behav. Biomed. Mater.* **2018**, *84*, 1–11. [CrossRef] [PubMed]
26. Pei, X.; Wu, L.; Zhou, C.; Fan, H.; Gou, M.; Li, Z.; Zhang, B.; Lei, H.; Sun, H.; Liang, J.; et al. 3D printed titanium scaffolds with homogeneous diamond-like structures mimicking that of the osteocyte microenvironment and its bone regeneration study. *Biofabrication* **2020**. [CrossRef]
27. de Wild, M.; Schumacher, R.; Mayer, K.; Schkommodau, E.; Thoma, D.; Bredell, M.; Kruse Gujer, A.; Gratz, K.W.; Weber, F.E. Bone regeneration by the osteoconductivity of porous titanium implants manufactured by selective laser melting: A histological and micro computed tomography study in the rabbit. *Tissue Eng. Part A* **2013**, *19*, 2645–2654. [CrossRef]
28. Karageorgiou, V.; Kaplan, D. Porosity of 3D biomaterial scaffolds and osteogenesis. *Biomaterials* **2005**, *26*, 5474–5491. [CrossRef] [PubMed]
29. Song, P.; Hu, C.; Pei, X.; Sun, J.; Sun, H.; Wu, L.; Jiang, Q.; Fan, H.; Yang, B.; Zhou, C.; et al. Dual modulation of crystallinity and macro-/microstructures of 3D printed porous titanium implants to enhance stability and osseointegration. *J. Mater. Chem B* **2019**, *7*, 2865–2877. [CrossRef]
30. Bornstein, M.M.; Hart, C.N.; Halbritter, S.A.; Morton, D.; Buser, D. Early loading of nonsubmerged titanium implants with a chemically modified sand-blasted and acid-etched surface: 6-month results of a prospective case series study in the posterior mandible focusing on peri-implant crestal bone changes and implant stability quotient (ISQ) values. *Clin. Implant. Dent. Relat Res.* **2009**, *11*, 338–347. [CrossRef]
31. Rumpler, M.; Woesz, A.; Dunlop, J.W.; van Dongen, J.T.; Fratzl, P. The effect of geometry on three-dimensional tissue growth. *J. R. Soc. Interface* **2008**, *5*, 1173–1180. [CrossRef] [PubMed]
32. Otsuki, B.; Takemoto, M.; Fujibayashi, S.; Neo, M.; Kokubo, T.; Nakamura, T. Pore throat size and connectivity determine bone and tissue ingrowth into porous implants: Three-dimensional micro-CT based structural analyses of porous bioactive titanium implants. *Biomaterials* **2006**, *27*, 5892–5900. [CrossRef] [PubMed]
33. Deng, F.; Liu, L.; Li, Z.; Liu, J. 3D printed Ti6Al4V bone scaffolds with different pore structure effects on bone ingrowth. *J. Biol Eng.* **2021**, *15*, 4. [CrossRef] [PubMed]
34. Zhao, D.; Huang, Y.; Ao, Y.; Han, C.; Wang, Q.; Li, Y.; Liu, J.; Wei, Q.; Zhang, Z. Effect of pore geometry on the fatigue properties and cell affinity of porous titanium scaffolds fabricated by selective laser melting. *J. Mech. Behav. Biomed. Mater.* **2018**, *88*, 478–487. [CrossRef] [PubMed]
35. Zhang, S.; Cheng, X.; Yao, Y.; Wei, Y.; Han, C.; Shi, Y.; Wei, Q.; Zhang, Z. Porous niobium coatings fabricated with selective laser melting on titanium substrates: Preparation, characterization, and cell behavior. *Mater. Sci. Eng. C Mater. Biol. Appl.* **2015**, *53*, 50–59. [CrossRef]
36. Ishimoto, T.; Kawahara, K.; Matsugaki, A.; Kamioka, H.; Nakano, T. Quantitative Evaluation of Osteocyte Morphology and Bone Anisotropic Extracellular Matrix in Rat Femur. *Calcif. Tissue Int.* **2021**. [CrossRef]
37. Ishimoto, T.; Yamada, K.; Takahashi, H.; Takahata, M.; Ito, M.; Hanawa, T.; Nakano, T. Trabecular health of vertebrae based on anisotropy in trabecular architecture and collagen/apatite micro-arrangement after implantation of intervertebral fusion cages in the sheep spine. *Bone* **2018**, *108*, 25–33. [CrossRef]

Article

Effect of Laser Scanning Speed on the Microstructure and Mechanical Properties of Laser-Powder-Bed-Fused K418 Nickel-Based Alloy

Zhen Chen [1], Yongxin Lu [2,3,*], Fan Luo [2], Shuzhe Zhang [1], Pei Wei [1], Sen Yao [1] and Yongxin Wang [1,*]

1 State Key Laboratory of Manufacturing System Engineering, Xi'an Jiaotong University, Xi'an 710049, China; chenzhen2025@xjtu.edu.cn (Z.C.); zsz1989@stu.xjtu.edu.cn (S.Z.); weipei0529@163.com (P.W.); ysabyh@163.com (S.Y.)

2 School of Material Science and Engineering, Xi'an Shiyou University, Xi'an 710065, China; lg030609@163.com

3 School of Materials Science and Engineering, Northwestern Polytechnical University, Xi'an 710072, China

* Correspondence: luyongxin618@163.com (Y.L.); wangyongxin126@mail.xjtu.edu.cn (Y.W.)

Abstract: Laser powder bed fusion (LPBF) is a powder-bed-based metal additive manufacturing process with multiple influencing parameters as well as multi-physics interaction. The laser scanning speed, which is one of the essential process parameters of the LPBF process, determines the microstructure and properties of the components by adjusting the instantaneous energy input of the molten pool. This work presents a comprehensive investigation of the effects of the laser scanning speed on the densification behavior, phase evolution, microstructure development, microhardness, and tensile properties of K418 alloy prepared by laser powder bed fusion. When the scanning speed is 800 mm/s, the microstructure of the material is dominated by cellular dendrite crystals, with coarse grains and some cracks in the melting tracks. When the scanning speed is increased to 1200 mm/s, a portion of the material undergoes a cellular dendrite–columnar crystal transition, the preferred orientation of the grains is primarily (001), and internal defects are significantly reduced. When the scanning speed is further increased to 1600 mm/s, columnar crystals become the main constituent grains, and the content of high-angle grain boundaries (HAGBs) within the microstructure increases, refining the grain size. However, the scanning speed is too fast, resulting in defects such as unmelted powder, and lowering the relative density. The experimental results show that by optimizing the laser scanning speed, the microhardness of the LPBF-ed K418 parts can be improved to 362.89 ± 5.01 HV, the tensile strength can be elevated to 1244.35 ± 99.12 MPa, and the elongation can be enhanced to 12.53 ± 1.79%. These findings could help determine the best scanning speed for producing K418 components with satisfactory microstructure and tensile properties via LPBF. In addition, since the LPBF process is largely not constrained and limited by the complexity of the geometric shape of the part, it is expected to manufacture sophisticated and complex structures with hollow, porous, mesh, thin-walled, special-shaped inner flow channels and other structures through the topology optimization design. However, due to the relatively narrow LPBF process window, this study will benefit from LPBF in producing a lightweight, complex, and low-cost K418 product, greatly improving its performance, and promoting the use of LPBF technology in the preparation of nickel-based superalloys.

Keywords: laser powder bed fusion; K418 alloy; orientation; phase; microhardness

Citation: Chen, Z.; Lu, Y.; Luo, F.; Zhang, S.; Wei, P.; Yao, S.; Wang, Y. Effect of Laser Scanning Speed on the Microstructure and Mechanical Properties of Laser-Powder-Bed-Fused K418 Nickel-Based Alloy. *Materials* **2022**, *15*, 3045. https://doi.org/10.3390/ma15093045

Academic Editors: Swee Leong Sing and Wai Yee Yeong

Received: 5 March 2022
Accepted: 19 April 2022
Published: 22 April 2022

1. Introduction

Due to its considerable creep strength, thermal and cold fatigue resistance, and oxidation resistance, K418 alloy is commonly used to produce hot precision parts operating below 900 °C, such as turbine working blades, guide vanes, guides, and whole-cast turbines [1–3]. Extensive research has been conducted to date on the influence of casting [4], forging [5], and injection molding [6] on the microstructure and mechanical properties of the K418 alloy.

The literature has primarily concentrated on K418 alloys processed conventionally. With the increasing demand for precision components with complex structures in manufacturing, novel process techniques including laser powder bed fusion (LPBF) become worthwhile to investigate [7].

LPBF technology is a major technological approach in metallic additive manufacturing, distinguished by simple processing, the absence of tooling, high forming accuracy, short fabrication cycle, etc. It can efficiently produce parts with complex structures, and has found widespread application in the aerospace, medical, energy, and other industrial fields [8–10]. The metal melts and solidifies rapidly when LPBF-molding metal parts under the scanning of a high-energy laser beam, and the transition of the powder from solid to liquid and liquid to solid is extremely fast and difficult to control—especially the cooling rate of the melt pool, which can reach 10^8 K/s [11–13]. Therefore, the uniqueness of the process of LPBF technology makes the durability of the material far superior to that of conventional materials. For example, Li et al. [14] studied the influence of laser scanning speed on the microstructure, phase evolution, and nanohardness of LPBF-ed Ti-45Al-2Cr-5Nb. As the scanning speed increased slightly from 500 to 800 mm/s, the nanohardness of LPBF-ed TiAl alloys was much higher than that of cast parts.

However, the parameters of the LPBF process have a complicated pattern of effects on the qualities of the formed parts, and many defects can occur with improper forming processes [15–18]. Jiang et al. [15] used LPBF to evaluate the effect of energy density on the microstructural evolution and properties of new stainless steels. The findings revealed that high scanning speeds not only caused the appearance of unmelted powder on the surface of the samples, but also increased their porosity. As a result, choosing appropriate and suitable process parameters can improve the mechanical properties. Wang et al. [19] observed the effects of scanning speed on the melt pool morphology, grain changes, and tensile properties of IN718 alloy produced by LPBF. The findings revealed that as the scanning speed increased, the melt pool's depth-to-width ratio increased, the microstructure transformed from cellular to columnar crystals, and the tensile strength and elongation potentially exceeded 1014 $\pm$ 19 MPa and 19.04 $\pm$ 1.12%, respectively, at a laser scanning speed of 1300 mm/s.

As can be seen from the preceding literature, the scanning speed, as a critical process parameter in LPBF technology, has a significant impact on the microstructure and mechanical properties of LPBF-ed materials. In this work, K418 alloy was prepared via the LPBF technique, and the effects of scanning speed on the densification, phase composition, microstructural evolution, and mechanical behavior of LPBF specimens were thoroughly discussed. The purpose of this research is to provide more references for the additive manufacturing of nickel-based alloys.

2. Experimental Procedure

The K418 powder used in this work was purchased from Beijing AVIC Mighty Company. The chemical composition of the K418 powder was as follows (wt.%): 12.93 Cr, 5.96 Al, 4.26 Mo, 2.1 Nb, 0.82 Ti, 0.14 Fe, 0.13 C, and the remainder Ni. The powder had an essentially spherical morphology, with a size range of 15 to 53 μm. A stainless steel substrate with good wettability was chosen as the substrate material. The rated power of the single-mode CW fiber laser was 200 W, the laser spot diameter was approximately 60 μm, and the maximum forming size was 120 mm $\times$ 120 mm $\times$ 150 mm. Argon was employed as a protective gas to prevent the raw material from oxidizing during LPBF. This research focuses on a single variable—scanning speed; the LPBF setup parameters were determined after several trials to be laser power of 200 W, laser scanning speed of 800~1400 mm/s, layer thickness of 0.03 mm, and hatch spacing of 0.07 mm, as shown in Table 1. The scanning strategy used was a rotational scanning strategy, implying a rotation angle of 67° between the nth and n + 1st layers.

Table 1. Laser powder bed fusion process parameters.

Process Parameters	Laser Power (W)	Scanning Speed (mm/s)	Layer Thickness (mm)	Hatch Spacing (mm)
Value	200	800, 1000, 1200, 1400, 1600	0.03	0.07

After fabrication, all LPBF-ed K418 samples were ground and polished, and the relative density of the samples was first determined using an automatic densitometer founded on Archimedes' principle, and observed for surface defects using an optical microscope (Axio Vert. A1, Jena, Germany). The phase composition of the samples was characterized using an X-ray diffractometer (XRD-6000, Shimadzu Instrument, Kyoto, Japan) with a 20–100° diffraction angle and a scan rate of 10°/min. Following that, the samples were etched for 1 min in aqua regia ($HCl:HNO_3$) = 3:1 using the standard metallographic procedure, and the microstructure of the specimens was observed using a scanning electron microscope. The grain orientation, grain boundary distribution, and grain size of the selected typical samples were analyzed using an electron backscatter diffractometer. Finally, the microhardness of the specimens' surfaces was characterized using a microhardness tester (HXD-1000TMC, Shanghai, China), with a load of 200 g and a holding time of 10 s. Measurements were made on the surface of each specimen at 0.05 mm intervals, with 15 measurement points, and their average values were calculated. The tensile properties of the LPBF samples were tested using a tensile testing machine at a stretching rate of 0.5 mm/min, in accordance with ASTM D638 GB/T 228-2010. To ensure data accuracy, three samples were tested for each parameter. After the tensile experiments, scanning electron microscopy was used to characterize the tensile fracture morphology.

3. Results and Discussion

3.1. Densification Behavior

Figure 1 shows the densities of the LPBF samples of K418 alloy at various scanning speeds, as well as the corresponding metallographic images. It can be seen that the relative density of the LPBF-ed K418 specimens increased slowly at first, and then rapidly as the laser scanning speed increased. When the laser scanning speed was reduced to 800–1000 mm/s, the densities were approximately 97.6%, and the corresponding optical photographs show that there were tiny pores and cracks on the surfaces of the samples. At a laser scanning speed of 1200 mm/s, the densities reached 97.76%, and the optical photographs revealed the best surface morphology of the samples. As the scanning speed increased to 1400–1600 mm/s, the surface of the samples became more porous and less well-formed, and the densities decreased rapidly. As a result, when the scanning speed was too slow, the laser beam stayed in the molten pool for an extended period of time, causing the energy input in the molten pool to be too high, and resulting in turbulence and instability of the melt in the molten pool which, in turn, resulted in porosity and balling, along with other defects, thereby reducing the relative density [20]. However, when the scanning speed was too fast, the molten pool formation and cooling time were very short, the Marangoni convection strength in the molten pool was weak, and the melt was prone to Plateau–Rayleigh instability due to surface tension and capillary force. Discontinuous melt tracks with more unmelted or semi-melted metal particles result in holes and cracks on the surface, reducing the material's densification.

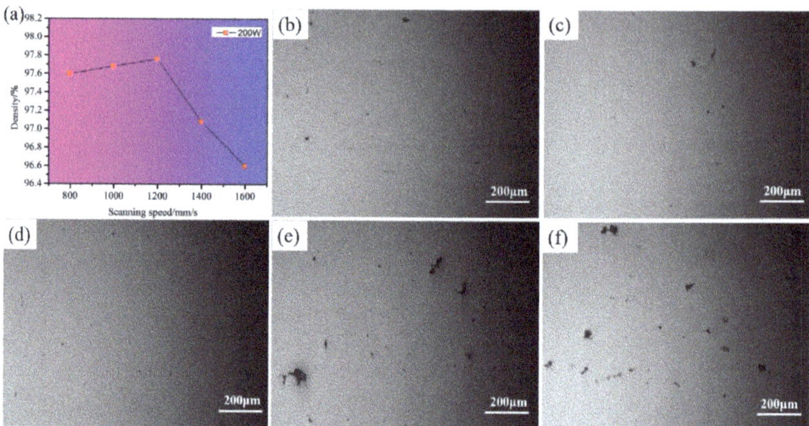

Figure 1. (**a**) Influence of different scanning speeds on relative density and (**b–f**) transverse section appearance: (**b**) 800 mm/s; (**c**) 1000 mm/s; (**d**) 1200 mm/s; (**e**) 1400 mm/s; (**f**) 1600 mm/s.

3.2. Phase Behavior

Figure 2 shows the XRD analysis of K418 samples prepared at different scanning speeds. The XRD results show that as the scanning speed increased, the changes in the γ phase and γ'/γ'' phase both became available in various grain orientations. As the key strengthening phase is the γ'/γ'' phase in Ni-based alloys, these are the strongest peak and the weakest peak, respectively. In addition, the sample grain size is proportional to the XRD diffraction peak's full width at half-maximum (*FWHM*), based on Scherrer and Wilson's equation formulae [21]:

$$FWHM = 0.9\mu/D\cos\theta \quad (1)$$

where *FWHM* is the diffraction peak, μ is the X-ray radiation wavelength, D denotes the mean grain size, and θ is the scanning angle. Therefore, on the strongest peak, the half-width of the γ'/γ'' phase diffraction peak at different scanning speeds can be calculated using X'Pert HighScore Plus software (Royal Dutch Philips, Amsterdam, The Netherlands), as illustrated in Table 2. When the laser scanning speed was 1200 mm/s, the half-width of γ'/γ'' phase diffraction peak was the largest, the average grain size of the material was reduced, and the properties of the material were enhanced [22,23]. In addition, no carbide peaks were detected in the XRD results, indicating a low level of internal carbide generation.

3.3. Microstructure Evolution

Figure 3 depicts the metallographic microstructure of K418 samples in the corrosion state as a function of scanning speeds. As the scanning speed increased, the morphology and arrangement of the molten pool changed. The laser input energy density was high at low scanning speeds, causing the temperature of the molten pool to rise too quickly, and the Marangoni convection intensity inside the molten pool to be too strong to produce turbulent flow, resulting in large fluctuations in the morphology of the forming surface and the laser energy absorbed by the K418 powder [24,25]. When the scanning speed was 1200 mm/s, the thermal flow of the molten pool was stable, the quality of the melt track quality improved, and cracks were reduced. However, as the scanning speed increased, there were defects such as unmelted powder and porosity on the surface of the melt pool, as well as the melt pool being short and irregular due to the faster laser shift speed, which cannot completely melt the powder.

Table 2. *FWHM* of γ'/γ'' phase diffraction peaks at different scanning speeds.

Scanning Speeds (mm/s)	800	1000	1200	1400	1600
FWHM	0.1968	0.4330	0.4723	0.1181	0.1378

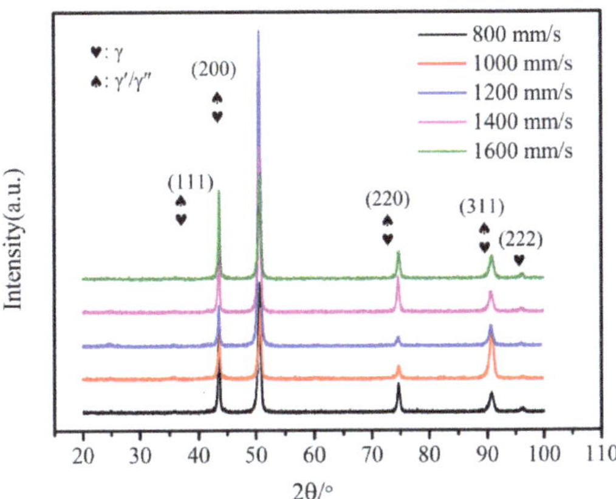

Figure 2. XRD analysis of K418 samples prepared at different scanning speeds.

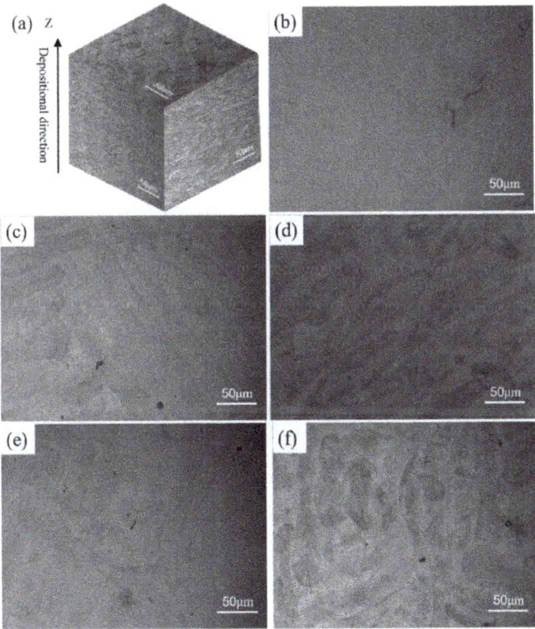

Figure 3. (**a**) The 3D metallographic diagram of k418, and (**b–f**) the transverse section metallography corresponding to different laser powers ((**b**) 800 mm/s; (**c**) 1000 mm/s; (**d**) 1200 mm/s; (**e**) 1400 mm/s; (**f**) 1600 mm/s).

Figure 4 illustrates SEM images of the K418 samples at various scanning speeds. At low scanning speeds, the specimen had a cellular dendritic microstructure inside the specimen, due to the increased residence time of the laser beam in a specific area, the restricted rate of energy loss in the melt pool, and the prolonged cooling time, promoting grain nucleation and growth. When the scanning speed was increased to 1200 mm/s, some of the microstructures in the specimen were converted to columnar crystals, which noticeably refined and grew continuously. However, increasing the scanning speed shortens the action time of the laser beam on the surface of the molten pool, accelerates heat conduction in the molten pool, and reduces the solidification time of the molten pool, providing suitable solidification conditions for the formation of columnar crystals. As a result, columnar crystals can be seen to preferentially grow along the deposition direction [26].

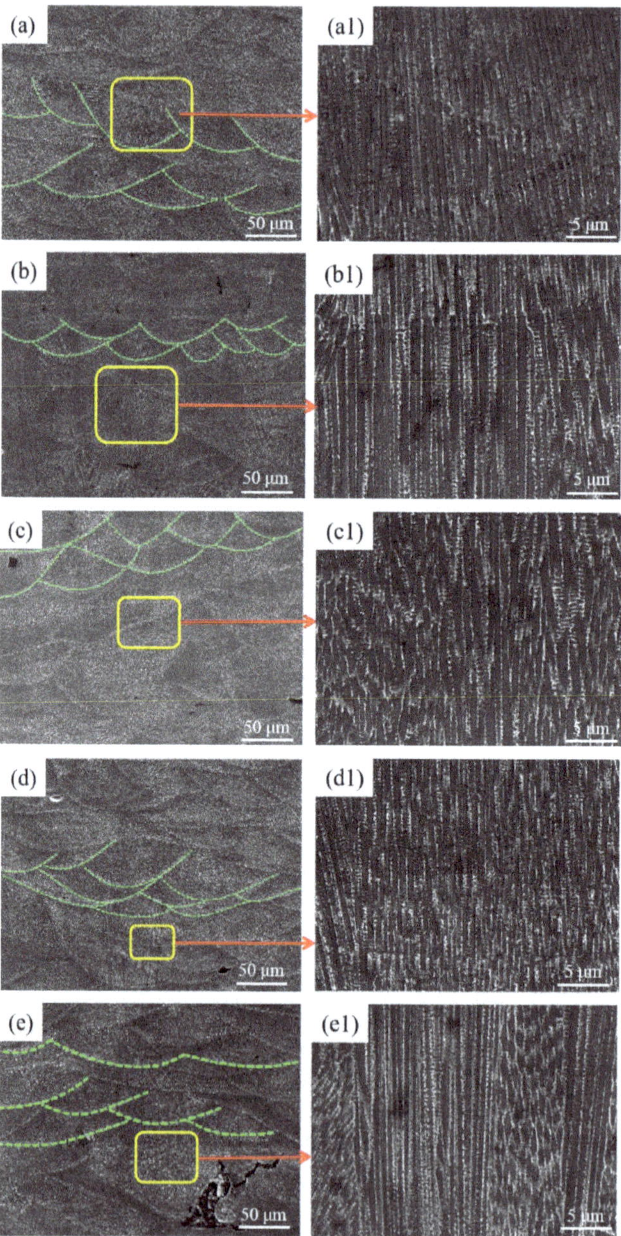

Figure 4. SEM image of the corresponding K418 sample under different scanning speeds: (**a**,**a1**) 800 mm/s; (**b**,**b1**) 1000 mm/s; (**c**,**c1**) 1200 mm/s; (**d**,**d1**) 1400 mm/s; (**e**,**e1**) 1600 mm/s.

Based on metallographic, scanning, and XRD analysis, it is clear that the microstructure transitions from cellular dendrites to columnar crystals, that the columnar crystals are obviously refined, and that the γ'/γ''-reinforced phase is formed at the boundary of the columnar crystals as the scanning speed increases. According to solidification theory, the cooling rate of the melt pool is positively correlated with the laser scanning speed during LPBF, and the phase transition of the microstructure during LPBF is strongly related to the cooling rate during solidification [24]. As a result, as the scanning speed increases from 800 mm/s to 1600 mm/s, the cooling rate tends to increase. Microstructure can be determined according to Equation (2) [27]:

$$\lambda_1 = A\left(G^{-0.55} \cdot V^{-0.28}\right) \tag{2}$$

where λ_1 is the primary dendrite spacing, A is a constant, G represents the temperature gradient, and V is the cooling rate. The primary dendrite spacing is inversely proportional to the cooling rate. Therefore, as the scanning speed increases, the interaction time between the laser beam and the powder decreases, the temperature of the center of the molten pool decreases, and the cooling rate increases, resulting in a decrease in primary dendrite spacing. Furthermore, when the scanning speed is too slow, the cooling rate of the molten pool is slow, there are more liquid phases in the molten pool, and the unmelted particles surrounding the molten pool are quickly dragged into the molten pool due to strong Marangoni convection and high surface tension, resulting in interlayer defects. Thus, residual stress is generated, resulting in the formation of interdendritic thermal cracks. When the scanning speed is too fast, the laser moves too quickly, the temperature of the molten pool is low, the cooling rate is high, and the surrounding powder particles in the molten pool are too late to melt completely, resulting in the occurrence of defects such as non-fusion and pores, as well as a large amount of accumulation in these defects. Furthermore, as the laser energy density increases, the LPBF-ed microstructure preferentially grows ultrafine columnar crystals rather than secondary dendrites, and the microstructure has a significant impact on the mechanical properties [28]. The XRD spectra in Figure 2 clearly show that the scanning speed influences the intensity of the diffraction peaks, implying that the grain orientation has shifted. As a result, the EBSD results at various scanning speeds differ significantly. It can be clearly seen from Figure 5 that the microstructure at 800 mm/s is composed of coarser cellular dendrite crystals with a balanced microstructure color distribution. Figure 6 depicts the grain orientation of a longitudinal section of K418 alloy at various scanning speeds. The weave intensity of the material is 2.93, as shown by its corresponding antipodal map (Figure 6a), indicating weak anisotropy. The red color inside the microstructure increases at a scanning speed of 1200 mm/s, implying that the crystals grow significantly in the first (001) orientation meritocracy. Furthermore, the inverse polarographic projection (Figure 6b) relationship reveals that the microstructure has a high weave strength, with an intensity factor of 6.72. The number of fine crystals within the microstructure increases as the scanning speed increases to 1600 mm/s, but the red area decreases slightly, indicating that the (001) directional selective orientation is weakened. The weakened orientation is also demonstrated by the weaving intensity in the corresponding anti-polar diagram in Figure 6c.

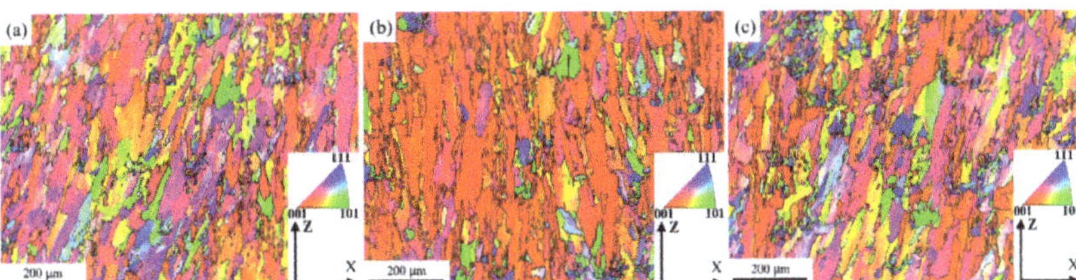

Figure 5. Grain orientation maps of K418 alloy at different scanning speeds: (**a**) 800 mm/s; (**b**) 1200 mm/s; (**c**) 1600 mm/s.

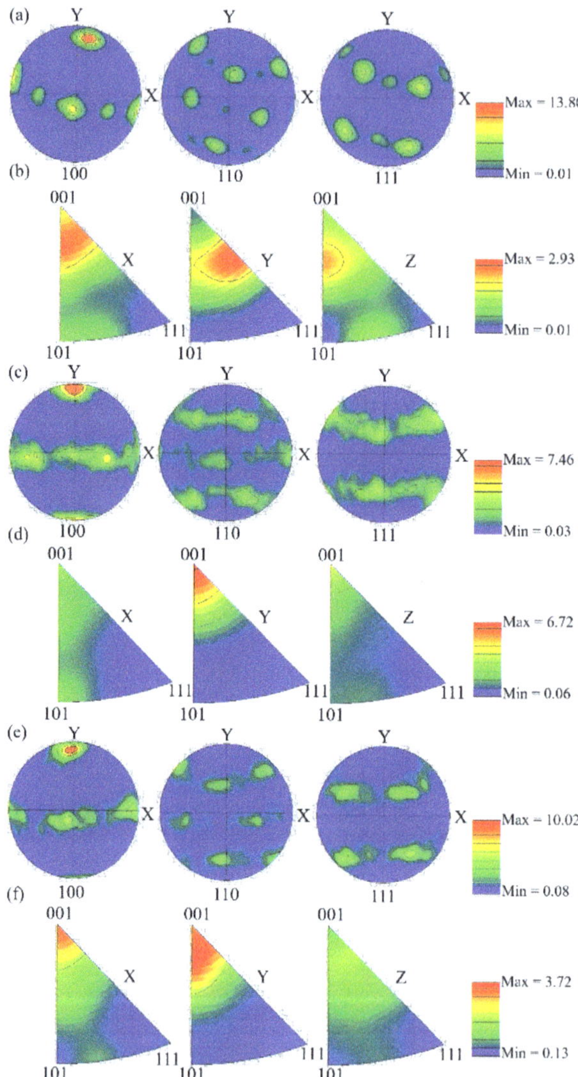

Figure 6. The polar diagrams ((**a**) 800 mm/s; (**c**) 1200 mm/s; (**e**) 1600 mm/s) and anti-polar diagrams ((**b**) 800 mm/s; (**d**) 1200 mm/s; (**f**) 1600 mm/s) of K418 alloy at different scanning speeds.

The difference in grain boundary orientation is also an important indicator for the microstructural characterization of the material. Figure 7 depicts the distribution of grain boundaries in K418 alloy at various laser scanning speeds. The microstructure of the specimen is dominated by high-angle grain boundaries (HAGBs). The preparation of samples layer by layer and row by row is well known to be an important feature of the LPBF technique. As the laser scans the next layer, the previously solidified melt pool is partially remelted, similar to the annealing heat treatment process, which inevitably results in recrystallization and, thus, a high percentage of HAGBs [29]. However, the migration and integration rate of subgrain boundaries decreases as the laser scanning speed increases, resulting in a reduction in HAGBs. As a result, as shown in Figure 7e, the grain size decreases as the scanning speed increases.

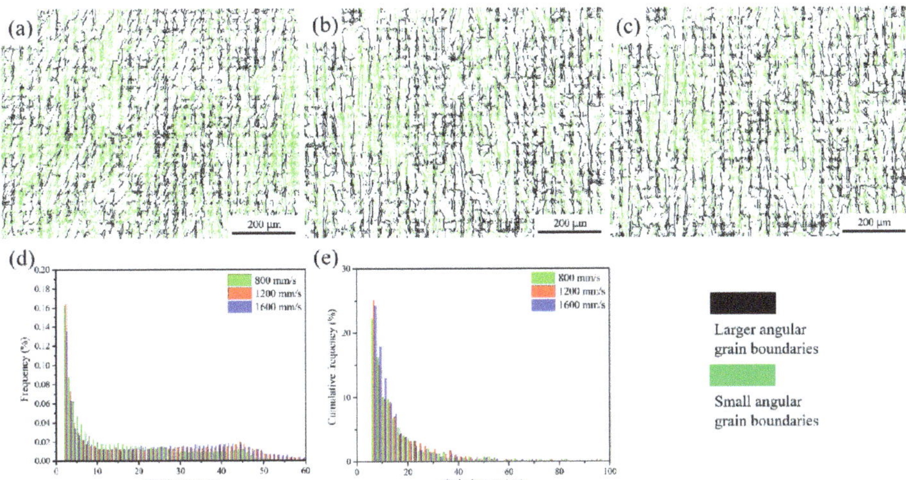

Figure 7. Grain boundary distribution of K418 alloy at different scanning speeds ((**a**) 800 mm/s; (**b**) 1200 mm/s; (**c**) 1600 mm/s), (**d**) grain boundary size statistics of K418 alloy at different scanning speeds, and (**e**) grain size statistics of K418 alloy at different scanning speeds.

Figure 8 depicts the Schmidt factor and distribution statistics for the K418 alloy at various scanning speeds. According to Schmidt's law, a smaller Schmidt factor results in a higher yield strength [30]. Figure 8 shows that the Schmidt factors are approximate at different scanning speeds and, thus, their yield strengths do not differ significantly. As the laser scanning speed increases from 800 mm/s to 1600 mm/s, the migration and integration rates of the subgrain boundaries decrease, causing the HAGB content to increase from 40.4% to 57.2% and the grain size to increase from 0.99 μm to 1.29 μm, as shown in Table 3.

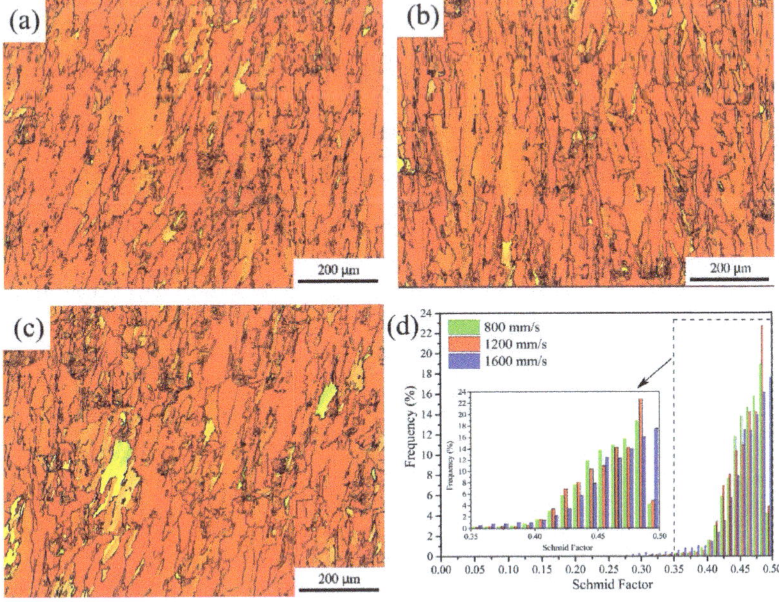

Figure 8. Schmidt factor distribution ((**a**) 800 mm/s; (**b**) 1200 mm/s; (**c**) 1600 mm/s) and (**d**) statistical diagram of K418 alloy at different scanning speeds.

Table 3. Statistics of EBSD test results under different scanning speeds.

Scanning Speeds (mm/s)	HAGBs	Average Grain Size	Substructured	Deformed
800	40.4%	0.99 μm	16.67%	75.29%
1200	52.5%	1.01 μm	13.07%	80.23%
1600	57.2%	1.29 μm	13.22%	75.97%

3.4. Microhardness

The microhardness of LPBF-ed K418 samples at various scanning speeds is depicted in Figure 9. When the scanning speed is low, the energy heat input is high, resulting in coarse internal grains of the microstructure and low hardness. As the scanning speed increases, the grains are refined, the melt pool morphology tends to be uniform, and the extremely fast cooling rate causes the matrix grains to grow too late, resulting in fine crystal reinforcement and high hardness. The linear relationship between Vickers hardness and grain size is defined by Hall and Petch as follows [31]:

$$H_v = H_0 + K_H D^{-\frac{1}{2}} \tag{3}$$

where H_0 and K_H are the corresponding constants. The equation clearly shows that hardness is positively related to grain size. Finer grain size results in higher hardness values. When the scanning speed is too fast, the laser energy input is relatively low, resulting in discontinuous melting tracks or unmelted powder, internal defects, and a reduction in the microhardness of the formed part.

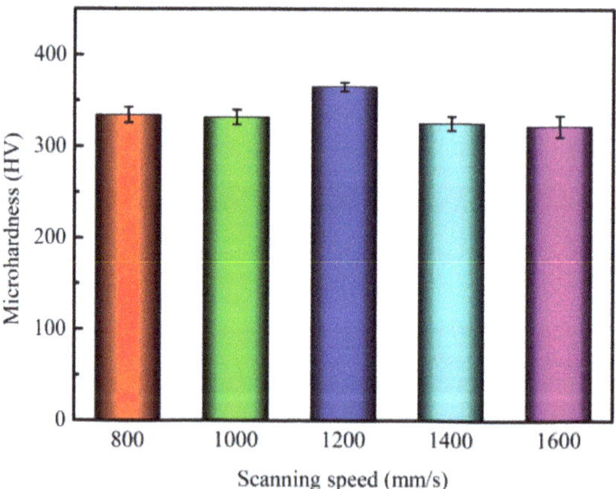

Figure 9. Influence of different scanning speeds on the microhardness of K418.

3.5. Tensile Property

Due to the extremely high temperature gradient and cooling rate (~10^8 K/s) of the LPBF processing, the grain size after solidification is at the micrometer scale. Simultaneously, the material tends to generate metastable phases that deviate from the equilibrium state due to the repeated thermal cycling process of rapid heating and rapid cooling. The microstructural evolution is largely determined by the temperature gradient in the molten pool, whereas the temperature gradient is influenced by the scanning speed, which affects the microstructural evolution characteristics by manipulating the local solidification conditions of the molten pool. The diffusion process is frequently constrained by the narrow solidification window, second-phase precipitation or component segregation is rare, and fine columnar crystals are obtained, inhibiting the growth of secondary dendrite arms and ensuring excellent mechanical properties of LPBF-ed K418 parts. On the other hand, since nickel has a face-centered cubic (FCC) crystal structure, the maximum heat flow and temperature gradient are generally consistent with the deposition direction during LPBF processing—that is, from the substrate along the direction of the build-up layer height increase. Therefore, the microstructure of LPBF-ed K418 alloy tends to exhibit <100> cubic-textured columnar grains, which influence the final mechanical properties. Figure 10 depicts the tensile properties of K418 samples at various scanning speeds, demonstrating that the tensile strength and elongation of the samples tend to increase and then decrease as the scanning speed increases, whereas the yield strength shows only a slight change. The tensile properties were the highest at 1200 mm/s, with the tensile strength reaching 1244.35 $\pm$ 99.12 MPa, the yield strength reaching 863.89 $\pm$ 132.71 MPa, and the elongation of the material reaching a maximum of 12.53 $\pm$ 1.79%. When the scanning speed is less than 1200 mm/s, the powder absorption time is longer, the melt pool cooling time is slower, and the subsequent molten powder is prone to sputtering, resulting in poor melt track bonding. When the scanning speed exceeds 1200 mm/s, the scanning track becomes discontinuous, the metal powder is unable to be completely melted, and defects such as pores and unmelted particles appear. When loaded, the defects act as a source of crack propagation, resulting in early failure and reduced ductility.

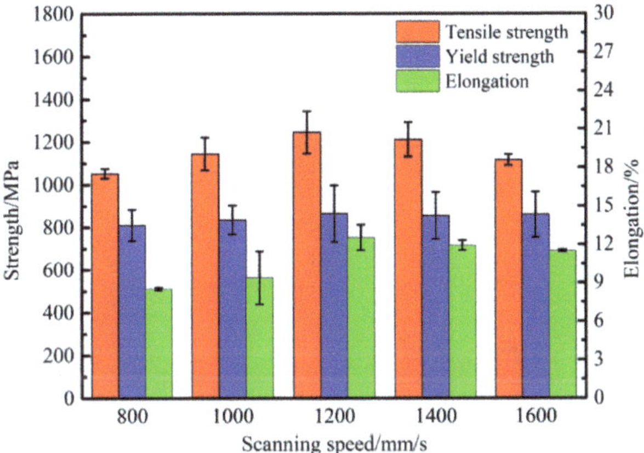

Figure 10. Influence of different scanning speeds on the tensile properties of K418.

Figure 11 depicts the SEM morphology of a tensile fracture at various scanning speeds. When the scanning speed is low, a large number of holes as well as cracks appear in the fracture morphology, resulting in lower strength of the material. When the scanning speed is increased to 1200 mm/s, the defects are obviously reduced, the fractures are distributed with a certain number of tough nests, and the typical feature-tearing ribs can also be clearly observed, as shown in Figure 11c. Then, as the scanning speed increases, a certain number of unmelted particles and pores appear at the fracture. Figure 12 depicts a schematic diagram of the defect distribution in the tensile fracture cross-section corresponding to different laser powers. According to Figure 12 and the macroscopic morphology of the tensile fracture, it can be concluded that the material's tensile properties are related to internal defects such as unmelted particles, pores, and cracks. At a low scanning speed (800–1000 mm/s), defects such as pores and cracks exist in the internal microstructure and grain boundaries, fracturing the specimen and reducing its tensile properties. However, due to the presence of numerous small uniform dendrites and fewer defects in the microstructure, the ideal tensile properties are obtained at a scanning speed of 1200 mm/s. Although the internal microstructure of the specimen was significantly refined when the scanning speed was increased (1400–1600 mm/s), the cooling rate was too fast due to the excessive scanning speed, and the powder melted incompletely, resulting in unmelted particles, pores and other defects, which generated stress concentration and, thus, reduced the tensile properties [32].

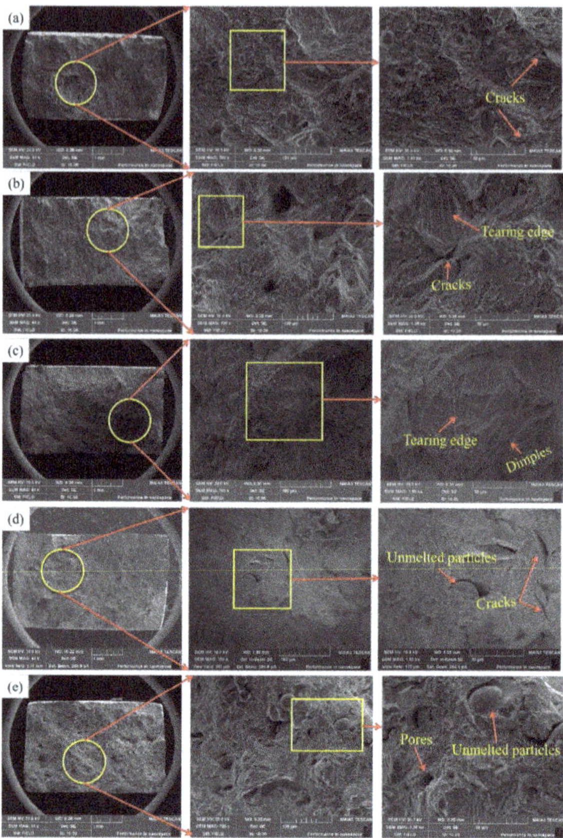

Figure 11. SEM morphology of tensile fractures corresponding to different laser powers ((**a**) 800 mm/s; (**b**) 1000 mm/s; (**c**) 1200 mm/s; (**d**) 1400 mm/s; (**e**) 1600 mm/s).

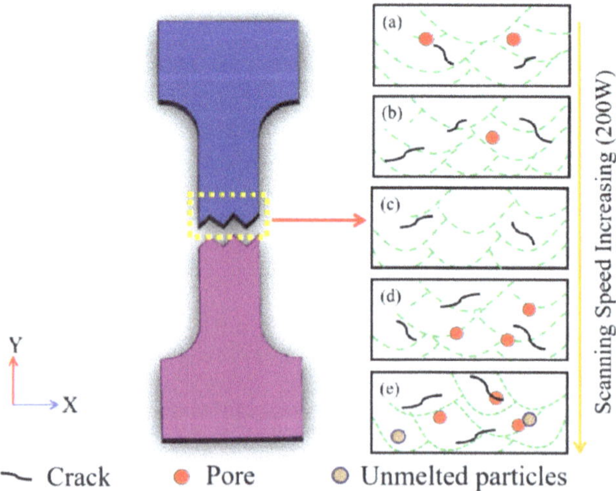

Figure 12. Defect distribution diagrams of tensile fractures' transverse sections corresponding to different laser powers ((**a**) 800 mm/s; (**b**) 1000 mm/s; (**c**) 1200 mm/s; (**d**) 1400 mm/s; (**e**) 1600 mm/s).

4. Conclusions

In this work, the effects of laser scanning speed on the relative density, microstructural characteristics, phase transformation, microhardness, and tensile properties of LPBF-ed K418 samples were investigated. The main findings are summarized as follows:

(1) There are numerous factors affecting the LPBF processing, and the scanning speed is one of the key process parameters influencing the microstructure and properties of the LPBF-ed K418 superalloy. A satisfactory tensile strength of 1244.35 $\pm$ 99.12 MPa and elongation of 12.53 $\pm$ 1.79% were obtained through process optimization.

(2) As the scanning speed increased from 800 mm/s to 1600 mm/s, the tensile strength and elongation of the material tended to first increase and then decrease, while the yield strength remained stable. The optimal comprehensive mechanical properties were obtained when the scanning speed was 1200 mm/s.

(3) The microstructure of LPBF-ed K418 exhibits directional preferential growth of columnar crystals along the (001) direction, which is dominated by high-angle grain boundaries (HAGBs), and the volume fraction of HAGBs decreases with increasing scanning speed.

(4) The optimized laser scanning speed can be used as a reference for the LPBF preparation of nickel-based superalloys, and it is expected that this study will promote the industrial application of nickel-based superalloys prepared via LPBF.

Author Contributions: Conceptualization, Y.L. and Z.C.; methodology, Z.C. and F.L.; software, F.L., P.W.; validation, F.L., Z.C. and S.Y.; formal analysis, F.L.; investigation, Y.L., S.Z.; resources, Z.C.; data curation, F.L.; writing—original draft preparation, Y.L. and F.L.; writing—review and editing, Z.C; visualization, F.L.; supervision, Z.C. and Y.W.; project administration, Z.C.; funding acquisition, Z.C.; All authors have read and agreed to the published version of the manuscript.

Funding: This research was supported by the first China special assistant Postdoctoral Science Foundation (Grant No. 2019TQ0259) and the National Science Fund for Distinguished Young Scholars (Grant No.52005391).

Conflicts of Interest: The authors declare no conflict of interest.

References

1. Pang, M.; Yu, G.; Wang, H.-H.; Zheng, C.-Y. Microstructure study of laser welding cast nickel-based superalloy K418. *J. Mater. Process. Technol.* **2008**, *207*, 271–275. [CrossRef]
2. Chen, C.J.; Cao, Q.; Zhang, M.; Chang, Q.M.; Zhang, S.C. Laser Repair Cladding of K418 Ni-Base Superalloy by CoCrNiW Powder. *Adv. Mater. Res.* **2011**, *291–294*, 1417–1420. [CrossRef]
3. Luo, Y.; Ma, T.; Shao, W.; Zhang, G.; Zhang, B. Effects of heat treatment on microstructures and mechanical properties of GH4169/K418 functionally graded material fabricated by laser melting deposition. *Mater. Sci. Eng. A* **2021**, *821*, 141601. [CrossRef]
4. Yang, F.; Wang, J.; Yu, J.; Zhou, Z.; Wang, B.; Tu, T.; Ren, X.; Deng, K.; Ren, Z. Microstructure and Mechanical Properties of Ni-based Superalloy K418 Produced by the Continuous Unidirectional Solidification Process. *J. Mater. Eng. Perform.* **2019**, *28*, 6483–6491. [CrossRef]
5. Mignanelli, P.; Jones, N.; Perkins, K.; Hardy, M.; Stone, H. Microstructural evolution of a delta containing nickel-base superalloy during heat treatment and isothermal forging. *Mater. Sci. Eng. A* **2015**, *621*, 265–271. [CrossRef]
6. Zhang, L.; Chen, X.; Li, D.; Chen, C.; Qu, X.; He, X.; Li, Z. A comparative investigation on MIM418 superalloy fabricated using gas- and water-atomized powders. *Powder Technol.* **2015**, *286*, 798–806. [CrossRef]
7. Liu, S.; Shin, Y.C. Additive manufacturing of Ti6Al4V alloy: A review. *Mater. Des.* **2019**, *164*, 107552. [CrossRef]
8. He, B.; Wu, W.; Zhang, L.; Lu, L.; Yang, Q.; Long, Q.; Chang, K. Microstructural characteristic and mechanical property of Ti6Al4V alloy fabricated by selective laser melting. *Vacuum* **2018**, *150*, 79–83. [CrossRef]
9. Li, N.; Huang, S.; Zhang, G.; Qin, R.; Liu, W.; Xiong, H.; Shi, G.; Blackburn, J. Progress in additive manufacturing on new materials: A review. *J. Mater. Sci. Technol.* **2019**, *35*, 242–269. [CrossRef]
10. Song, B.; Zhao, X.; Li, S.; Han, C.; Wei, Q.; Wen, S.; Liu, J.; Shi, Y. Differences in microstructure and properties between selective laser melting and traditional manufacturing for fabrication of metal parts: A review. *Front. Mech. Eng.* **2015**, *10*, 111–125. [CrossRef]
11. Sun, D.; Gu, D.; Lin, K.; Ma, J.; Chen, W.; Huang, J.; Sun, X.; Chu, M. Selective laser melting of titanium parts: Influence of laser process parameters on macro- and microstructures and tensile property. *Powder Technol.* **2019**, *342*, 371–379. [CrossRef]

12. Liu, C.; Tong, J.; Jiang, M.; Chen, Z.; Xu, G.; Liao, H.; Wang, P.; Wang, X.; Xu, M.; Lao, C. Effect of scanning strategy on microstructure and mechanical properties of selective laser melted reduced activation ferritic/martensitic steel. *Mater. Sci. Eng. A* **2019**, *766*, 138364. [CrossRef]
13. Saedi, S.; Moghaddam, N.S.; Amerinatanzi, A.; Elahinia, M.; Karaca, H.E. On the effects of selective laser melting process parameters on microstructure and thermomechanical response of Ni-rich NiTi. *Acta Mater.* **2018**, *144*, 552–560. [CrossRef]
14. Li, W.; Liu, J.; Zhou, Y.; Li, S.; Wen, S.; Wei, Q.; Yan, C.; Shi, Y. Effect of laser scanning speed on a Ti-45Al-2Cr-5Nb alloy processed by selective laser melting: Microstructure, phase and mechanical properties. *J. Alloy Compd.* **2016**, *688*, 626–636. [CrossRef]
15. Jiang, P.; Zhang, C.; Zhang, S.; Zhang, J.; Chen, J.; Chen, H. Additive manufacturing of novel ferritic stainless steel by selective laser melting: Role of laser scanning speed on the formability, microstructure and properties. *Opt. Laser Technol.* **2021**, *140*, 107055. [CrossRef]
16. Chen, Z.; Wei, P.; Zhang, S.; Lu, B.; Zhang, L.; Yang, X.; Huang, K.; Huang, Y.; Li, X.; Zhao, Q. Graphene reinforced nickel-based superalloy composites fabricated by additive manufacturing. *Mater. Sci. Eng. A* **2020**, *769*, 138484. [CrossRef]
17. Moshtaghi, M.; Safyari, M. Effect of dwelling time in VIM furnace on chemical composition and mechanical properties of a Ni-Fe-Cr alloy. *Vacuum* **2019**, *169*, 108890. [CrossRef]
18. Was, G.S. Grain-Boundary Chemistry and Intergranular Fracture in Austenitic Nickel-Base Alloys—A Review. *Corrosion* **1990**, *46*, 319–330. [CrossRef]
19. Wang, H.-Y.; Wang, B.-B.; Wang, L.; Cui, R.; Luo, L.-S.; Su, Y.-Q. Impact of laser scanning speed on microstructure and mechanical properties of Inconel 718 alloys by selective laser melting. *China Foundry* **2021**, *18*, 170–179. [CrossRef]
20. Amirjan, M.; Sakiani, H. Effect of scanning strategy and speed on the microstructure and mechanical properties of selective laser melted IN718 nickel-based superalloy. *Int. J. Adv. Manuf. Technol.* **2019**, *103*, 1769–1780. [CrossRef]
21. Wang, X.; Carter, L.N.; Pang, B.; Attallah, M.M.; Loretto, M.H. Microstructure and yield strength of SLM-fabricated CM247LC Ni-Superalloy. *Acta Mater.* **2017**, *128*, 87–95. [CrossRef]
22. Raghavan, S.; Zhang, B.; Wang, P.; Sun, C.-N.; Nai, M.L.S.; Li, T.; Wei, J. Effect of different heat treatments on the microstructure and mechanical properties in selective laser melted INCONEL 718 alloy. *Mater. Manuf. Processes* **2016**, *32*, 1588–1595. [CrossRef]
23. Wang, R.; Chen, C.; Liu, M.; Zhao, R.; Xu, S.; Hu, T.; Shuai, S.; Liao, H.; Ke, L.; Vanmeensel, K.; et al. Effects of laser scanning speed and building direction on the microstructure and mechanical properties of selective laser melted Inconel 718 superalloy. *Mater. Today Commun.* **2021**, *30*, 103095. [CrossRef]
24. Chen, Z.; Xiang, Y.; Wei, Z.; Wei, P.; Lu, B.; Zhang, L.; Du, J. Thermal dynamic behavior during selective laser melting of K418 superalloy: Numerical simulation and experimental verification. *Appl. Phys. A* **2018**, *124*, 313. [CrossRef]
25. Tan, G.H.; Wang, S.L.; Ming, W.U.; Sun, W.J.; Chen, Y.H.; Li-Ming, K.E. Effect of Cracks in Electron Beam Welding of K418 Superalloy and Its Microstructure and Properties. *J. Netshape Form. Eng.* **2019**, *170*, 107697. [CrossRef]
26. Xiong, W.; Hao, L.; Li, Y.; Tang, D.; Cui, Q.; Feng, Z.; Yan, C. Effect of selective laser melting parameters on morphology, microstructure, densification and mechanical properties of supersaturated silver alloy. *Mater. Des.* **2019**, *170*, 107697. [CrossRef]
27. Yang, H.; Meng, L.; Luo, S.; Wang, Z. Microstructural evolution and mechanical performances of selective laser melting Inconel 718 from low to high laser power. *J. Alloy Compd.* **2020**, *828*, 154473. [CrossRef]
28. Chen, Z.; Chen, S.; Wei, Z.; Zhang, L.; Wei, P.; Lu, B.; Zhang, S.; Xiang, Y. Anisotropy of nickel-based superalloy K418 fabricated by selective laser melting. *Prog. Nat. Sci.* **2018**, *28*, 496–504. [CrossRef]
29. Qiu, C.; Adkins, N.J.E.; Attallah, M.M. Selective laser melting of Invar 36: Microstructure and properties. *Acta Mater.* **2016**, *103*, 382–395. [CrossRef]
30. Li, W.; Yang, Y.; Li, M.; Liu, J.; Cai, D.; Wei, Q.; Yan, C.; Shi, Y. Enhanced mechanical property with refined microstructure of a novel γ-TiAl/TiB2 metal matrix composite (MMC) processed via hot isostatic press. *Mater. Des.* **2018**, *141*, 57–66. [CrossRef]
31. Ye, J.; Han, B.Q.; Lee, Z.; Ahn, B.; Nutt, S.R.; Schoenung, J. A tri-modal aluminum based composite with super-high strength. *Scr. Mater.* **2005**, *53*, 481–486. [CrossRef]
32. Cherry, J.A.; Davies, H.M.; Mehmood, S.; Lavery, N.P.; Brown, S.G.R.; Sienz, J. Investigation into the effect of process parameters on microstructural and physical properties of 316L stainless steel parts by selective laser melting. *Int. J. Adv. Manuf. Technol.* **2015**, *76*, 869–879. [CrossRef]

 *materials*

MDPI

Article

Practical Approach to Eliminate Solidification Cracks by Supplementing AlMg4.5Mn0.7 with AlSi10Mg Powder in Laser Powder Bed Fusion

Constantin Böhm *, Martin Werz * and Stefan Weihe

Materials Testing Institute (MPA), University of Stuttgart, Pfaffenwaldring 32, D-70569 Stuttgart, Germany; stefan.weihe@mpa.uni-stuttgart.de

* Correspondence: constantin.boehm@mpa.uni-stuttgart.de (C.B.); martin.werz@mpa.uni-stuttgart.de (M.W.)

Abstract: The range of available aluminum alloy powders for laser powder bed fusion (LPBF) is restricted to mainly Al–Si based alloys. Currently aluminum alloy powders, designed for lightweight application, based on Al–Mg (5000 series), Al–Si–Mg (6000 series), or Al–Zn–Mg (7000 series), cannot be processed by LPBF without solidification cracks. This has an impact on the potential of LPBF for lightweight applications. In fusion welding, solidification cracks are eliminated by using filler materials. This study aims to transfer the known procedure to LPBF, by supplementing EN AW-5083 (AlMg4.5Mn0.7) with AlSi10Mg. EN AW-5083 and two modifications (+7 wt.% and +15 wt.% AlSi10Mg) were produced by LPBF and analyzed. It was found that, in EN AW-5083, the solidification cracks have a length $\geq$200 μm parallel to the building direction. Furthermore, the solidification cracks can already be eliminated by supplementing 7 wt.% AlSi10Mg. The microstructure analysis revealed that, by supplementing AlSi10Mg, the melt pool boundaries become visible, and the grain refines by 40% relative to the base alloy. Therefore, adding a low melting point phase and grain refinement are the mechanisms that eliminate solidification cracking. This study illustrates a practical approach to eliminate solidification cracks in LPBF.

Keywords: laser powder bed fusion; solidification crack; hot crack; aluminum alloy powder; EN AW-5083; AA5083; AlSi10Mg; selective laser melting

Citation: Böhm, C.; Werz, M.; Weihe, S. Practical Approach to Eliminate Solidification Cracks by Supplementing AlMg4.5Mn0.7 with AlSi10Mg Powder in Laser Powder Bed Fusion. *Materials* **2022**, *15*, 572. https://doi.org/10.3390/ma15020572

Academic Editors: Swee Leong Sing and Wai Yee Yeong

Received: 26 November 2021
Accepted: 10 January 2022
Published: 13 January 2022

1. Introduction

Solidification cracks are severe defects that limit the processability of the commercially-available aluminum alloy powders for laser powder bed fusion (LPBF), also known as selective laser melting (SLM) [1]. This hinders the potential of lightweight design with LPBF.

The LPBF process is physically similar to fusion welding. In LPBF, a part is created by iteratively melting metal powder along multiple tracks, with a focused laser beam in a single plane, and adding new powder layers to build up the three-dimensional part. The part is built up on a substrate plate, where the powder is added layer-by-layer by a powder recoater. In contrast to fusion welding, single or multiple seams are welded to join parts.

The LPBF process has numerous influence factors that all impact the quality of the part [2]. Overall, as reviewed in [3], the research in LPBF focusses on residual stresses and distortion, the influence of the process on the surface quality, and their influence on the mechanical properties of the LPBF-produced parts, as well as the interaction of the laser beam with the metal powder [4–6]. Another important research subject is the density of the part. Research groups focus on understanding the mechanisms of pore formation [7], in order to find optimal process parameters for multiple materials [8,9]. Furthermore, processing defects, such as splattering [10], balling [11], and disproportionate evaporation of volatile elements [2,12,13], which can degrade the quality of the LPBF-produced part, are addressed in current research. Another research area deals with the influence of the properties of the metal powder on the finished part, as well as the process stability [14–16].

In contrast to the process-related research, the LPBF machines are equally subject of a research and development effort [17].

Additional effort is invested into developing high-strength aluminum alloy powders for LPBF, as summarized by Leirmo [1], where solidification cracks must be overcome to ensure their processability with LPBF. Various strategies are used to eliminate or reduce solidification cracks in fusion welding [18]. Solidification cracks can be reduced in fusion welding (i) by controlling the solidification structure by supplementing grain refiner [19,20] or adjusting the process parameters to increase the grain refinement [21] or using specially-designed alloys [22] and (ii) by using favorable welding conditions by adjusting the process parameters to reduce the stress on the weld seam [23] or adapting the joint configuration [18].

These strategies were already applied to eliminate or reduce solidification cracks during the LPBF process. For example, the powder alloy composition of AA7075 (AlZnMg5.5Cu) was controlled by supplementing elemental silicon [24] or AlSi10Mg powder [25] to create a powder–powder mixture. Mechanically mixing metal powders is a novel technique used in LPBF research that enables flexible alloy design. This technique was performed for various metal systems in LPBF [8,24–30]. In all these studies [8,24–30], the metal powders were mixed dry for several minutes to hours. There is yet no uniform procedure for mixing the powders. Furthermore, the effects of the mixing parameters on the homogeneity of the resulting powder is still ongoing research [30]. Moreover, the LPBF process parameters were adjusted in [13,31–33], and the base plate heating was adjusted in [12] to eliminate or reduce solidification cracks. In addition, the solidification structure was controlled to reduce solidification cracks by adding grain refiner [13,32,34–36].

Pre-alloying powder alloys [13,32] and/or coating with nanoparticles [34,36] requires expensive or sophisticated technology and trained staff. In contrast, mixing the base alloy with a supplement powder to modify and control the alloy composition appears to be a practical technique. For example, Aversa et al. [25] supplemented 50 wt.% of AlSi10Mg to AA7075 to eliminate solidification cracks, and Montero-Sistiaga et al. [24] supplemented 4 wt.% elemental silicon to AA7075. However, supplementing too much silicon results in brittle phases that can reduce the tensile ductility. For the 50 wt.% AA7075 and 50 wt.% AlSi10Mg alloy, Aversa et al. [25] found an elongation of break of 2.7%, whereas for AA7075-T6, a typical value is 10% [37]. Therefore, an objective for this research field is to minimize the amount of supplemented AlSi10Mg.

In order to calculate the minimal required AlSi10Mg addition to a base powder alloy with a solidification crack criterion [38,39], the mechanisms that lead to the elimination of solidification cracks need to be understood first. Furthermore, experimental data for low additions of AlSi10Mg is required to validate solidification crack criteria. Up to now, there is still a knowledge gap in current literature.

The 5000 series aluminum alloys show good weldability, good formability, and high-strength, as well as good corrosion resistance [37,40]. One example of the 5000 series is EN AW-5083. The alloy composition is listed in Table 1. Its high magnesium content reduces the density of the alloy, as well as increases its weldability and solid solution strength. Therefore, it is a widely-used alloy for lightweight applications [37,40]. However, solidification cracks form when processing this alloy with LPBF [13]. How solidification cracks in EN AW-5083 can be avoided by supplementing AlSi10Mg has not yet been investigated in literature. There is a lack of experimental data.

Table 1. Alloy composition of EN AW-5083 (AlMg4.5Mn0.7), according to DIN EN 573-3 [41].

	Si (wt.%)	Fe (wt.%)	Cu (wt.%)	Mn (wt.%)	Mg (wt.%)	Cr (wt.%)	Al (wt.%)
EN AW-5083 (AlMg4.5Mn0.7)	0.4	0.4	0.1	0.4–1.0	4.0–4.9	0.05–0.25	Balance

The objective of this study is to understand to what extent the selected process parameters can reduce the solidification cracking susceptibility in EN AW-5083 and how even small supplements of AlSi10Mg (+7 wt.% and +15 wt.% AlSi10Mg) can reduce the solidification cracking susceptibility of EN AW-5083. This study provides essential experimental data and insights for computer-aided alloy design.

2. Materials and Methods

In order to investigate the influence of the amount of supplemented powder and process parameters, cubical specimens ($10 \times 10 \times 10$ mm^3) were manufactured with LPBF and analyzed metallographically, as seen in Figure 1. All specimens were produced with a LPBF system by AconityMINI (Aconity3D GmbH, Herzogenrath, Germany).

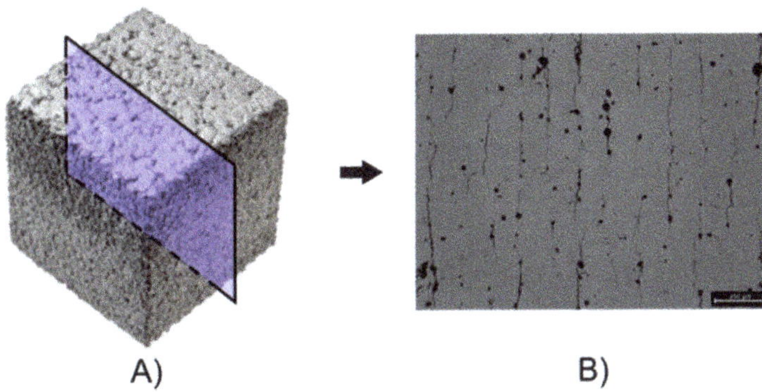

Figure 1. (**A**) Cubical specimens, $10 \times 10 \times 10$ mm^3, were produced with LPBF and cut perpendicular to the building direction. (**B**) The cross sections were cut, polished, and captured with a light microscope, in order to reveal the solidification cracks parallel to the building direction.

The used scan strategy was unidirectional alternating by 90° in every layer. For all conducted experiments, the powder layer thickness was t = 30 µm, and no substrate heating was used. The scan velocity (v = 500–1500 mm/s) was varied because it is a main influence on the solidification cracking sensitivity, as shown in [38]. Furthermore the focal diameter was adjusted, in order to increase the size of the weld pool, thereby decreasing the expected solidification crack sensitivity [18]. The hatch distance was adjusted according to the focal diameter to get the required overlap of the melt tracks. Table 2 summarizes the process parameters used. Table 2 also highlights the volumetric energy density (VED) = $P/(v \cdot h \cdot d_f)$. Overall, the laser power was adjusted to reach a value of VED, which led to dense parts made of AlSi10Mg in prior LPBF experiments conducted at the institute.

Table 2. Applied process parameters for the LPBF experiments. For all experiments, no base plate heating was used, and the layer thickness was kept constant at 30 µm.

	Laser Power P (W)	Scan Velocity v (mm/s)	Focal Diameter d_f (µm)	Hatch Distance h (µm)	Volumetric Energy Density VED (J/mm^3)
A	150	500	100	80	125
B	250	1000	100	80	104
C	250	1500	100	80	69
D	250	500	200	150	111
E	325	1000	200	150	72
F	350	1500	200	150	52

For each combination of process parameters, one specimen was produced and investigated further.

Two alloy modifications were prepared, in order to investigate the effect of supplementing AlSi10Mg on the solidification cracks. The alloy composition of the base alloys, supplement AlSi10Mg powder, and two modifications are listed in Table 3. The amount of AlSi10Mg was chosen, in order to get double (+7 wt.% AlSi10Mg) and fourfold (+15 wt.% AlSi10Mg) the maximum allowed silicon content in EN AW-5083, according to DIN EN 573-3 [41] (Table 1). The goal was to keep the amount of supplemented AlSi10Mg as low as possible because there is no available data in this region.

Table 3. Alloy composition of the base alloy, supplement AlSi10Mg powder, and two modifications.

	Si (wt.%)	Mg (wt.%)	Mn (wt.%)	Al (wt.%)	Source
EN AW-5083 (AlMg4.5Mn0.7)	0.14	4.0	0.67	Balance	Data sheet
AlSi10Mg	9.84	0.32		Balance	Data sheet
MOD1: +7 wt.% AlSi10Mg	0.82	3.74	0.62	Balance	Calculated
MOD2: +15 wt.% AlSi10Mg	1.60	3.45	0.57	Balance	Calculated

The modifications were prepared by weighing and manually mixing them, as shown in Figure 2. The base alloy EN AW-5083 (supplied by Rosswag GmbH, Pfinztal, Germany) and supplement AlSi10Mg (supplied by TLS Technik GmbH & Co. Spezialpulver KG, Bitterfeld-Wolfen, Germany) were mixed in a sealed container by manually shaking and rotating it along its longitudinal axis for 30 s. The mixing procedure was repeated five times for each powder mixture. After mixing, the resulting powder blend was inspected by naked eye, in order to qualitatively evaluate the homogeneity of the blend. This evaluation was possible due to the optical difference of AlSi10Mg, which appears darker due to the high silicon content, compared to the base alloy EN AW-5083 (AlMg4.5Mn0.7), which appears brighter under ambient lighting, as shown in Figure 2.

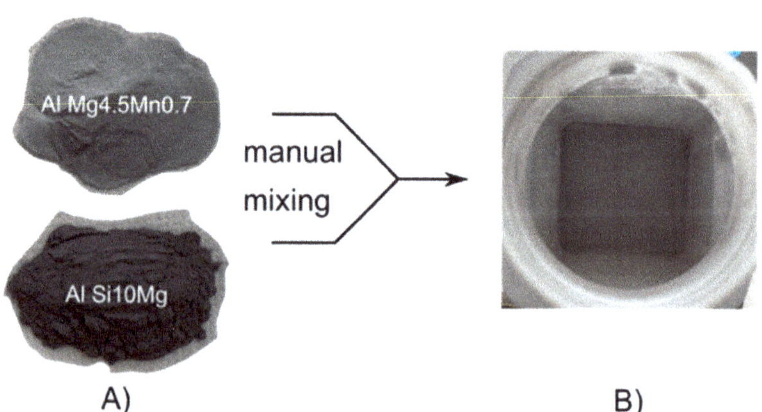

Figure 2. (**A**) Base alloy EN AW-5083 (AlMg4.5Mn0.7) was mixed with the supplement AlSi10Mg powder. (**B**) Homogeneous mixture by manually shaking the two powders in a sealed container.

The powder size distribution for EN AW-5083 was D10 = 26.36 μm, D50 = 41.37 μm, and D90 = 62.16 μm for the respective percentile quantile. For AlSi10Mg it was D10 = 14.53 μm, D50 = 30.13 μm, and D90 = 44.14 μm. The powder supplier measured and provided this data with a dynamic image analysis.

The resulting silicon content (Table 3) in the modifications exceeded the allowed silicon content of the EN AW-5083 standard, where a maximum of 0.4 wt.% is specified. The composition of the modifications is comparable to the EN AW-5026 alloy, which has an allowed silicon content of 0.55–1.4 wt.% and magnesium content of 3.9–4.9 wt.% [41]. Due to the increased silicon content, the modifications are considered to be hardenable by Mg_2Si precipitations.

The LPBF-produced specimens of the base alloy and two modifications were cut, polished, and analyzed with light and electron microscopes. The specimens were polished wet with SiC abrasive paper, with grit sizes in the following sequence: P180, P320, P400, P800, P1200 (European *P*-grade). Afterwards, the specimens were further polished in a sequence with diamond suspension with the following sizes: 15 μm, 6 μm, and 1 μm. Lastly, the specimens were vibration polished with SiO_2 of size 0.05 μm for approximately two hours for the inspection with the light microscope and approximately 24 h for the inspection with the electron microscope.

The analysis with the light microscope (Leica DM8000M, Leica, Wetzlar, Germany) was conducted for all prepared specimens. The light microscope images were used to approximate the density/porosity of the part. For the calculation of the density, the image processing software ImageJ v1.52a [42] was used. First, the greyscale light microscope image was adjusted to a black-and-white image, with the function *Threshold* of ImageJ. Next, the function *Analyze Particles* was used to calculate the area fraction of the pores. The automatic settings of ImageJ were used for all image processing steps. One image per specimen was analyzed in this way.

The analysis of the microstructure was done with a scanning electron microscope (SEM) (Auriga SEM System, Carl Zeiss AG, Oberkochen, Germany). The analysis was conducted for the base alloy and two modifications prepared with the process parameters B (Table 2). For the microstructure analysis, an electron back scattering diffraction (EBSD), as well as an energy-dispersive X-ray spectroscopy (EDX), was performed. The EDX signal was measured during the EBSD measurement. Therefore, it is to be interpreted qualitatively. The microstructure analysis with SEM was performed with an acceleration voltage of 20 kV at a magnification of ×100 of an 800 × 800 μm region, with a step size of 2 μm. A DigiView detector (EDAX, Mahwah, NJ, USA) was used for EBSD, and an Octane Pro detector (EDAX, Mahwah, NJ, USA) was used for EDX.

3. Results

3.1. Metallographic Analysis of EN AW-5083 (AlMg4.5Mn0.7)

The process parameters were varied, in order to evaluate their influence on the solidification cracks in EN AW 5083 (AlMg4.5Mn0.7). Of particular interest was the variation of the scan velocity v because it influences the solidification and cooling rates [18]. Both are influence factors on the solidification crack susceptibility [38]. Figure 3 depicts the polished cross sections of the cubic specimens, viewed with the light microscope. The used process parameters are listed in Table 2. The density was calculated by image processing and varied between 95.4–97.5%, as summarized in Table 4.

In all images of Figure 3, cracks parallel to the building direction are present. The crack length is between 100 μm to 300 μm. The horizontal spacing between the cracks is between 50–150 μm. In the chosen process window, the cracks could not be eliminated. However, in Figure 3A,D, the cracks occur qualitatively lesser in frequency and length, compared to the Figure 3C,F.

This seconds the initial assumption that the cracks are solidification cracks. With an increase in the *VED*, the temperature gradient and cooling rate is reduced, which subsequently leads to a reduction of the solidification crack susceptibility [38]. If liquation cracks were present, an increase in the *VED* would lead to a higher crack frequency, due to an increased heat-affected zone [18].

Figure 4 depicts images, captured with SEM, of the fracture surface of an EN AW-5083 specimen. The sample was cut and then forcefully torn apart, in order to reveal the surface of the solidification cracks. Figure 4C shows the residual fracture surface and the freely-solidified dendritic structure. The ductile forced fracture surface is a result of the laboratory procedure to beak the specimen. However, the freely-solidified dendritic structure in Figure 4D is a result of the solidification during the LPBF process.

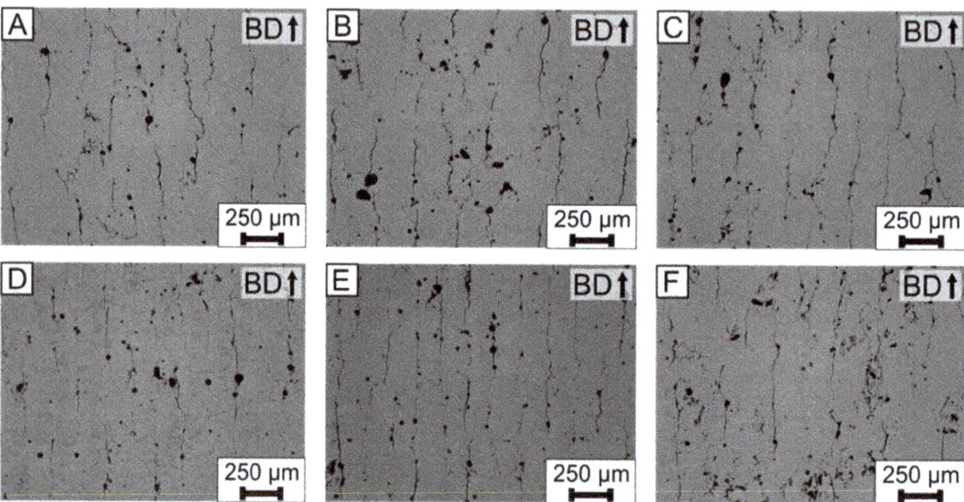

Figure 3. Light microscope images of the polished cross sections for the base alloy EN AW-5083 (AlMg4.5Mn0.7). (**A–F**) Process parameters are listed in Table 2 and correspond to the letter of each cross section. Furthermore, the results of the density measurements, which were done with image processing, are listed in Table 4.

Table 4. Results of the density measurements of the EN AW-5083 specimens, done with image processing. The process parameters used are listed in Table 2.

Process Parameter	Density (%)
A	97.49
B	96.03
C	96.92
D	97.39
E	96.88
F	95.44

As a conclusion, the freely-solidified dendritic structure and increase in the *VED*, leading to a reduction in the crack frequency, are evidence for solidification cracks. Furthermore, the variation of the process parameters in the given process window (*VED* = 52 J/mm^3–25 J/mm^3) could not eliminate the solidification cracks in the LPBF-produced EN AW-5083 (AlMg4.5Mn0.7) specimens. Kouraytem et al. [43] showed that pores, such as those present in Figure 3, can lead to solidification cracks. Therefore, there could be a combination of process parameters, where no pores or solidification cracks are present. However, this combination of process parameters is currently unknown for EN AW-5083.

3.2. Modifying EN AW-5083 (AlMg4.5Mn0.7) with AlSi10Mg

3.2.1. Modification I: AlMg4.5Mn0.7 +7 wt.% AlSi10Mg

The process parameters were varied, in order to evaluate their influence on the solidification cracks in the modified alloy MOD1 (AlMg4.5Mn0.7 +7 wt.% AlSi10Mg). Figure 5 depicts the polished cross sections of the cubic specimens, viewed with the light microscope. The process parameters used are listed in Table 2. The resulting density (Table 5) of the specimens was measured with image processing and range between 95.3–99.1%.

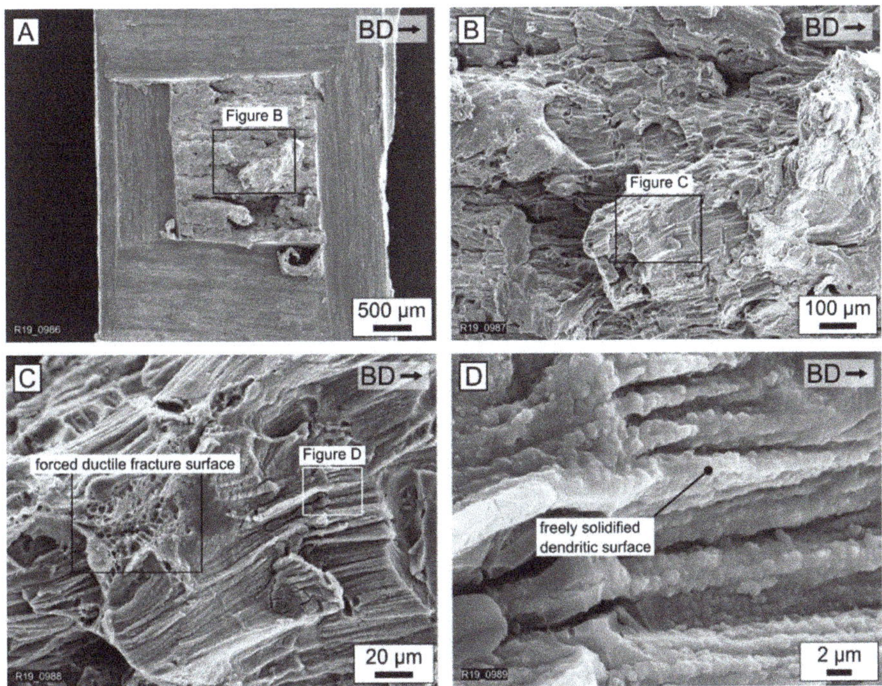

Figure 4. SEM images of the fracture surface of the base alloy EN AW-5083 (AlMg4.5Mn0.7). (**A**) Indicating the procedure for the forced fracture in the laboratory, (**B**) magnification highlighting the melt pools, (**C**) magnification highlighting the forced ductile fracture surface and dendritic structure, and (**D**) magnification of freely-solidified dendritic surface, typically for solidification cracks.

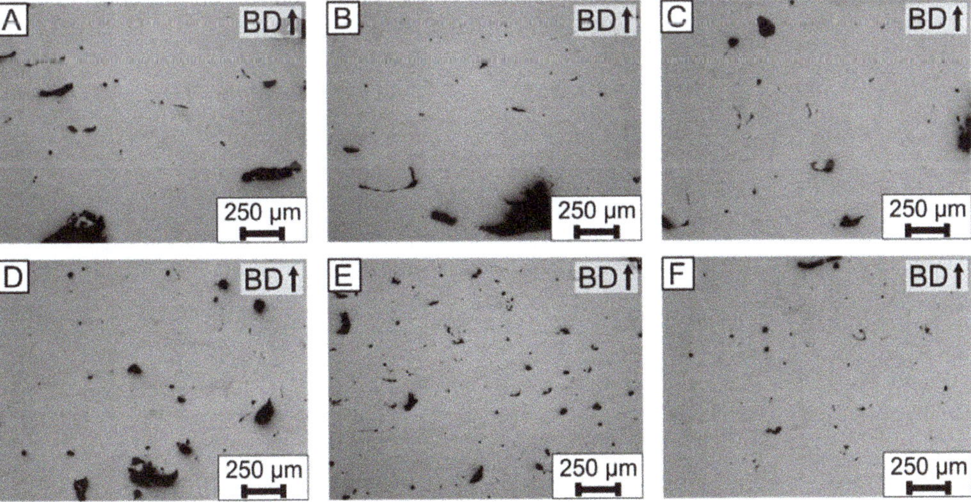

Figure 5. Light microscope images of the polished cross sections for the MOD1 +7 wt.% AlSi10Mg (Table 3). (**A–F**) Process parameters are listed in Table 2 and correspond to the letter of each cross section. The results of the density measurements, which were calculated with image processing, are listed in Table 5.

Table 5. Results of the density measurements of Modification I specimens, done with image processing. The process parameters used are listed in Table 2.

Process Parameter	Density (%)
A	97.67
B	95.26
C	99.11
D	98.56
E	98.31
F	98.96

In all images of Figure 5, no solidification cracks are visible for all combinations of the process parameters. However, there are various processing defects, such as lack of fusion or porosities. Porosities and lack of fusion can be reduced by adjusting the process parameters, for example by increasing the *VED*, as proposed by Weingarten et al. [7] and Ghasemi-Tabasi et al. [9].

Even though pores can initiate solidification cracks [43], no solidification cracks were observed for the modification I with a residual porosity. In conclusion, supplementing AlSi10Mg can eliminate the solidification cracks.

3.2.2. Modification II: AlMg4.5Mn0.7 +15 wt.% AlSi10Mg

The process parameters were varied, in order to evaluate their influence on the solidification cracks in the modified alloy MOD2 (AlMg4.5Mn0.7 +15 wt.% AlSi10Mg). Figure 6 depicts the polished cross sections of the cubic specimens, viewed with the light microscope. The process parameters used are listed in Table 2. Table 6 lists the results of the density measurements, which were done with image processing. The density of the MOD2 specimens range between 96.6–99.1% for the used process parameters.

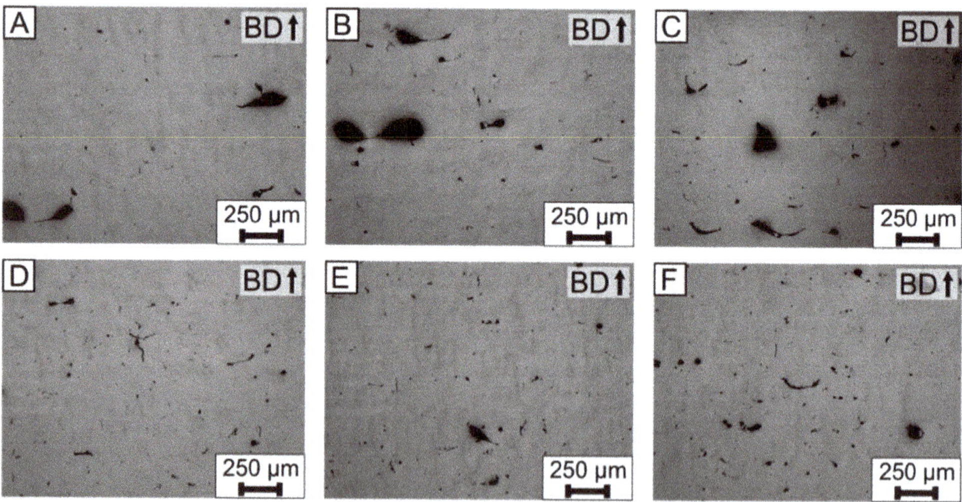

Figure 6. Light microscope images of the polished cross sections for the MOD2 +15 wt.% AlSi10Mg (Table 3). (**A–F**) Process parameters are listed in Table 2 and correspond to the letter of each cross section. Table 6 lists the results of the density measurements, which were calculated with image processing.

In all images of Figure 6, no solidification cracks are visible for any combination of the process parameters, similar to the results of Modification I (Figure 5). However, single microcracks, with different orientations, are visible in Figure 6B,D,E. Furthermore, there are various processing defects, such as lack of fusion or porosities. Lack of fusion and porosities can be reduced by adjusting the process parameters further.

Table 6. Results of the density measurements of Modification II specimens, done with image processing. The process parameters used are listed in Table 2.

Process Parameter	Density (%)
A	98.51
B	96.60
C	98.61
D	99.10
E	98.89
F	97.61

For both modifications, solidification cracks were eliminated, even though there were pores. As shown by Kouraytem et al. [43], pores can lead to solidification cracks.

3.3. Microstructure Analysis

In this section, the microstructure is studied in more detail, in order to highlight the mechanisms that eliminate the solidification cracks. The effect of supplementing AlSi10Mg on the grain structure is demonstrated. Furthermore, the distribution of the alloying elements, compared to the base powder alloy, is analyzed qualitatively.

3.3.1. EBSD Measurement

Figure 7 depicts the EBSD analysis of the base alloy and its two modifications. The analyzed specimens were processed with the process parameters B in Table 2. The colors indicate the different grain orientations. All images are aligned according to their building direction (BD).

Figure 7. Electron back scattering diffraction (EBSD) for the (**A**) base alloy EN AW-5083, (**B**) MOD1 +7 wt.% AlSi10Mg, and (**C**) MOD2 +15 wt.% AlSi10Mg (Table 3). The process parameters for all specimens were p = 250 W, v = 1000 mm/s, and d_f = 100 µm (Table 2).

In the base alloy EN AW-5083, the grains grow parallel to the building direction. The grains in the base alloy reach lengths between 300–500 µm. There are no melt pool boundaries visible. The grains outgrow multiple layers.

In the modification +7 wt.% AlSi10Mg, the grains grow parallel to the building direction. However, they are smaller, compared to the ones of the base alloy. The melt pool boundaries are visible for the modification +7 wt.% AlSi10Mg. Similar observations can be drawn from the modification +15 wt.% AlSi10Mg.

The EBSD analysis revealed that the base alloy has elongated grains that could lead to the increased solidification crack susceptibility. Supplementing the base alloy with AlSi10Mg leads to a qualitative grain refinement.

The grain diameter and area fraction of these grains can be calculated from the EBSD analysis. Figure 8 compares the area fraction of a specific grain diameter for the different alloys in this study with results from literature [13]. The x-axis in Figure 8 represents the grain diameter (µm) in a logarithmic scale. The y-axis represents the area fraction (%) of a specific grain diameter.

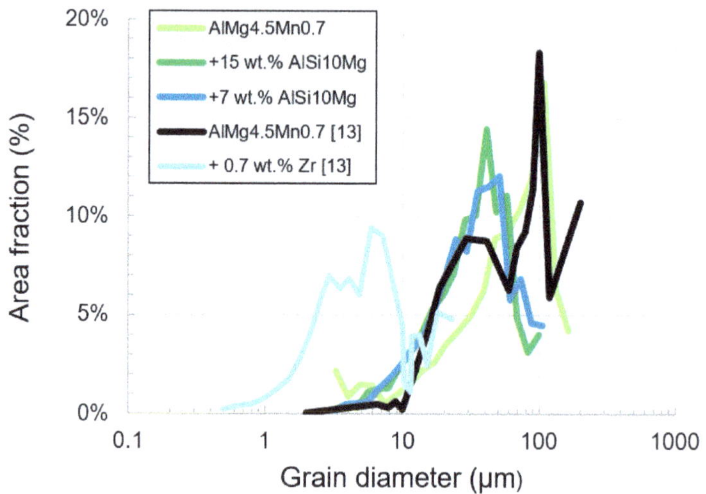

Figure 8. Histogram of the area fraction over the grain diameter. The data was extracted from the EBSD measurement Figure 7.

The grain diameter with the maximum area fraction for both base alloys is approximately 100 µm. Modifying the base alloy reduces the grain diameter with the maximum area fraction to 40 µm (+15 wt.% AlSi10Mg), 50 µm (+7 wt.% AlSi10Mg), and 6 µm (+0.7 wt.% Zr). Furthermore, by modifying the alloy, the mean grain diameter is reduced, as listed in Table 7. Relative to the base alloy, the mean grain diameter is reduced by 40% by modifying it with +7 wt.% or +15 wt.% AlSi10Mg. Modifying the base alloy with zirconium reduces the mean grain diameter, relative to the base alloy by 91% [13]. Zirconium has a higher potential for grain refinement than AlSi10Mg because it adds more nucleation particles to the alloy.

Table 7. Calculated mean grain diameter from the EBSD results for the different alloys.

Alloy	Mean Grain Diameter (µm)	Source
AlMg4.5Mn0.7	69	This study.
+7 wt.% AlSi10Mg	41	This study.
+15 wt.% AlSi10Mg	39	This study.
AlMg4.5Mn0.7	82	[13]
+0.7 wt.% Zr	7	[13]

Supplementing the alloy with AlSi10Mg increases the fraction of dissolved alloying elements in the melt, which subsequently restricts the grain growth and promotes a grain refinement. The grain refinement decreases the solidification cracks significantly. In other studies, grain refiners, such as titanium or zirconium, were supplemented to the base alloy to reduce the solidification cracks [34,36].

3.3.2. Qualitative EDX Measurement

Figure 9 depicts the results of the qualitative EDX measurement. The EDX signal was recorded during the EBSD measurement. In Figure 9, the color bars qualitatively indicate the amount of silicon present in the cross section. The color bars are not comparable with each other. A different color scheme was chosen to avoid misinterpretation. The goal of the EDX measurement was to get an indication of how homogeneously the silicon is distributed in the analyzed cross section.

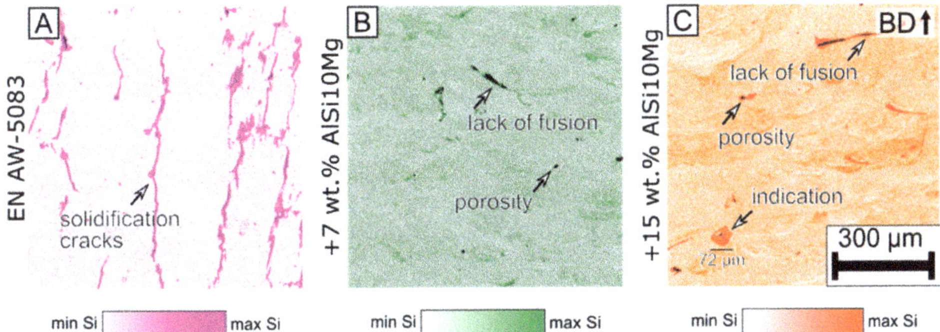

Figure 9. Qualitative energy-dispersive X-ray spectroscopy (EDX) was done during EBSD measurement for the (**A**) base alloy EN AW-5083, (**B**) MOD1 +7 wt.% AlSi10Mg, and (**C**) MOD2 +15 wt.% AlSi10Mg (Table 3). The process parameters for all specimens were p = 250 W, v = 1000 mm/s, and d_f = 100 µm. (Table 2).

In the +7 wt.% (Figure 9B) and +15 wt.% AlSi10Mg (Figure 9C) modification, some melt pool boundaries are more pronounced than others. This shows that the silicon is not equally distributed. Furthermore, in the +15 wt.% AlSi10Mg modification (Figure 9C), a size 72 µm indication is highlighted where the silicon signal is increased. It is hypothesized that the indication is a not fully dissolved AlSi10Mg powder particle.

In the base alloy, an increased silicon signal is measured at the location of the solidification cracks. During the polishing process, the SiO_2-rich polishing emulsion accumulates within the solidification cracks. This leads to an increase in the signal. Therefore, the results close to cracks should be disregarded. Similar behavior is observed for lack of fusion and pores.

The qualitative EDX measurements suggest that no pronounced segregation is present in the modifications.

4. Discussion

4.1. Mechanisms to Reduce Solidification Cracks

In fusion welding, solidification cracks can be reduced by *(i)* controlling the solidification structure and *(ii)* using favorable welding conditions [18]. In this study, these mechanisms were investigated *(i)* by adding supplement powder and *(ii)* changing the process parameters.

In Figure 3, the base alloy was processed with varying process parameters with no base plate heating. It is expected that, with an increase in *VED*, the temperature gradient and cooling rate decreased [18], which leads to reduction in the solidification crack susceptibility, according to Rappaz et al. [38]. Only a minor change in the solidification crack frequency was observed in Figure 3. Overall, varying *VED* = 52 J/mm^3 and 125 J/mm^3 could not lead to a significant reduction in the solidification crack length and frequency.

In comparison, Zhou et al. [13] found that, at *VED* = 224 J/mm^3 and 513 J/mm^3 (P = 200 W and 350 W, v = 100 mm/s and 400 mm/s, t = 30 µm, h = 130 µm), no cracks were detected for the EN AW-5083 (AlMg4.5Mn0.7) alloy. However, Zhou et al. [13] measured a significant reduction in the magnesium concentration, due to a disproportionate evaporation during the LPBF process, as well as densities between 93% and 99% at *VED* = 224 J/mm^3 and 513 J/mm^3. Furthermore, at higher scan velocities, v > 400 mm/s solidification cracks could not be avoided. In conclusion, Zhou et al. [13] showed that processing EN AW-5083 without solidification cracks is possible, at the cost of (a) changing the magnesium concentration from 3.27 wt.% in the powder to about 1.5 wt.% in the specimen, (b) lower densities, and (c) slower build up rates ($v < 400$ mm/s). Even though solidification cracks were avoided with $VED = 224\,\mathrm{J/mm^3}$ and $513\,\mathrm{J/mm^3}$, the LPBF-produced specimens cannot be specified as EN AW-5083 anymore, due to the disproportionate evaporation of magnesium. It is concluded that *(i)* adjusting the parameters P, v, and d_f to reduce the solidification cracks is not a valid strategy to eliminate solidification cracks in EN AW-5083 or AA5083. This agrees with the findings of this study.

For the experiments shown in Figures 5 and 6, the base alloy is supplemented with +7 wt.% and +15 wt.% AlSi10Mg. The modifications eliminate the solidification cracks in the complete process window. Kouraytem et al. [43] showed that pores, similar to the ones present in Figures 5 and 6, can cause solidification cracks. Therefore, the results suggest that the benefits of supplementing AlSi10Mg cannot be decreased by the pores observed in these cross sections. The results suggests that the benefit of supplementing AlSi10Mg is independent of the combination of the process parameters. Furthermore, the findings suggest that eliminating solidifications cracks in LPBF is a metallurgical problem. This allows to further adjust the combination of process parameters to avoid the common processing defects, such as pores, lack of fusion, and even the change in the magnesium concentration, as shown by [13], in the supplemented base alloy without expecting solidification cracks.

As it is known from fusion welding, adding silicon leads to a low melting point phase that reduces the solidification crack susceptibility [18,44]. Furthermore, it was found that supplementing AlSi10Mg increases the grain refinement and texture. With supplementing +7 wt.% AlSi10Mg, the melt pool boundaries are visible and restrict the grains to overgrow multiple layers, as can be seen in Figure 7. In this experiment, the grain refinement effect was saturated already at +7 wt.% AlSi10Mg and mean grain diameter is reduced by 40% relative to the base alloy (Table 7). It is hypothesized that either all the nucleation particles are already activated or the process parameters are limiting further grain refinement. In summary, supplementing AlSi10Mg increases the grain growth restriction [20].

Zhou et al. [13] produced a pre-alloyed powder alloy with an addition of 0.7 wt.% zirconium to AA5083. This leads to a significant grain refinement of 90%, relative to the base alloy mean grain size (Table 7). Mehta et al. [32] showed that this approach is also applicable for AA6061. Martin et al. [34] added zirconium nano particles to the powder alloy to promote grain refinement and eliminate the solidification cracks in AA7075. These findings further support the hypothesis that solidification cracks are a metallurgical problem.

Modifying a base alloy with a supplement powder, such as AlSi10Mg, reduces the solidification cracks because of two mechanisms. First, the grain refinement is increased and, second, a low melting-point silicon rich phase is added.

4.2. Feasibility of the Mixing Strategy

A homogeneous microstructure leads to homogeneous material properties. Therefore, the added silicon from the AlSi10Mg is preferably distributed homogeneously, so that all solidification cracks are eliminated. Figure 9 depicts the qualitative EDX measurements. The qualitative EDX measurements suggest that there is a segregation present. However, overall, the silicon is distributed sufficiently enough because the solidification cracks were avoided in all cross sections, as shown in Figures 5 and 6. Adding a silicon-rich supplement powder to an aluminum alloy was already done successful in previous studies: Aversa et al. supplemented 50 wt.% AA7075 with 50% AlSi10Mg [25], and Montero-Sistiaga et al. supplemented up to 4 wt.% elemental silicon to AA7075 [24].

In conclusion, it is a feasible process strategy to supplement the base alloy by manually mixing it with AlSi10Mg, in order to eliminate solidification cracks. As shown by Skelton et al. [30], the size distribution of the metal powders can have an effect on the homogeneity of the blend. However, some questions still remain unanswered. What is the influence of the mixing strategy on the material properties? What are the limits of mixing that lead to an inhomogeneous silicon distribution? These questions are known in the field of particuology and addressed in [45]. The influence of the mixture procedure on the quality of the part, namely the porosity, lack of fusion, or the segregation, should be addressed and researched in-depth.

5. Conclusions

Solidification cracks hinder the development of new aluminum alloy powders for LPBF. In this study, a practical approach to overcome solidification cracks was presented. Up to now, there is no known experimental data on supplementing AlSi10Mg powder to the base alloy powder EN AW-5083 (AlMg4.5Mn0.7). Most research focusses on the aluminum alloy powder AA7075.

It was possible to eliminate the solidification cracks by supplementing +7 wt.% AlSi10Mg powder to EN AW-5083 powder. The powders were mixed before the specimens were manufactured with LPBF. The microstructural analysis showed that supplementing AlSi10Mg leads to a grain refinement of 40%, relative to the mean grain size of the base alloy. Subsequently, the grain refinement caused by supplementing AlSi10Mg helps to eliminate the solidification cracks, besides the known effect of adding a silicon-rich phase.

By varying the process parameters between VED = 52 J/mm^3 and 125 J/mm^3, the solidification cracks in the base alloy EN AW-5083 could not be eliminated. The results suggest that eliminating solidification cracks in EN AW-5083 during LPBF is a metallurgical problem and requires a modification of the base alloy.

In literature, pre-alloying or adding nanoparticles are both procedures that are used to avoid solidification cracks. In comparison to those, the presented approach is considered practical because of the availability of AlSi10Mg as a supplementing powder. The approach can be easily transferred to other powder alloys and supplement powders.

Furthermore, this study highlights the mechanisms to eliminate the solidification cracks, which are prerequisites for modelling the phenomenon. Additionally, this study provides valuable experimental data, which can also be used for modelling and calculating the required amount of supplement powder. Numerous models are available to describe the solidification crack susceptibility. They can be used as a starting point for computer-aided alloy design.

Author Contributions: Conceptualization, C.B. and M.W.; methodology, C.B.; formal analysis, C.B.; writing—original draft preparation, C.B.; writing—review and editing, S.W. and M.W.; funding acquisition, S.W. All authors have read and agreed to the published version of the manuscript.

Funding: This research was funded by the Federal Ministry of Education and Research (BMBF, Germany), funding number 13XP5115A (Acronym: DiStAl), in the scope of the initiative MATERIALD1GITIAL https://www.materialdigital.de/. This publication was supported by the Open Access Publishing Fund of the University of Stuttgart.

Institutional Review Board Statement: Not applicable.

Informed Consent Statement: Not applicable.

Data Availability Statement: The data presented in this study are available on request from the corresponding author.

Acknowledgments: We would like to express our gratitude to Rosswag GmbH for providing the powder alloy.

Conflicts of Interest: The authors declare no conflict of interest.

References

1. Leirmo, J.L. High Strength Aluminium Alloys in Laser-Based Powder Bed Fusion—a Review. *Procedia CIRP* **2021**, *104*, 1747–1752. [CrossRef]
2. Spierings, A.B. Dissertationsschrift Spierings. Available online: https://www.research-collection.ethz.ch/handle/20.500.11850/253924 (accessed on 9 January 2022).
3. Aboulkhair, N.T.; Simonelli, M.; Parry, L.; Ashcroft, I.; Tuck, C.; Hague, R. 3D printing of Aluminium alloys: Additive Manufacturing of Aluminium alloys using selective laser melting. *Prog. Mater. Sci.* **2019**, *106*, 100578. [CrossRef]
4. Ye, J.; Khairallah, S.A.; Rubenchik, A.M.; Crumb, M.F.; Guss, G.; Belak, J.; Matthews, M.J. Energy Coupling Mechanisms and Scaling Behavior Associated with Laser Powder Bed Fusion Additive Manufacturing. *Adv. Eng. Mater.* **2019**, *21*, 1900185. [CrossRef]
5. Leis, A.; Weber, R.; Graf, T. Process Window for Highly Efficient Laser-Based Powder Bed Fusion of AlSi10Mg with Reduced Pore Formation. *Materials* **2021**, *14*, 5255. [CrossRef]
6. Boley, C.D.; Khairallah, S.A.; Rubenchik, A.M. Calculation of laser absorption by metal powders in additive manufacturing. *Appl. Opt.* **2015**, *54*, 2477–2482. [CrossRef]
7. Weingarten, C.; Buchbinder, D.; Pirch, N.; Meiners, W.; Wissenbach, K.; Poprawe, R. Formation and reduction of hydrogen porosity during selective laser melting of AlSi10Mg. *J. Mater. Processing Technol.* **2015**, *221*, 112–120. [CrossRef]
8. Ewald, S.; Kies, F.; Hermsen, S.; Voshage, M.; Haase, C.; Schleifenbaum, J.H. Rapid Alloy Development of Extremely High-Alloyed Metals Using Powder Blends in Laser Powder Bed Fusion. *Materials* **2019**, *12*, 1706. [CrossRef]
9. Ghasemi-Tabasi, H.; Jhabvala, J.; Boillat, E.; Ivas, T.; Drissi-Daoudi, R.; Logé, R.E. An effective rule for translating optimal selective laser melting processing parameters from one material to another. *Addit. Manuf.* **2020**, *36*, 101496. [CrossRef]
10. Guo, Q.; Zhao, C.; Escano, L.I.; Young, Z.; Xiong, L.; Fezzaa, K.; Everhart, W.; Brown, B.; Sun, T.; Chen, L. Transient dynamics of powder spattering in laser powder bed fusion additive manufacturing process revealed by in-situ high-speed high-energy x-ray imaging. *Acta Mater.* **2018**, *151*, 169–180. [CrossRef]
11. Li, R.; Liu, J.; Shi, Y.; Wang, L.; Jiang, W. Balling behavior of stainless steel and nickel powder during selective laser melting process. *Int. J. Adv. Manuf. Technol.* **2012**, *59*, 1025–1035. [CrossRef]
12. Kaufmann, N.; Imran, M.; Wischeropp, T.M.; Emmelmann, C.; Siddique, S.; Walther, F. Influence of Process Parameters on the Quality of Aluminium Alloy EN AW 7075 Using Selective Laser Melting (SLM). *Phys. Procedia* **2016**, *83*, 918–926. [CrossRef]
13. Zhou, L.; Hyer, H.; Park, S.; Pan, H.; Bai, Y.; Rice, K.P.; Sohn, Y. Microstructure and mechanical properties of Zr-modified aluminum alloy 5083 manufactured by laser powder bed fusion. *Addit. Manuf.* **2019**, *28*, 485–496. [CrossRef]
14. Vock, S.; Klöden, B.; Kirchner, A.; Weißgärber, T.; Kieback, B. Powders for powder bed fusion: A review. *Prog. Addit. Manuf.* **2019**, *4*, 383–397. [CrossRef]
15. Pleass, C.; Jothi, S. Influence of powder characteristics and additive manufacturing process parameters on the microstructure and mechanical behaviour of Inconel 625 fabricated by Selective Laser Melting. *Addit. Manuf.* **2018**, *24*, 419–431. [CrossRef]
16. Gokcekaya, O.; Ishimoto, T.; Todo, T.; Wang, P.; Nakano, T. Influence of powder characteristics on densification via crystallographic texture formation: Pure tungsten prepared by laser powder bed fusion. *Addit. Manuf. Lett.* **2021**, *1*, 100016. [CrossRef]
17. Khorasani, A.; Gibson, I.; Veetil, J.K.; Ghasemi, A.H. A review of technological improvements in laser-based powder bed fusion of metal printers. *Int. J. Adv. Manuf. Technol.* **2020**, *108*, 191–209. [CrossRef]
18. Kou, S. *Welding Metallurgy*, 2nd ed.; Wiley-Interscience: Hoboken, NJ, USA, 2003; ISBN 0471434914.
19. Tang, Z.; Vollertsen, F. Influence of grain refinement on hot cracking in laser welding of aluminum. *Weld. World* **2014**, *58*, 355–366. [CrossRef]
20. Schempp, P.; Rethmeier, M. Understanding grain refinement in aluminium welding: Henry Granjon Prize 2015 winner category B: Materials behaviour and weldability. *Weld. World* **2015**, *59*, 767–784. [CrossRef]
21. Böhm, C.; Hagenlocher, C.; Wagner, J.; Graf, T.; Weihe, S. Analytical Description of the Criterion for the Columnar-To-Equiaxed Transition During Laser Beam Welding of Aluminum Alloys. *Met. Mater. Trans. A* **2021**, *52*, 2720–2731. [CrossRef]
22. Weller, D.; Hagenlocher, C.; Weber, R.; Graf, T. Influence of the solidification path of AlMgSi aluminium alloys on the critical strain rate during remote laser beam welding. *Sci. Technol. Weld. Join.* **2020**, *25*, 101–105. [CrossRef]
23. Hagenlocher, C.; Weller, D.; Weber, R.; Graf, T. Analytical Description of the Influence of the Welding Parameters on the Hot Cracking Susceptibility of Laser Beam Welds in Aluminum Alloys. *Met. Mater. Trans. A* **2019**, *50*, 5174–5180. [CrossRef]
24. Montero-Sistiaga, M.L.; Mertens, R.; Vrancken, B.; Wang, X.; van Hooreweder, B.; Kruth, J.-P.; van Humbeeck, J. Changing the alloy composition of Al7075 for better processability by selective laser melting. *J. Mater. Processing Technol.* **2016**, *238*, 437–445. [CrossRef]
25. Aversa, A.; Marchese, G.; Manfredi, D.; Lorusso, M.; Calignano, F.; Biamino, S.; Lombardi, M.; Fino, P.; Pavese, M. Laser Powder Bed Fusion of a High Strength Al-Si-Zn-Mg-Cu Alloy. *Metals* **2018**, *8*, 300. [CrossRef]
26. Fischer, M.; Joguet, D.; Robin, G.; Peltier, L.; Laheurte, P. In situ elaboration of a binary Ti–26Nb alloy by selective laser melting of elemental titanium and niobium mixed powders. *Mater. Sci. Eng. C* **2016**, *62*, 852–859. [CrossRef]
27. Ewald, S.; Köhnen, P.; Ziegler, S.; Haase, C.; Schleifenbaum, J.H. Precise control of microstructure and mechanical properties of additively manufactured steels using elemental carbon powder. *Mater. Lett.* **2021**, *295*, 129788. [CrossRef]
28. Shoji Aota, L.; Bajaj, P.; Zschommler Sandim, H.R.; Aimé Jägle, E. Laser Powder-Bed Fusion as an Alloy Development Tool: Parameter Selection for In-Situ Alloying Using Elemental Powders. *Materials* **2020**, *13*, 3922. [CrossRef]

29. Bradford, R.L.; Cao, L.; Klosterman, D.; Herman, F.; Forman, L.; Browning, C. A metal–metal powder formulation approach for laser additive manufacturing of difficult-to-print high-strength aluminum alloys. *Mater. Lett.* **2021**, *300*, 130113. [CrossRef]
30. Skelton, J.M.; Sullivan, E.J.; Fitz-Gerald, J.M.; Floro, J.A. Efficacy of elemental mixing of in situ alloyed Al-33wt%Cu during laser powder bed fusion. *J. Mater. Processing Technol.* **2022**, *299*, 117379. [CrossRef]
31. Stopyra, W.; Gruber, K.; Smolina, I.; Kurzynowski, T.; Kuźnicka, B. Laser powder bed fusion of AA7075 alloy: Influence of process parameters on porosity and hot cracking. *Addit. Manuf.* **2020**, *35*, 101270. [CrossRef]
32. Mehta, A.; Zhou, L.; Huynh, T.; Park, S.; Hyer, H.; Song, S.; Bai, Y.; Imholte, D.D.; Woolstenhulme, N.E.; Wachs, D.M.; et al. Additive manufacturing and mechanical properties of the dense and crack free Zr-modified aluminum alloy 6061 fabricated by the laser-powder bed fusion. *Addit. Manuf.* **2021**, *41*, 101966. [CrossRef]
33. Nie, X.; Chen, Z.; Qi, Y.; Zhang, H.; Zhang, C.; Xiao, Z.; Zhu, H. Effect of defocusing distance on laser powder bed fusion of high strength Al–Cu–Mg–Mn alloy. *Virtual Phys. Prototyp.* **2020**, *15*, 325–339. [CrossRef]
34. Martin, J.H.; Yahata, B.D.; Hundley, J.M.; Mayer, J.A.; Schaedler, T.A.; Pollock, T.M. 3D printing of high-strength aluminium alloys. *Nature* **2017**, *549*, 365–369. [CrossRef]
35. Zhuravlev, E.; Milkereit, B.; Yang, B.; Heiland, S.; Vieth, P.; Voigt, M.; Schaper, M.; Grundmeier, G.; Schick, C.; Kessler, O. Assessment of AlZnMgCu alloy powder modification for crack-free laser powder bed fusion by differential fast scanning calorimetry. *Mater. Des.* **2021**, *204*, 109677. [CrossRef]
36. Heiland, S.; Milkereit, B.; Hoyer, K.-P.; Zhuravlev, E.; Kessler, O.; Schaper, M. Requirements for Processing High-Strength AlZnMgCu Alloys with PBF-LB/M to Achieve Crack-Free and Dense Parts. *Materials* **2021**, *14*, 7190. [CrossRef] [PubMed]
37. Ostermann, F. *Anwendungstechnologie Aluminium*; Springer Vieweg: Wiesbaden, Germany, 2014; ISBN 978-3-662-43806-0.
38. Rappaz, M.; Drezet, J.-M.; Gremaud, M. A new hot-tearing criterion. *Met. Mater. Trans. A* **1999**, *30*, 449–455. [CrossRef]
39. Kou, S. A criterion for cracking during solidification. *Acta Mater.* **2015**, *88*, 366–374. [CrossRef]
40. Hirsch, J. Aluminium in Innovative Light-Weight Car Design. *Mater. Trans.* **2011**, *52*, 818–824. [CrossRef]
41. *DIN EN 573-3:2013-12: Aluminium and Aluminium Alloys—Chemical Composition and Form of Wrought Products—Part 3: Chemical Composition and Form of Products*; German Version EN 573Y3; Beuth Publishing DIN: Berlin, Germany, 2013.
42. Rasband, W.; Image, J. U.S. National Institutes of Health: Bethesda, MD, USA. Available online: http://imagej.nih.gov/ij2011 (accessed on 9 January 2022).
43. Kouraytem, N.; Chiang, P.-J.; Jiang, R.; Kantzos, C.; Pauza, J.; Cunningham, R.; Wu, Z.; Tang, G.; Parab, N.; Zhao, C.; et al. Solidification crack propagation and morphology dependence on processing parameters in AA6061 from ultra-high-speed x-ray visualization. *Addit. Manuf.* **2021**, *42*, 101959. [CrossRef]
44. Liu, J.; Kou, S. Susceptibility of ternary aluminum alloys to cracking during solidification. *Acta Mater.* **2017**, *125*, 513–523. [CrossRef]
45. Bridgwater, J. Mixing of powders and granular materials by mechanical means—A perspective. *Particuology* **2012**, *10*, 397–427. [CrossRef]

Article

TiNi-Based Bi-Metallic Shape-Memory Alloy by Laser-Directed Energy Deposition

Yitao Chen [1,*], Cesar Ortiz Rios [2], Braden McLain [1], Joseph W. Newkirk [2] and Frank Liou [1]

[1] Department of Mechanical and Aerospace Engineering, Missouri University of Science and Technology, Rolla, MO 65409, USA; btmywv@mst.edu (B.M.); liou@mst.edu (F.L.)

[2] Department of Materials Science and Engineering, Missouri University of Science and Technology, Rolla, MO 65409, USA; codqc@mst.edu (C.O.R.); jnewkirk@mst.edu (J.W.N.)

* Correspondence: yc4gc@mst.edu

Abstract: In this study, laser-directed energy deposition was applied to build a Ti-rich ternary Ti–Ni–Cu shape-memory alloy onto a TiNi shape-memory alloy substrate to realize the joining of the multifunctional bi-metallic shape-memory alloy structure. The cost-effective Ti, Ni, and Cu elemental powder blend was used for raw materials. Various material characterization approaches were applied to reveal different material properties in two sections. The as-fabricated Ti–Ni–Cu alloy microstructure has the TiNi phase as the matrix with Ti_2Ni secondary precipitates. The hardness shows no high values indicating that the major phase is not hard intermetallics. A bonding strength of 569.1 MPa was obtained by tensile testing, and digital image correlation reveals the different tensile responses of the two sections. Differential scanning calorimetry was used to measure the phase-transformation temperatures. The austenite finishing temperature of higher than 80 °C was measured for the Ti–Ni–Cu alloy section. For the TiNi substrate, the austenite finishing temperature was tested to be near 47 °C at the bottom and around 22 °C at the upper substrate region, which is due to the repeated laser scanning that acts as annealing on the substrate. Finally, the multiple shape-memory effect of two shape-memory alloy sides was tested and identified.

Keywords: additive manufacturing; directed energy deposition; shape-memory alloys; joining of metals; elemental powders

Citation: Chen, Y.; Ortiz Rios, C.; McLain, B.; Newkirk, J.W.; Liou, F. TiNi-Based Bi-Metallic Shape-Memory Alloy by Laser-Directed Energy Deposition. *Materials* **2022**, *15*, 3945. https://doi.org/10.3390/ma15113945

Academic Editors: Wai Yee Yeong and Swee Leong Sing

Received: 29 March 2022
Accepted: 27 May 2022
Published: 1 June 2022

1. Introduction

Joining different metal alloys into a single bi-metallic component can realize a wide variety of combinations with excellent material properties [1–3], and the joining of bi-metals is necessary in many industrial environments where different properties are needed at different locations [3]. The directed energy deposition (DED) additive manufacturing (AM) process has become an important approach to realize various types of metal joining and metal repair since strong metallurgical bonding can be achieved [4–6]. So far, DED has been used in various applications including metal joining and part repairing [2,6–8]. Also, since the AM process has a high degree of freedom for the spatial distribution of both geometry and material compositions, it can be much more flexible in developing more novel alloy structures [9,10]. Typical works using DED to join similar or dissimilar alloys include steels, Ti alloys, Ni-based superalloys, and Cu alloys. Sahasrabudhe et al. [2] joined SS410 with Ti-6Al-4V to obtain a high application temperature at the Ti-alloy end and the economic corrosion-resistive steel end. Jones et al. [11] applied DED for the manufacturing of steel-Inconel bi-metallic structures. Zhang et al. [12] used DED to join steels and Cu in order to obtain both high strengths from steel and high thermal conductivity from Cu. Functionally graded structures and interlayers can also be used for joining two dissimilar alloy parts by DED, which benefits from the flexible nature of the additive process [13–15].

Shape-memory alloy (SMA) is a type of smart metallic material that is able to demonstrate different shapes at different temperatures and memorize its original shape through austenite-martensite phase transformation [16,17]. TiNi alloy is a popular metallic material that can demonstrate shape-memory effects [17–19]. For TiNi alloys, it has been found that their phase-transformation temperature (TT) can be changed by adjusting the chemical composition and processing methods [17,20]. Therefore, the TTs of TiNi alloy can cover a wide temperature range for multiple functions [20]. Also, if different types of TiNi SMAs can coexist in a single component, it will obtain more unique and attractive multiple shape-memory behaviors. For example, Khan et al. [21] flexibly applied a laser to TiNi alloys to generate a discrepancy between laser-affected areas and created location-dependent TTs within a single part. Nematollahi et al. [22] applied selective laser melting (SLM), a powder-bed-based AM process, to build a graded TiNi SMA by changing the processing parameters at two different sections. For DED, by taking advantage of the flexible capability in similar or dissimilar joining, SMAs with multiple shape-memory behaviors can also be combined using DED to demonstrate more functions. Most of the aforementioned DED processes used for metal joining focused on improving the comprehensive mechanical, structural, and thermal properties. However, the joining effect and shape-memory behavior after joining two different types of TiNi-based SMAs using DED have not been widely reported.

For DED, both the feedstock SMA materials and the substrate SMA materials can have a wide range of selections. In this paper, we applied the DED approach to join two different TiNi-based parts into a bi-metallic SMA in order to exhibit multiple shape-memory behaviors. A Ti-rich ternary Ti–Ni–Cu SMA was deposited on a near equiatomic binary TiNi alloy to realize the bi-metallic SMAs, which were joined by two SMA parts with different element types, element compositions, and processing methods. A small amount of a third element can be applied to create more applications [23], such as by adding Cu to replace a small portion of Ni, Ti–Ni–Cu SMA can obtain narrower hysteresis and potential applications in the biomedical field [24,25]. The Ti–Ni–Cu ternary SMA was fabricated by the powder mixture of Ti, Ni, and Cu elemental powders. The elemental powder mixture was used as an alternative form of raw materials other than pre-alloyed powders [26,27]. The SMA with an as-mixed atomic composition of Ti–45at.%Ni–5at.%Cu was fabricated by elemental powders in [28,29] with titanium as the substrate material. In this work, more details will be studied on using elemental powder DED to build Ti–Ni–Cu/TiNi bi-metallic alloys. The joining effect at the interface, the bonding strength, and the multiple functional behaviors from different sections of SMA will all be evaluated.

2. Materials and Methods

In this work, a Ti-rich Ti–Ni–Cu alloy single-wall structure was deposited on a TiNi substrate by a DED processing system mainly including a laser system, a CNC-controlled moving stage, and a powder feeder. Near spherical Cp-Ti powder (AP&C, particle size 53~150 μm), Ni powder (Atlantic Equipment Engineers Inc., Upper Saddle River, NJ, USA, particle size −100/+325 mesh), and Cu powder (Royal Metal Powders Inc., Maryville, TN, USA, average particle size 110 μm [12]) were mixed with the atomic composition of Ti/Ni/Cu = 53/44/3 to create a Ti-rich composition that is different from the TiNi substrate. We use Ti–Ni–Cu alloy to represent this Ti–44at.%Ni–3at.%Cu ternary SMA that was deposited and studied in this paper. The powder blend was homogenized using a Turbula T2F mixer for 0.5 hr. The substrate material is near-equiatomic TiNi binary SMA bar material purchased from Kellogg's Research Labs (New Boston, NH, USA), which was cut to size, 25 mm × 10 mm × 5 mm. The Ti–Ni–Cu alloy single-wall deposition strategy on the TiNi substrate is illustrated in Figure 1a. The substrate was fixed by a vise on the moving stage, and the single wall structure was deposited on the surface of 25 mm × 5 mm of the substrate. An IPG Photonics CW fiber laser with a 1064 nm wavelength was used in this work to create a melt pool on the substrate. The spot size was adjusted and fixed at approximately 2.5 mm. A powder nozzle was used to feed powders into the melt pool and the nozzle was 10 mm above the substrate surface. The moving stage followed a single track and multilayer toolpath. The toolpath consists of uniform +Y direction and −Y direction movements with no dwell time for multiple cycles. Between each change in the Y direction, the stage dropped down vertically by one increment, which represents the layer thickness. The travel speed for the stage in the +Y and −Y directions was 250 mm/min. During the traveling guided by the toolpath, the Ti, Ni, and Cu powder mixture was delivered into the melt pool through the powder nozzle by the powder feeder supplied by Powder Motion Labs. Solid layers build up along the Z direction as seen in Figure 1a,b, which is called the building direction (BD). The laser power was set at 600 W at the first layer and 400 W for all the remaining layers. A total of 60 layers were deposited, and the dimension of the DED Ti–Ni–Cu alloy along the Y direction (long direction) is 25 mm, whereas the total deposition height of the Ti–Ni–Cu alloy is 10 mm. The entire process was carried out within an argon atmosphere to minimize the oxidation effect.

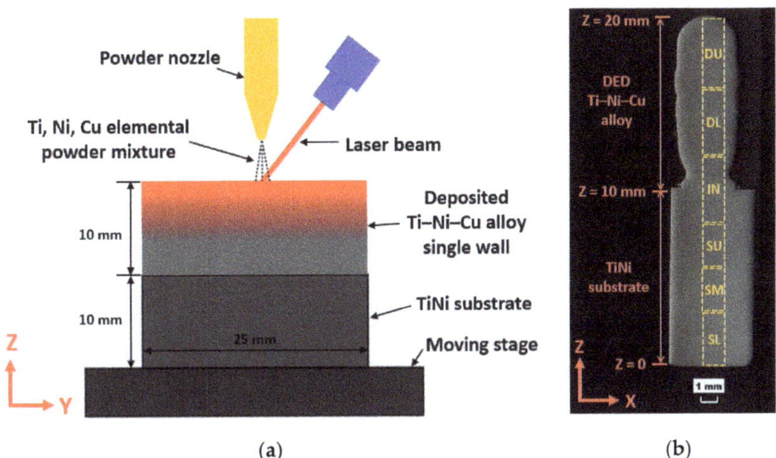

Figure 1. (**a**) The schematic of joining bi-metallic SMA using DED in this work. (**b**) Image of the XZ-plane cross-section of the bi-metallic structure including the TiNi substrate and the DED Ti–Ni–Cu alloy. The bottom line of the substrate has been marked as Z = 0. The Z height of the interface and the top point are 10 mm and 20 mm, respectively.

The cross-section of the 10 mm tall Ti–Ni–Cu alloy single wall joined with the 10 mm tall TiNi substrate was cut off by wire-EDM and was prepared for material characterization, including microstructural, functional, and mechanical testing and joining evaluation. A sample from the cross-section along the XZ-plane of the bi-metallic structure was prepared by following the metallographic specimen preparation steps using a Buehler EcoMet 250 grinder-polisher in order to observe the microstructure through microscopes. The sample was mounted by epoxy and ground using SiC sandpapers from 320 grit to 1200 grit. After grinding, the sample was then polished by 9 μm, 6 μm, 3 μm, 1 μm diamond suspension, and 0.05 μm colloidal alumina. After polishing, the sample was etched using Kroll's reagent (2 mL HF, 6 mL HNO_3, and 92 mL distilled water) by swabbing the surface for 30 s. Optical microscopic imaging was done on the sample by a Hirox KH-8700 optical microscope (OM) (Hirox Co., Ltd., Tokyo, Japan). High-resolution electron microscopic imaging and energy-dispersive spectroscopy (EDS) analysis were performed by a Thermo Fisher Scientific Helios Hydra CX scanning electron microscope (SEM) (Thermo Fisher Scientific, Waltham, MA, USA) and a Helios NanoLab 600 SEM (Thermo Fisher Scientific, Waltham, MA, USA).

Figure 1b is an OM image of the well-polished cross-section of the bi-metallic SMA structure. To determine the locations, the bottom point of the TiNi substrate is marked as Z = 0. Correspondingly, the interfacial line and the top point of the DED Ti–Ni–Cu alloy single wall are marked as Z = 10 mm and Z = 20 mm, respectively. Vickers hardness values of both the DED Ti–Ni–Cu alloy part and the TiNi substrate were measured by a Struers Duramin-5 Vickers hardness tester. According to the Z coordinate shown in Figure 1b,

Vickers hardness testing was conducted at a height from Z = 1 mm to Z = 19 mm with a 1 mm interval. For each height, three indentations were taken, and the average hardness of each height was calculated. Each indentation used a load of 1.96 N and a dwell time of 10 s.

The bonding strength between the Ti–Ni–Cu alloy part and the TiNi substrate was measured by uniaxial tensile testing via an Instron 5969 universal testing machine (Instron, Norwood, MA, USA) and analyzed by the digital image correlation (DIC) technique. Figure 2a shows the sketch of the miniature tensile specimen that was developed to study mechanical behaviors of small size materials. The gauge length of the specimen is 3 mm, and the thickness is 1 mm. More information can be found in a previous study [30]. A 10 kN load cell was applied for the testing of the miniature tensile specimen. The crosshead speed was controlled to maintain the strain rate at 0.003 mm/mm/s during the tensile test. Figure 2b displays the camera setup for DIC during the tensile testing. The following fracture surface analysis was also conducted using the Helios Hydra CX SEM.

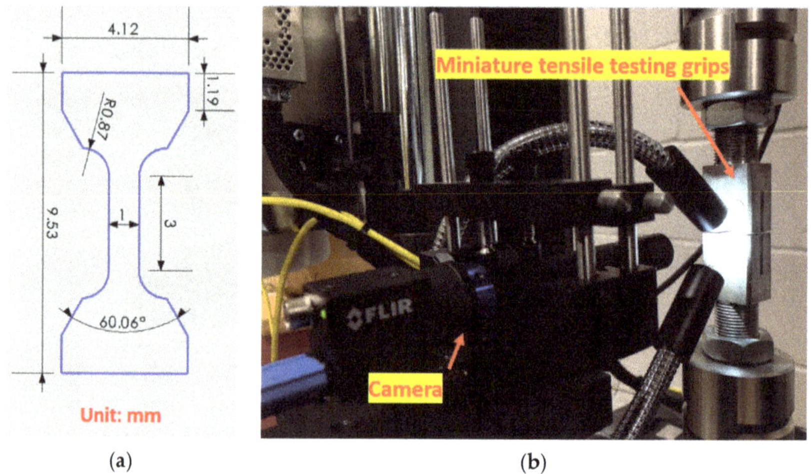

Figure 2. (**a**) The sketch of the miniature tensile specimen design. (**b**) The camera setup for DIC.

Differential scanning calorimetry (DSC) was used for studying the TTs of various sections of the as-deposited Ti–Ni–Cu/TiNi bi-metallic structure using a TA Instruments Q2000 differential scanning calorimeter (New Castle, DE, USA) with a ramping heating/cooling rate of 10 °C/min. Apart from DSC, the shape-memory effect of the as-deposited bi-metallic SMA was tested by recording the shape recovery on a hot plate with changing temperatures.

3. Results and Discussions

3.1. Microstructural Characterization

Figure 3 shows the micrographs of the as-deposited Ti–Ni–Cu alloy. Figure 3a is an OM image of the interface between the DED Ti–Ni–Cu alloy and the TiNi substrate. A clear interfacial line can be observed and no gas pores and cracks are found near the interfacial line, which indicates a good interaction between the powder flow of the first layer and the substrate materials within the melt pool. Figure 3b is obtained from the Ti–Ni–Cu alloy section, where darker phases distribute within the matrix and form a columnar microstructure. With the assistance of SEM/EDS, higher magnification images are acquired and shown in Figure 3c,d. In Figure 3c, EDS points analysis was performed within both the matrix (Point 1) and the dark minor phase (Point 2). The atomic compositions from the EDS analysis of Point 1 and Point 2 are listed in Figure 3c. It can be seen that in the dark phases, the atomic composition of Ti is twice the atomic composition of (Ni + Cu). Therefore, the dark minor phases are Ti_2Ni intermetallics, which are marked in Figure 3b. The composition of Point 1 indicates that the matrix is Ti-rich TiNi with a Ti atomic composition of about

51.5 at.% and a small number of Cu atoms to replace the Ni atoms. The high magnification image of Figure 3d clearly shows the martensite twinning structure and dark Ti_2Ni particles. Large-area EDS analysis was also performed at heights of Z = 10 mm, 12 mm, 14 mm, 16 mm, and 18 mm of the Ti–Ni–Cu alloy section. The average atomic composition of the five heights measured by EDS is Ti (52.1 ± 0.5) Ni (43.3 ± 0.9) Cu (4.6 ± 0.4). The fluctuation of Ti, Ni, and Cu compositions among the five heights is much lower than the former work that deposited Ti–45at.%Ni–5at.%Cu alloy on the titanium substrates [29]. The dilution from the TiNi substrate in this work has less influence on the as-deposited section. The SEM image of the interfacial region is demonstrated in Figure 4a. Figure 4b–d are the EDS mapping of the Ti, Ni, and Cu elements. It can be seen that the Cu signal is much weaker below the interfacial line; however, there is not a significant change in the signal of Ti and Ni between the area above and below the interface, since the compositions of Ti and Ni within the Ti–Ni–Cu alloy and TiNi substrate are similar. The Ni signal shows a gradual and slight decrease from the TiNi substrate area to the Ti–Ni–Cu alloy area since the atomic composition of Ni in the substrate is a little higher than the Ti–Ni–Cu alloy.

3.2. Hardness Distribution

Vickers hardness distribution from Z = 1 mm to Z = 19 mm with a 1 mm interval is plotted in Figure 5. Figure 5 shows the average hardness of the three indentations at each Z height. It can be seen that the Vickers hardness distribution from bottom to top shows typical low hardness (200~300 HV0.2) of the entire cross-sectional area along the Z direction of the bi-metallic structure. It was reported in [31] that the hardness of the intermetallic Ti_2Ni phase could be as high as 700 HV. Thus, the low hardness reflects that the major phase of this bi-metallic structure is the TiNi phase, and the Ti_2Ni intermetallic phase is the dispersive secondary phase in the as-deposited Ti–Ni–Cu alloy section due to the rich atomic composition of the Ti element. Almost all of the average hardness values of the DED Ti–Ni–Cu alloy (above Z = 10 mm) are within the range of 200~250 HV0.2. Within the substrate, from Z = 1 mm to Z = 9 mm, an obvious difference in the average hardness value can be observed between the lower half of the substrate and the upper half of the substrate. The average hardness values of the lower half of the substrate (Z < 6 mm) are closer to 300 HV0.2. The hardness also obtains a higher value at the location of Z = 10 mm, which could be due to the local mixing of the Ti, Ni, and Cu elements between the TiNi substrate and the first layer of DED Ti–Ni–Cu alloy that might result in a local small composition deviation in the TiNi matrix compared to the original TiNi substrate.

To obtain more information related to the Z-height-dependent hardness value, the sample sectioning plan for the following DSC analysis is designed according to the dashed boxes in Figure 1b. From bottom to top, they are the lower part of the substrate (SL, Z height: 0~3 mm), the middle part of the substrate (SM, Z height: 3~5.5 mm), the upper part of the substrate (SU, Z height: 5.5~8 mm), the interface including both the substrate and the bottom region of the DED Ti–Ni–Cu alloy (IN, Z height: 8~12 mm), the lower part of the DED Ti–Ni–Cu alloy (DL, Z height: 12~16 mm), and the upper part of the DED Ti–Ni–Cu alloy (DU, Z height: 16~20 mm). In this way, the phase-transition behaviors of samples SL and SM represent the lower half region of the substrate with higher average hardness, and sample SU reflects the upper half region of the substrate with lower average hardness.

3.3. Phase Transformation

DSC results in all six different sections: DU, DL, IN, SU, SM, and SL, according to Figure 1b, are demonstrated in Figure 6 from Figure 6a–f, respectively. Each DSC curve consists of a heating and cooling cycle with peaks that indicate the austenite formation during heating and the martensite formation during cooling. Typical characterization temperatures, including martensite starting temperature (M_s), martensite finishing temperature (M_f), austenite starting temperature (A_s), and austenite finishing temperature (A_f), are determined by the tangent method [32] as shown in the DSC curve of sample DU in Figure 6a. Austenite peak temperature (A_p) and martensite peak temperature (M_p) are also included.

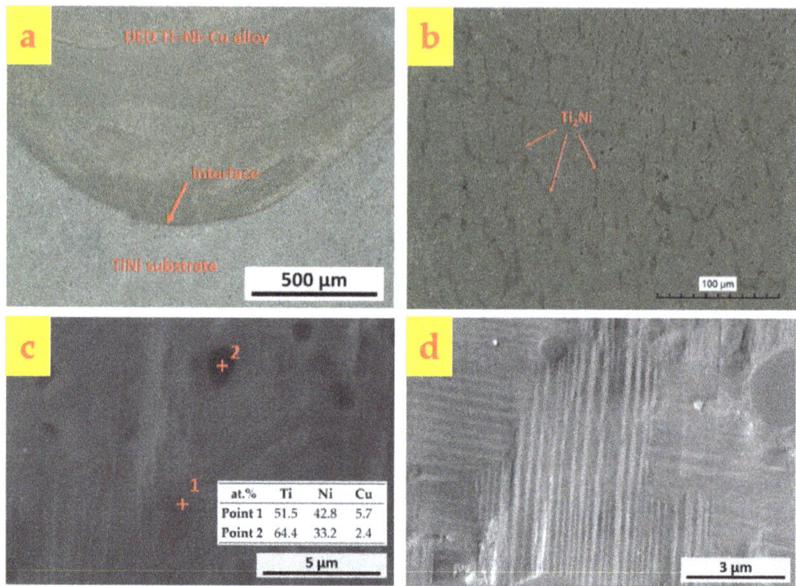

Figure 3. Microstructural features of the as-deposited Ti–Ni–Cu SMA. (**a**) Interfacial region between the DED Ti–Ni–Cu SMA and the TiNi substrate. (**b**) Microstructure of TiNi matrix and Ti_2Ni. (**c**) EDS point analysis of TiNi matrix and Ti_2Ni. (**d**) Twinning structure of martensitic TiNi phase.

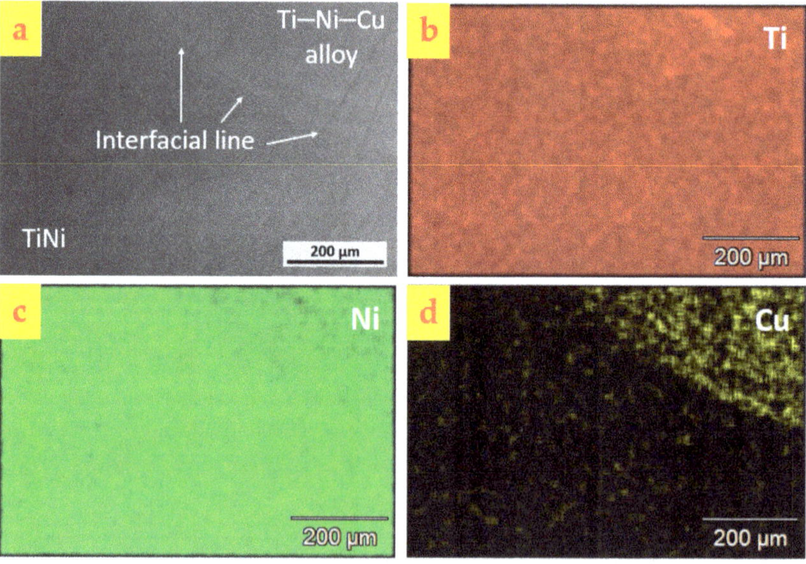

Figure 4. (**a**) Interface between the DED Ti–Ni–Cu alloy and the TiNi substrate. (**b**) Element mapping of Ti. (**c**) Element mapping of Ni. (**d**) Element mapping of Cu. Notice that the intensity of the Cu signal from the Ti–Ni–Cu alloy above the interfacial line is much stronger than the area within the TiNi substrate below the interfacial line.

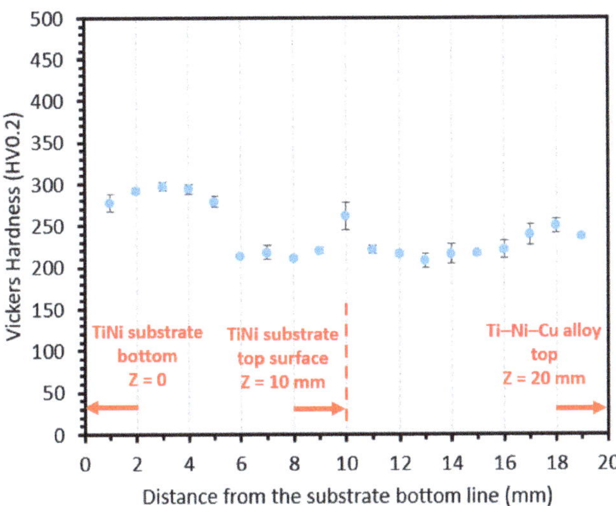

Figure 5. Vickers hardness distribution from the bottom of the TiNi substrate to the top of the DED Ti–Ni–Cu alloy. Notice that the lack of high hardness value from the hard intermetallics indicates that the major phase is the TiNi phase. Relatively higher hardness values can be observed at the lower section of the substrate from Z = 1 mm to Z = 5 mm and at the point of Z = 10 mm near the interface.

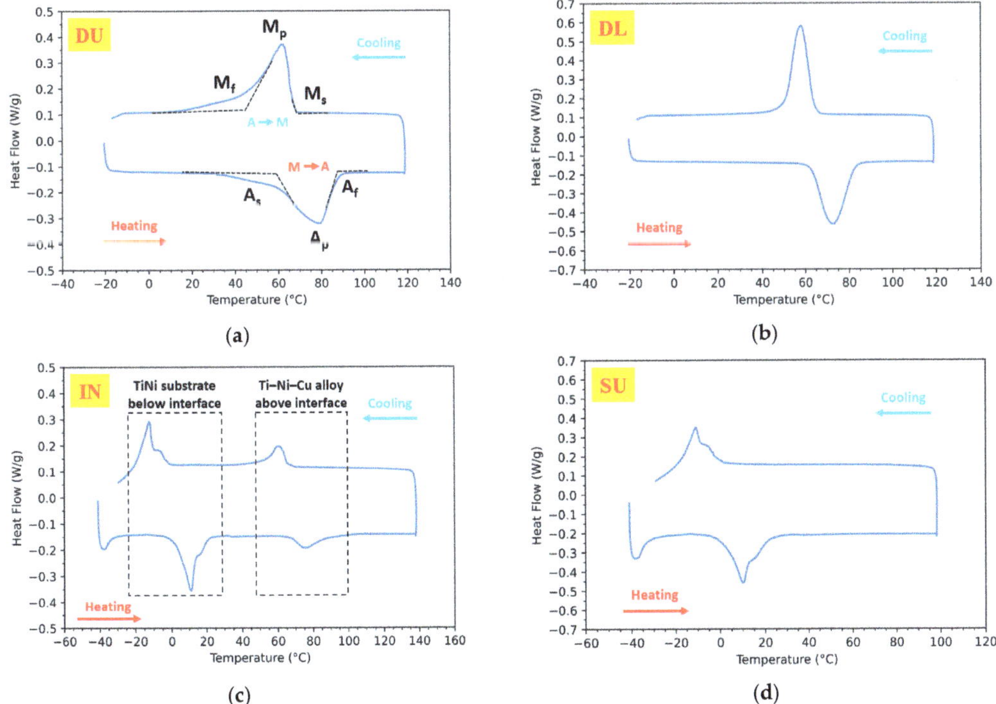

Figure 6. *Cont.*

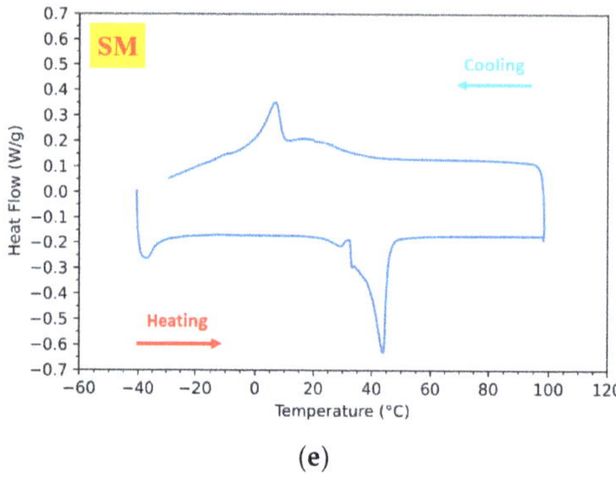

(e)

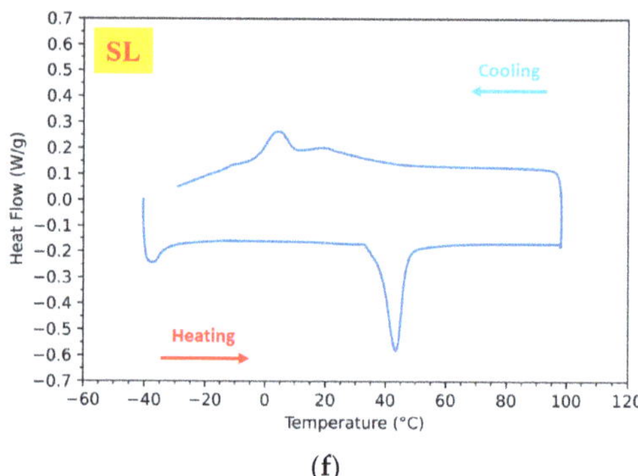

(f)

Figure 6. DSC heating–cooling curves of the six regions (marked in Figure 1b) of the as-deposited bi-metallic SMA. (**a**) The upper part of the DED Ti–Ni–Cu alloy (DU). The characterization temperatures: austenite starting temperature (A_s), austenite peak temperature (A_p), austenite finishing temperature (A_f), martensite starting temperature (M_s), martensite peak temperature (M_p), and martensite finishing temperature (M_f) are labeled. (**b**) The lower part of the DED Ti–Ni–Cu alloy (DL). (**c**) The interfacial region (IN). Notice that there are two distinctive phase-transformation peaks during heating and cooling from Ti–Ni–Cu alloy and TiNi substrate, which are marked by dashed boxes. (**d**) The upper part of the substrate (SU). (**e**) The middle part of the substrate (SM). (**f**) The lower part of the substrate (SL).

Table 1 summarizes the A_f of all sections. Figure 6 shows that the Ti-rich upper Ti–Ni–Cu alloy shows a higher austenite finishing temperature A_f. As for the sections DU and DL, the TiNi matrix is highly Ti-rich and the A_f reaches higher than 80 °C. The DSC curve of the IN sample with the interface has two distinctive peaks both during heating and during cooling in Figure 6c, which are marked by dashed boxes. The high-temperature peaks come from the Ti-rich Ti–Ni–Cu alloy portion above the interface, whereas the low-temperature peaks are due to the equiatomic TiNi substrate below the interface. So, after the joining by DED, both phase-transformation behaviors of the two sections near the interface can still be clearly illustrated in DSC. For the substrate, at the upper part of the substrate, which includes the lower half of IN and the section of SU, the values of A_f are as low as 21.2 °C and 22.8 °C, whereas the A_f values of the SM and SL are 46.5 °C and 47.7 °C.

Table 1. Summary of the A_f value of all sections within the bi-metallic SMA structure and the corresponding Z heights.

Sample Section	A_f (°C)
DU	88.3
DL	82.9
IN (above interface)	90.9
IN (below interface)	21.2
SU	22.8
SM	46.5
SL	47.7

It can also be seen that all of the samples from the Ti–Ni–Cu alloy section exhibit one-step phase transformation in both the heating and cooling processes. For the substrate, during cooling the upper half of the substrate, IN and SU, exhibit features close to one-step martensite transformation, whereas the cooling curves of the lower part including SM and SL are closer to double overlapped peaks during cooling, which could be related to the intermediate R-phase [33,34] that initially exists in the substrate. For the near-equiatomic TiNi substrate used in this work, the upper part went through repeated high-temperature annealing from the laser, which tends to exhibit lower TTs and one-step austenite-martensite transition after annealing [33,35]. The lower part of the substrate, SM and SL, which are far from the deposited part, underwent a relatively lower heating temperature; the TTs are not highly affected by the laser-annealing processes. It was also reported that the formation of the R-phase can be favored by secondary precipitates such as Ni_4Ti_3 and dislocation substructures [36]. Therefore, the higher hardness values of the lower substrate from Z = 1 mm to Z = 5 mm are likely from the existence of precipitates and dislocation substructures. In contrast, the high-temperature annealing effect at the upper substrate reduces the hardness due to the reduction in those factors.

The thermal hysteresis ΔT is defined by the temperature difference between A_p and M_p [37]. The ΔT value of the DED Ti–Ni–Cu alloy section was calculated for the whole Ti–Ni–Cu alloy section. The ΔT values of DU, DL, and the upper part of IN ranges from 14.4 °C to 17.2 °C, which is narrower than reported additively manufactured Ti-rich TiNi binary SMAs such as in [38]. This then shows the effects of the addition of the Cu element in using the AM process to fabricate TiNi-based SMAs.

3.4. Tensile Behavior

The DIC technique was applied using GOM Correlate software to plot the stress–strain curve of the bi-metal under a uniaxial load using the function of the virtual extensometer. A testing sample was extracted vertically across the interface with the half-gauge section belonging to the DED Ti–Ni–Cu alloy single wall and the other half gauge located within the TiNi substrate according to the design in Figure 2a, as seen in the dashed rectangular box in the lower right corner of Figure 7. Thus, the axial direction of the sample and the tensile loading direction are parallel to the Z axis shown in Figure 1. The tensile stress–strain curve is shown in Figure 7, which demonstrates the typical stages of SMA. In order to obtain a good alignment on the miniature tensile specimen, a preload was applied at the beginning of the uniaxial tensile testing. The first near-linear section mainly corresponds to the initial elastic response. After the initial elastic stage, the curve starts a near-horizontal stress plateau, which mainly comes from the stress-induced martensite formation of the austenite-dominated top part of the TiNi substrate (the lower half of the IN section) [39]. After the plateau, the curve goes into the elastic stage of the detwinned martensitic structure, followed by plastic deformation after the elastic region [39]. It can be seen from the stress–strain curve of the sample in Figure 7 that the sample obtains an ultimate tensile strength σ_{UTS} of 569.1 MPa. The total tensile strain at the fracture is about 8.2%. The local axial strain (ε) within the gauge section when the total tensile strain reaches 0, 1%, 2%, 3%, 4%, 5%, 6%, 7%, and 8% are mapped and shown in Figure 8. From the entire evolution of the strain map, it can be observed that as the tensile load increases, the gauge shows two distinctive sections with higher local axial strain in the TiNi substrate (approximately 10%) and lower local axial strain in the DED Ti–Ni–Cu alloy (approximately 4~5%). Anomalous points start to appear in the strain map near the interface when the total tensile strain reaches 8% due to crack formation. Therefore, the mechanical behavior at different sections of the bi-metallic SMA can be clearly tracked using the DIC technique.

The SEM images of the details on the fracture surface of the sample are shown in Figure 9. Figure 9a,b shows the ductile regions where the dimple-like structure dominates, whereas the feature in Figure 9c,d shows low ductility. Figure 9c includes both local dimples and cleavage regions, and in the higher magnification image of Figure 9d, it also has perpendicular patterns marked by perpendicular short, dashed lines, which may correspond to the martensite twinning structure within the Ti–Ni–Cu alloy single-wall section shown in Figure 3d. Also, the dark particles can be found in Figure 9d. Figure 9e–h represent the SEM/EDS mapping of the long, dashed rectangular area in Figure 9d. The Ti, Ni, Cu element mapping of Figure 9e indicates that the dark particle could be Ti_2Ni, which may act as a stress concentrator near the interface that initiates the cracking under the tensile loading.

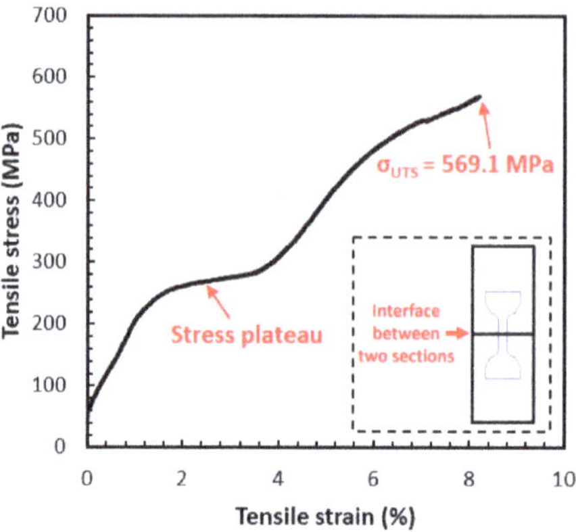

Figure 7. The stress–strain curve of the DED as-deposited bi-metallic structure. The miniature tensile sample was extracted with the interfacial line located at the center of the gauge section.

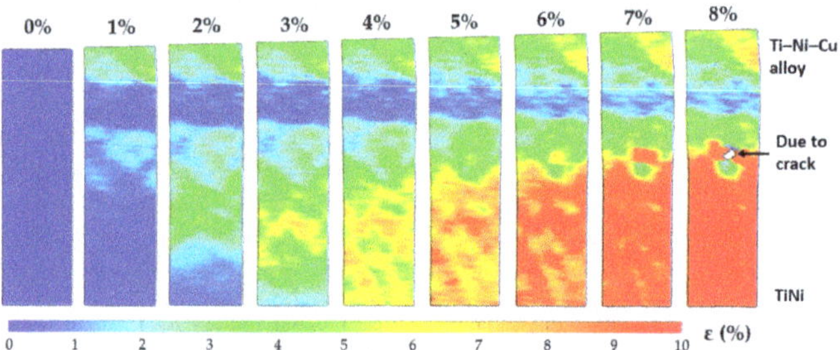

Figure 8. The strain map of the local axial strain within the gauge section at the moments when the total tensile strain equals 0, 1%, 2%, 3%, 4%, 5%, 6%, 7%, and 8%.

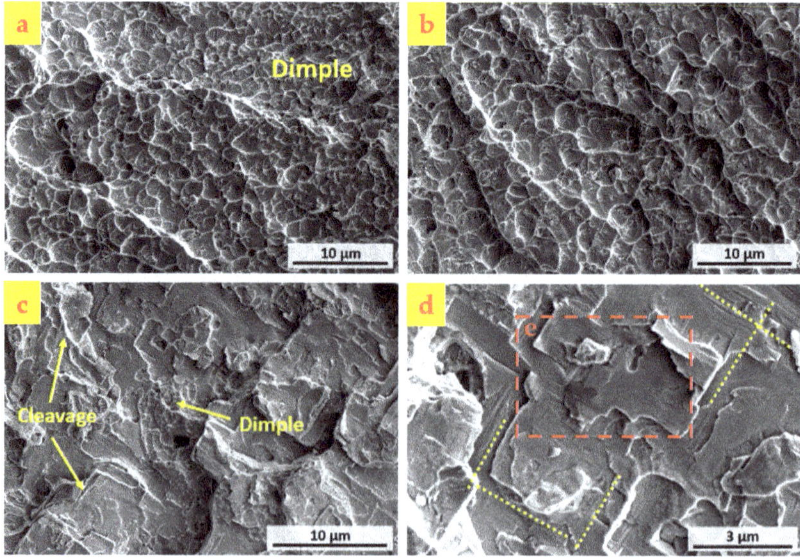

Figure 9. *Cont.*

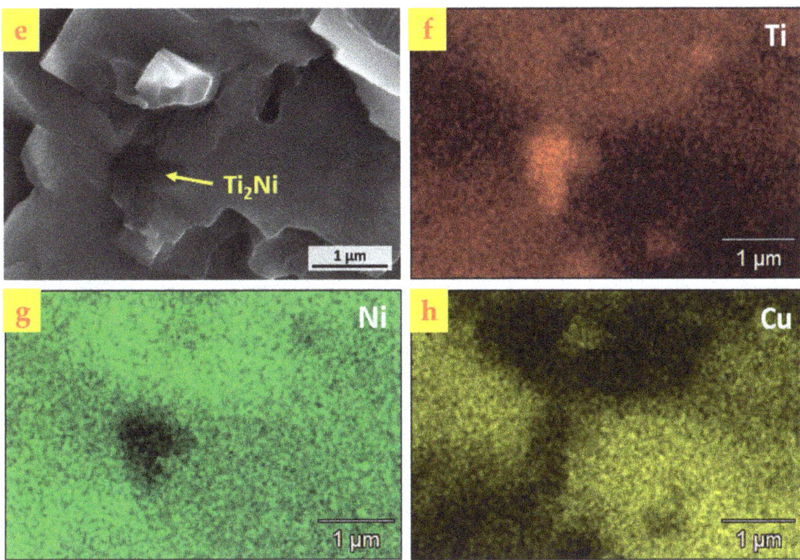

Figure 9. (**a**,**b**) Dimple-like ductile area of the fracture surface. (**c**,**d**) Brittle area of the fracture surface. (**e**) High magnification image of (**d**) in long, dashed rectangular box. (**f**) Element mapping of Ti in (**e**). (**g**) Element mapping of Ni in (**e**). (**h**) Element mapping of Cu in (**e**).

3.5. Demonstration of Multiple Shape-Memory Effect

The multiple shape-memory behavior is demonstrated by a cross-section slice from the as-deposited bi-metallic SMA. Both the DED Ti–Ni–Cu alloy side and the TiNi substrate side are bent compared to the original straight shape. After bending, the bi-metallic SMA piece was then placed on a hot plate at 70 °C, which is higher than (A_f + 20 °C) of the TiNi substrate, as seen in Figure 10a. Figure 10b–d capture the shape change of the bi-metallic SMA on the hot plate at 70 °C after 10 s, 15 s, and 30 s, respectively. It can be noticed that at 70 °C, the curved TiNi substrate side gradually changes back to the straight status. In comparison, the Ti–Ni–Cu alloy side keeps the curved shape since the temperature of 70 °C is not able to complete the martensite-austenite transition of the Ti–Ni–Cu alloy section. As the time reaches 60 s, the hot plate temperature starts to rise and finally reaches 120 °C. Figure 10e shows that at 120 °C, which is higher than the (A_f + 20 °C) of the DED Ti–Ni–Cu alloy part, the Ti–Ni–Cu alloy side also recovers to the original straight state. Therefore, the multiple shape-memory behaviors at two sides of the bi-metallic structure are identified.

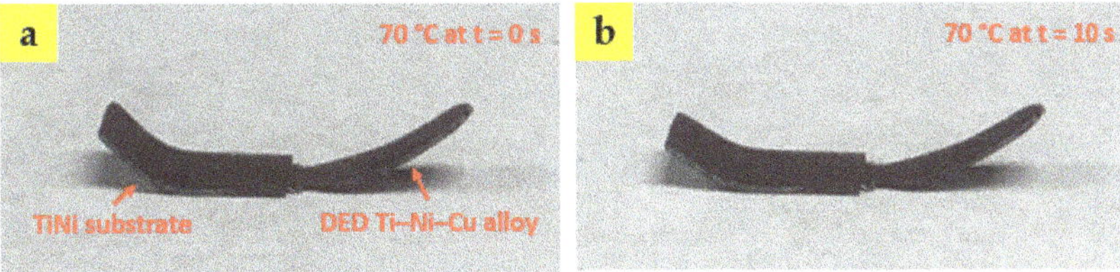

Figure 10. *Cont.*

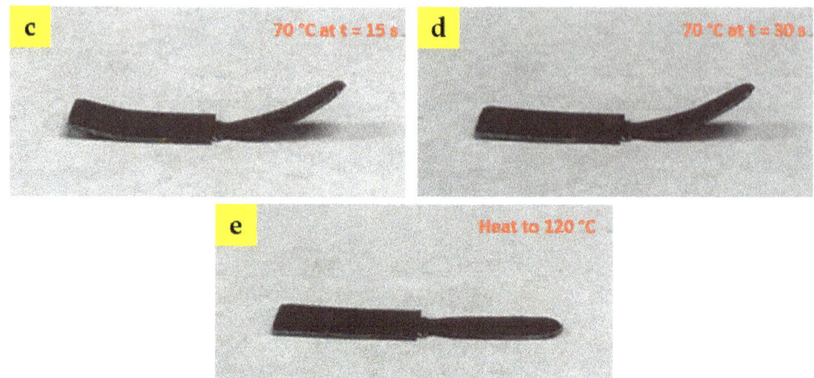

Figure 10. Demonstration of the SMA using a hot plate at 70 °C and 120 °C: (**a**) The bi-metallic SMA sample was bent at both the TiNi substrate side and the DED Ti–Ni–Cu alloy side and placed at 70 °C at t = 0 s. (**b**) t = 10 s. (**c**) t = 15 s. (**d**) t = 30 s. Notice that the TiNi substrate has almost recovered to the original status, whereas the Ti–Ni–Cu alloy side keeps the curved shape. (**e**) The hot plate was then heated from 70 °C to 120 °C. Finally, the Ti–Ni–Cu alloy side recovers.

4. Conclusions

In this work, a Ti-rich Ti–Ni–Cu ternary SMA with an as-mixed atomic composition of Ti–44at.%Ni–3at.%Cu was fabricated on a near-equiatomic commercial TiNi binary SMA substrate, which shows the capability of the elemental powder DED process to join two different types of SMA sections and build SMAs with multiple shape-memory behaviors. The findings from studying the microstructural, mechanical, and functional behaviors of the as-fabricated bi-metallic SMA are summarized below.

Microscopic imaging revealed the dense metallurgical bonding between the as-deposited Ti–Ni–Cu alloy part and the TiNi substrate with no gas pores or cracks being seen at the interface. The TiNi matrix phase and Ti_2Ni minor phase are identified by EDS, and the martensite twinning structure is observed within the TiNi matrix phase of the as-deposited Ti–Ni–Cu alloy section.

Vickers hardness distribution from bottom to top shows the typical low hardness (200~300 HV 0.2) of SMAs across the entire cross-section. All major phases are TiNi phases rather than Ti_2Ni hard intermetallic phases, and Ti_2Ni is the main secondary phase in the Ti–Ni–Cu alloy section.

Tensile testing combined with the DIC technique on a miniature tensile sample shows a bonding strength of 569.1 MPa and the fracture occurs at 8.2% total tensile strain. DIC strain mapping indicates the difference in local axial strain distribution between two sections of different SMA types, which are approximately 4~5% in the Ti–Ni–Cu alloy side and approximately 10% in the TiNi substrate.

DSC analysis at different sections shows that the Ti-rich DED Ti–Ni–Cu alloy shows higher A_f. For the near-equiatomic TiNi substrate, the lower half of the substrate maintains a higher A_f and more obvious R-phase features than the upper half of the substrate since the upper part went through higher temperature annealing from laser power, which tends to exhibit lower TTs and a single austenite-martensite transition without an R-phase.

The shape-memory behavior is tested by an as-deposited bi-metallic slice at 70 °C and 120 °C. At 70 °C, shape recovery occurs only at the TiNi substrate side, whereas when the temperature reaches 120 °C, both sides perform the shape recovery. Future studies will include the heat treatment effect on the microstructure and functional properties of DED bi-metallic SMAs.

Author Contributions: Conceptualization, Y.C.; methodology, Y.C.; software, Y.C. and B.M.; formal analysis, Y.C., C.O.R. and J.W.N.; investigation, Y.C., C.O.R. and B.M.; writing—original draft preparation, Y.C.; writing—review and editing, Y.C., C.O.R., B.M., J.W.N. and F.L.; supervision, J.W.N. and F.L.; project administration, F.L.; funding acquisition, F.L. All authors have read and agreed to the published version of the manuscript.

Funding: This research was partially funded by the U.S. Department of Energy STTR Contract DE-SC0018879, the National Science Foundation Grants CMMI 1625736, and the Material Research Center at Missouri S&T. Their financial support is greatly appreciated.

Institutional Review Board Statement: Not applicable.

Informed Consent Statement: Not applicable.

Data Availability Statement: Not applicable.

Acknowledgments: The authors would like to thank the Intelligent Systems Center, Material Research Center, and Department of Chemistry Shared Instruments Lab at Missouri S&T for all the research support.

Conflicts of Interest: The authors declare no conflict of interest.

References

1. Cai, W.; Daehn, G.; Vivek, A.; Li, J.; Khan, H.; Mishra, R.S.; Komarasamy, M. A state-of-the-art review on solid-state metal joining. *J. Manuf. Sci. Eng.* **2019**, *141*, 031012. [CrossRef]
2. Sahasrabudhe, H.; Harrison, R.; Carpenter, C.; Bandyopadhyay, A. Stainless steel to titanium bimetallic structure using LENS™. *Addit. Manuf.* **2015**, *5*, 1–8. [CrossRef]
3. Rajesh Kannan, A.; Mohan Kumar, S.; Pramod, R.; Pravin Kumar, N.; Siva Shanmugam, N.; Palguna, Y. Microstructure and mechanical properties of wire arc additive manufactured bi-metallic structure. *Sci. Technol. Weld. Join.* **2020**, *26*, 47–57. [CrossRef]
4. Feenstra, D.R.; Molotnikov, A.; Birbilis, N. Effect of energy density on the interface evolution of stainless steel 316L deposited upon INC 625 via directed energy deposition. *J. Mater. Sci.* **2020**, *55*, 13314–13328. [CrossRef]
5. Piscopo, G.; Iuliano, L. Current research and industrial application of laser powder directed energy deposition. *Int. J. Adv. Manuf. Technol.* **2022**, *119*, 6893–6917. [CrossRef]
6. Ahn, D.G. Directed energy deposition (DED) process: State of the art. *Int. J. Precis. Eng. Manuf.-Green Technol.* **2021**, *8*, 703–742. [CrossRef]
7. Saboori, A.; Aversa, A.; Marchese, G.; Biamino, S.; Lombardi, M.; Fino, P. Application of directed energy deposition-based additive manufacturing in repair. *Appl. Sci.* **2019**, *9*, 3316. [CrossRef]
8. Carroll, B.E.; Otis, R.A.; Borgonia, J.P.; Suh, J.O.; Dillon, R.P.; Shapiro, A.A.; Hofmann, D.C.; Liu, Z.K.; Beese, A.M. Functionally graded material of 304L stainless steel and inconel 625 fabricated by directed energy deposition: Characterization and thermodynamic modeling. *Acta Materialia* **2016**, *108*, 46–54. [CrossRef]
9. Feenstra, D.R.; Banerjee, R.; Fraser, H.L.; Huang, A.; Molotnikov, A.; Birbilis, N. Critical review of the state of the art in multi-material fabrication via directed energy deposition. *Curr. Opin. Solid State Mater. Sci.* **2021**, *25*, 100924. [CrossRef]
10. Mosallanejad, M.H.; Niroumand, B.; Aversa, A.; Saboori, A. In-situ alloying in laser-based additive manufacturing processes: A critical review. *J. Alloy. Compd.* **2021**, *872*, 159567. [CrossRef]
11. Jones, N.F.; Beuth, J.L.; de Boer, M.P. Directed energy deposition joining of Inconel 625 to 304 stainless steel with direct bonding. *J. Mater. Res.* **2021**, *36*, 3701–3712. [CrossRef]
12. Zhang, X.; Li, L.; Pan, T.; Chen, Y.; Zhang, Y.; Li, W.; Liou, F. Additive manufacturing of copper-tool steel dissimilar joining: Experimental characterization and thermal modeling. *Mater. Charact.* **2020**, *170*, 110692. [CrossRef]
13. Khodabakhshi, F.; Farshidianfar, M.H.; Bakhshivash, S.; Gerlich, A.P.; Khajepour, A. Dissimilar metals deposition by directed energy based on powder-fed laser additive manufacturing. *J. Manuf. Processes* **2019**, *43*, 83–97. [CrossRef]
14. Reichardt, A.; Dillon, R.P.; Borgonia, J.P.; Shapiro, A.A.; McEnerney, B.W.; Momose, T.; Hosemann, P. Development and characterization of Ti-6Al-4V to 304L stainless steel gradient components fabricated with laser deposition additive manufacturing. *Materials Design* **2016**, *104*, 404–413. [CrossRef]
15. Banait, S.M.; Paul, C.P.; Jinoop, A.N.; Kumar, H.; Pawade, R.S.; Bindra, K.S. Experimental investigation on laser directed energy deposition of functionally graded layers of Ni-Cr-B-Si and SS316L. *Opt. Laser Technol.* **2020**, *121*, 105787. [CrossRef]
16. Jani, J.M.; Leary, M.; Subic, A.; Gibson, M.A. A review of shape memory alloy research, applications and opportunities. *Mater. Des.* **2014**, *56*, 1078–1113. [CrossRef]
17. Otsuka, K.; Ren, X. Physical metallurgy of Ti–Ni-based shape memory alloys. *Prog. Mater. Sci.* **2005**, *50*, 511–678. [CrossRef]
18. Gloanec, A.L.; Bilotta, G.; Gerland, M. Deformation mechanisms in a TiNi shape memory alloy during cyclic loading. *Mater. Sci. Eng. A* **2013**, *564*, 351–358. [CrossRef]
19. Liu, Y.; Shan, L.; Shan, J.; Hui, M. Experimental study on temperature evolution and strain rate effect on phase transformation of TiNi shape memory alloy under shock loading. *Int. J. Mech. Sci.* **2019**, *156*, 342–354. [CrossRef]

20. Lah, A.Š.; Fajfar, P.; Lavrič, Z.; Bukošek, V.; Rijavec, T. Preparation of Shape Memory NiTiNOL Filaments for Smart Textiles. *Tekstilec* **2016**, *59*, 168–174.
21. Khan, M.I.; Pequegnat, A.; Zhou, Y.N. Multiple memory shape memory alloys. *Adv. Eng. Mater.* **2013**, *15*, 386–393. [CrossRef]
22. Nematollahi, M.; Safaei, K.; Bayati, P.; Elahinia, M. Functionally graded NiTi shape memory alloy: Selective laser melting fabrication and multi-scale characterization. *Mater. Lett.* **2021**, *292*, 129648. [CrossRef]
23. Liang, Y.; Jiang, S.; Zhang, Y.; Hu, L.; Zhao, C. Microstructure evolution and deformation mechanism of NiTiFe shape memory alloy based on plane strain compression and subsequent annealing. *Mater. Chem. Phys.* **2018**, *215*, 112–120. [CrossRef]
24. Chang, S.H.; Chiu, W.C. Selective leaching and surface properties of Ti50Ni50−xCux (x = 0–20 at.%) shape memory alloys for biomedical applications. *Appl. Surf. Sci.* **2015**, *324*, 106–113. [CrossRef]
25. Jhou, W.-T.; Wang, C.; Ii, S.; Chiang, H.-S.; Hsueh, C.-H. TiNiCuAg shape memory alloy films for biomedical applications. *J. Alloy. Compd.* **2018**, *738*, 336–344. [CrossRef]
26. Chen, Y.; Zhang, X.; Parvez, M.M.; Liou, F. A review on metallic alloys fabrication using elemental powder blends by laser powder directed energy deposition process. *Materials* **2020**, *13*, 3562. [CrossRef]
27. Schwendner, K.I.; Banerjee, R.; Collins, P.C.; Brice, C.A.; Fraser, H.L. Direct laser deposition of alloys from elemental powder blends. *Scr. Mater.* **2001**, *45*, 1123–1129. [CrossRef]
28. Shiva, S.; Palani, I.A.; Paul, C.P.; Mishra, S.K.; Singh, B. Investigations on phase transformation and mechanical characteristics of laser additive manufactured TiNiCu shape memory alloy structures. *J. Mater. Process. Technol.* **2016**, *238*, 142–151. [CrossRef]
29. Chen, Y.; Zhang, X.; Parvez, M.M.; Newkirk, J.W.; Liou, F. Fabricating TiNiCu Ternary Shape Memory Alloy by Directed Energy Deposition via Elemental Metal Powders. *Appl. Sci.* **2021**, *11*, 4863. [CrossRef]
30. Karnati, S.; Axelsen, I.; Liou, F.F.; Newkirk, J.W. Investigation of tensile properties of bulk and SLM fabricated 304L stainless steel using various gage length specimens. In Proceedings of the 2016 International Solid Freeform Fabrication Symposium. University of Texas at Austin, Austin, TX, USA, 8–10 August 2016.
31. Ren, H.S.; Xiong, H.P.; Pang, S.J.; Chen, B.; Wu, X.; Cheng, Y.Y.; Chen, B.Q. Microstructures and mechanical properties of transient liquid-phase diffusion-bonded Ti3Al/TiAl joints with TiZrCuNi interlayer. *Metall. Mater. Trans. A* **2016**, *47*, 1668–1676. [CrossRef]
32. De Araújo, C.J.; Da Silva, N.J.; Da Silva, M.M.; Gonzalez, C.H. A comparative study of Ni–Ti and Ni–Ti–Cu shape memory alloy processed by plasma melting and injection molding. *Mater. Des.* **2011**, *32*, 4925–4930. [CrossRef]
33. Xu, L.; Wang, R. The effect of annealing and cold-drawing on the super-elasticity of the Ni-Ti shape memory alloy wire. *Mod. Appl. Sci.* **2010**, *4*, 109. [CrossRef]
34. Zhao, C.; Liang, H.; Luo, S.; Yang, J.; Wang, Z. The effect of energy input on reaction, phase transition and shape memory effect of NiTi alloy by selective laser melting. *J. Alloy. Compd.* **2020**, *817*, 153288. [CrossRef]
35. López-Medina, M.; Hernández-Navarro, F.; Flores-Zúñiga, H.; Soto-Parra, D.E. Reversible elastocaloric effect related to B2–R transformation in Ni50. 5Ti49. 5 alloy. *J. Appl. Phys.* **2021**, *129*, 115104. [CrossRef]
36. Carroll, M.C.; Somsen, C.; Eggeler, G. Multiple-step martensitic transformations in Ni-rich NiTi shape memory alloys. *Scr. Mater.* **2004**, *50*, 187–192. [CrossRef]
37. Shahmir, H.; Nili-Ahmadabadi, M.; Huang, Y.; Langdon, T.G. Evolution of microstructure and hardness in NiTi shape memory alloys processed by high-pressure torsion. *J. Mater. Sci.* **2014**, *49*, 2998–3009. [CrossRef]
38. Li, S.; Hassanin, H.; Attallah, M.M.; Adkins, N.J.; Essa, K. The development of TiNi-based negative Poisson's ratio structure using selective laser melting. *Acta Mater.* **2016**, *105*, 75–83. [CrossRef]
39. Gall, K.; Tyber, J.; Brice, V.; Frick, C.P.; Maier, H.J.; Morgan, N. Tensile deformation of NiTi wires. *J. Biomed. Mater. Res. Part A Off. J. Soc. Biomater. Jpn. Soc. Biomater. Aust. Soc. Biomater. Korean Soc. Biomater.* **2005**, *75*, 810–823. [CrossRef]

MDPI

Article

Multi-Fidelity Surrogate-Based Process Mapping with Uncertainty Quantification in Laser Directed Energy Deposition

Nandana Menon †, Sudeepta Mondal † and Amrita Basak *

Department of Mechanical Engineering, The Pennsylvania State University, University Park, State College, PA 16802, USA; nfm5316@psu.edu (N.M.); sudeepta979@gmail.com (S.M.)
* Correspondence: aub1526@psu.edu
† These authors contributed equally to this work.

Abstract: A multi-fidelity (MF) surrogate involving Gaussian processes (GPs) is used for designing temporal process maps in laser directed energy deposition (L-DED) additive manufacturing (AM). Process maps are used to establish relationships between the melt pool properties (e.g., melt pool depth) and process parameters (e.g., laser power and scan velocity). The MFGP surrogate involves a high-fidelity (HF) and a low-fidelity (LF) model. The Autodesk Netfabb® finite element model (FEM) is selected as the HF model, while an analytical model developed by Eagar-Tsai is chosen as the LF one. The results show that the MFGP surrogate is capable of successfully blending the information present in different fidelity models for designing the temporal forward process maps (e.g., given a set of process parameters for which the true depth is not known, what would be the melt pool depth?). To expand the newly-developed formulation for establishing the temporal inverse process maps (e.g., to achieve the desired melt pool depth for which the true process parameters are not known, what would be the optimal prediction of the process parameters as a function of time?), a case study is performed by coupling the MFGP surrogate with Bayesian Optimization (BO) under computational budget constraints. The results demonstrate that MFGP-BO can significantly improve the optimization solution quality compared to the single-fidelity (SF) GP-BO, along with incurring a lower computational budget. As opposed to the existing methods that are limited to developing steady-state forward process maps, the current work successfully demonstrates the realization of temporal forward and inverse process maps in L-DED incorporating uncertainty quantification (UQ).

Keywords: laser-directed energy deposition; melt pool; process maps; multi-fidelity Gaussian process; Bayesian Optimization; uncertainty quantification

Citation: Menon, N.; Mondal, S.; Basak, A. Multi-Fidelity Surrogate-Based Process Mapping with Uncertainty Quantification in Laser Directed Energy Deposition. *Materials* **2022**, *15*, 2902. https://doi.org/10.3390/ma15082902

Academic Editor: Swee Leong Sing and Wai Yee Yeong

Received: 9 February 2022
Accepted: 8 April 2022
Published: 15 April 2022

Publisher's Note: MDPI stays neutral with regard to jurisdictional claims in published maps and institutional affiliations.

1. Introduction

Laser-directed energy deposition (L-DED) offers tremendous opportunities in manufacturing metallic components because of its ability to fabricate three-dimensional near-net-shape parts with location-specific materials and microstructures [1]. Additionally, L-DED has been used to repair components made of different metallic materials such as steel [2], nickel [3], and titanium [4] alloys. In L-DED, powder or wire feedstocks are delivered at the desired location. The feedstock is melted by a high-power laser heat source as shown in Figure 1a. Process parameters such as the laser power, laser/substrate relative velocity (traverse speed or scan speed), and feedstock feed rate can be varied to achieve the desired deposit quality. The melt pool (as shown in Figure 1b,c), formed by the high-power laser source, plays a critical role in controlling the final microstructure of the L-DED part [5]. For example, investigations have revealed that high energy density produced larger melt pools in L-DED [6].

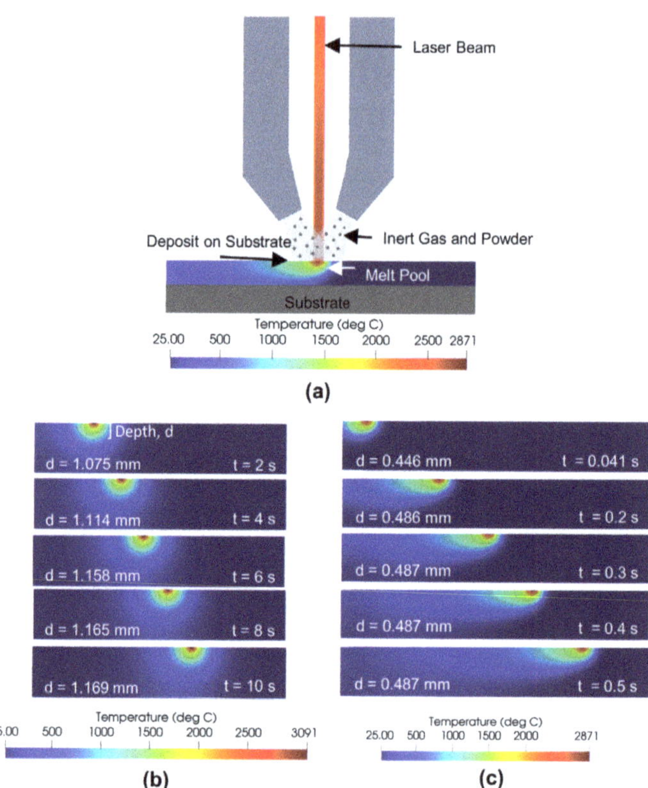

Figure 1. (**a**) A schematic of the L-DED process. (**b**) The melt pool evolution during the L-DED process for a representative power, P = 1000 W, and scan velocity, v = 1 mm/s showing a near-symmetric melt pool. (**c**) A second representative melt pool evolution at P = 1000 W, v = 25 mm/s showing the capability of Netfabb® in capturing an asymmetric melt pool evolution. The melt pool reaches steady-state much earlier in (**c**) than (**b**) due to higher v.

The melt pool shape and sizes are also affected by environmental conditions and part geometry in L-DED. For example, both simulations and experiments have proven that the presence of surfactants was a critical factor in dictating the shape of the melt pool as the surfactants were found to affect the surface tension [7]. The scan pattern was also found to affect the shape and size of the melt pool due to heat transfer across adjacent tracks and/or layers. Due to these parameters, the melt pool size may vary both in time and space during the L-DED process. Such a variation can detrimentally affect microstructures and, therefore, the mechanical properties of the part. Hence, the impact of toolpath on the melt pool geometry was also investigated [8].

Experimental investigation for evaluating the impact of these different process parameters on the melt pool properties is rather expensive. Moreover, since the thermophysical properties of the metallic materials are drastically different from one another, such an investigation has to be performed for each material of interest. Computational modeling can provide a much-required alternative for estimating the process parameters in L-DED. However, the process parameter development would still require a thorough exploration of the parameter space involving a design-of-experiment (DoE) approach. Computational investigation of such a DoE can be expensive if high-fidelity models are used for predicting the melt pool properties. Hence, it is critical to develop efficient methods for generating process maps that would efficiently estimate the melt pool properties as a function of the process parameters.

Despite recent advances in computational modeling-based process mapping of L-DED additive manufacturing (AM) processes, much of the existing work is focused on developing forward maps, i.e., estimating and predicting the steady-state melt pool characteristics (e.g., melt pool depth and width) as functions of process parameters (e.g., energy density, power, scan speed, and scan pattern, amongst others) or dimensionless numbers [9]. While such studies are extremely useful in understanding the impact of processing parameters on the melt pool properties, they suffer from four critical drawbacks:

- They are computationally expensive when high-fidelity (HF) models are used and less capable when low-fidelity (LF) models are used [10].
- They are deterministic and cannot define the prediction uncertainties when the simulation data is not available for the process parameters of interest [11].
- They are typically used to generate steady-state forward maps although the evolution of melt pool is a transient phenomenon as depicted in Figure 1b,c.
- They are seldom combined with computationally efficient optimization routines toward solving the inverse problem.

To address these limitations, one possibility is to blend the information obtained from the HF and LF simulation models using machine learning tools to develop a multi-fidelity (MF) [12] surrogate. The MF surrogate, thereafter, can be used as a proxy for the prediction of the melt pool properties as a function of process parameters at discrete time instants facilitating the design of temporal process maps. MF surrogates have been demonstrated to efficiently incorporate the information present in a hierarchy of varied fidelity models [13,14] to develop inexpensive estimates of the properties of interest. Such surrogates have been used frequently within the framework of co-kriging [15], where an LF model output acts as an auxiliary data source to enhance the prediction of an HF model [16]. This method is particularly useful when the HF model is computationally very expensive and difficult to evaluate. A modification of this approach, known as recursive co-kriging has been recently developed by Le Gratiet et al. [17,18]. This approach has eventually been used by Perdikaris et al. [19] to tackle the 'curse of dimensionality' while dealing with large datasets. One of the advantages of using MF surrogates is that they can be utilized as black box optimizers. Such optimizers have been used in a wide variety of engineering design problems [20–23]. However, such strategies have sparsely been explored in AM, particularly for optimizing the melt pool geometry.

Inspired by the existing research gap, this paper starts by implementing a multi-fidelity Gaussian process (MFGP) surrogate to design the forward process maps for the L-DED AM process at several discrete time instants of the simulation, thereby enabling the design of temporal process maps. Two different melt pool evolution models, namely, the Autodesk Netfabb® finite element model (FEM) [24] and the Eagar-Tsai [25] analytical model are chosen as the HF and LF ones, respectively. The MFGP surrogate predicts the melt pool depths with uncertainty for process parameters when the true depth is not known. The results demonstrate that the surrogate is capable of predicting the melt pool depth in L-DED at a level of accuracy comparable to a representative HF model, but at a fraction of its computational cost. Once the forward map is designed, the MFGP surrogate is integrated with a Bayesian Optimization (BO) algorithm to design an inverse process map. To demonstrate the efficacy of the inverse map, a case study is performed to obtain the desired melt pool depth under computational budget constraints. The optimal process parameters are found using an approach based on active learning (AL) [26]. Being a Bayesian machine learning framework, the posterior predictions are probabilistic, and are described by an output distribution (as opposed to single-valued estimates obtained from conventional deterministic regression models), which offers a principled methodology of uncertainty quantification (UQ) [27]. The performance of multi-fidelity Bayesian Optimization (MFGP-BO) is compared with single-fidelity Bayesian Optimization (SFGP-BO). SFGP-BO utilizes the HF model as the sole information source. The results demonstrate that MFGP-BO can not only find the optimal process parameters faster than SFGP-BO but also with an improved quality (defined as the percent deviation from the desired melt pool depth).

The paper is arranged in four sections, including the present one. The methodology section outlines the thermal models for melt pool prediction, MF surrogate development, and Bayesian Optimization. The next section presents the results, followed by a conclusion section outlining the applications and future perspective of the research. All simulations reported in this paper are conducted on an Intel® Xeon® Gold 6230 CPU processor with 128 GB of RAM.

2. Methods

2.1. Thermal Models for Melt Pool Predictions

Several different simulation models are available in the open literature for the prediction of melt pool properties in L-DED. Eagar and Tsai [25] improved Rosenthal's [28] analytical model by replacing the point heat-source with a Gaussian distribution. The Eagar-Tsai model has been widely employed to conduct rapid simulations for a wide range of materials. An analytical model involving metal fluid flow and mass transfer in addition to heat transfer in L-DED was developed by Gan et al. [29]. The solidus and liquidus thermal properties were considered, thereby, capturing the Marangoni convection which showed a strong influence on the melt pool shape [30]. This model was experimentally validated for single-layer IN718 deposits on AISI 1045 carbon steel [31]. Huang et al. [32] developed a comprehensive analytical model, from Rosenthal's 3D temperature distribution, by coupling both mass and energy flows. This model accommodated for the attenuation of laser power due to the change in clad geometry and powder-gas stream interactions. The model was experimentally validated for single-layer depositions of IN625. The effect of Marangoni flow was incorporated by using a modified thermal conductivity parameter.

While analytical models for L-DED provide valuable melt pool-related information, they include several simplifications. Computational capabilities using finite element models (FEMs) can address these simplifications. Anderson et al. [33] simulated L-DED of nickel-base superalloy CMSX-4® using a model developed by DebRoy [34]. However, it did not consider the variation of material properties with temperature. Kamara et al. [35] used a commercial FEM suite, Ansys Fluent to simulate the transient evolution of melt pool for IN718 powder-fed L-DED. Other commercial software are also available. Autodesk's Netfabb® [24] is a software application tailored for AM that has reduced the computation time significantly through adaptive meshing-based FEM techniques [36]. Netfabb® uses a nonlinear Newton–Raphson-based code and solves for the transient behavior of the L-DED process.

MF surrogates are developed, in this paper, to supplement the available HF predictions with inexpensive LF predictions. Combining data from multiple sources naively may result in biased predictions which do not accurately reflect the physics. Hence, the construction of a reliable MF surrogate depends on the careful selection of its constituent HF and LF models. Based on the existing literature [9], an analytical model (e.g., Eagar-Tsai), due to its many simplifications, shows lower computational cost. On the contrary, an FEM model (e.g., Netfabb®), shows higher computational expense but incorporates several modeling parameters (e.g., temperature-dependent thermophysical properties). These two models can define the hierarchical fidelity levels since they are defined by the same governing equations that explain the underlying physical phenomenon, provided they are numerically comparable in their discretized domains (i.e, both models have similar grid sizes). This observation forms the basis for selecting the Eagar-Tsai and Netfabb® models as LF and HF models, respectively, in this paper.

2.1.1. Eagar-Tsai's Model

The LF model used in this paper is the analytical model developed by Eagar and Tsai [25] which solves for the three-dimensional temperature distribution produced by a traveling distributed heat source moving on a semi-infinite plate. This model is a modification of the Rosenthal's model where a Gaussian heat distribution is used instead of a point source. Figure 2 explains the coordinate system used in the Eagar-Tsai model. The heat source is traveling with a uniform speed of v in the x-direction, and is assumed to be a 2D surface Gaussian. The temperature $T(x_c, y_c, z_c, t)$, at a particular location (x_c, y_c, z_c) and time t is calculated as:

$$T(x_c, y_c, z_c, t) - T_0 = \frac{\alpha_L P}{\pi \rho_p c_p (4\pi a_p)^{1/2}} \int_0^t \frac{dt'(t-t')^{-1/2}}{2a_p(t-t') + \sigma_L{}^2} e^{-\frac{(x_c - vt')^2 + y_c^2}{4a_p(t-t') + 2\sigma_L{}^2} - \frac{z_c^2}{4a_p(t-t')}} \quad (1)$$

Here, T_0 is the initial temperature of the substrate, α_L is the absorptivity of the laser beam, P is the power, ρ_p, k_p, and c_p are the density, thermal conductivity, and specific heat capacity of the material, respectively, $a_p \triangleq \frac{k_p}{\rho_p c_p}$ is the thermal diffusivity of the material, t' is the dummy integration variable, σ_L is the distribution parameters, and v is the scan velocity. Similar to Rosenthal's equation, the Eagar-Tsai model also makes several assumptions: (i) absence of heat transfer due to convection and radiation; (ii) constant thermal properties for the material; (iii) quasi-steady state semi-infinite medium, and absence of any phase change.

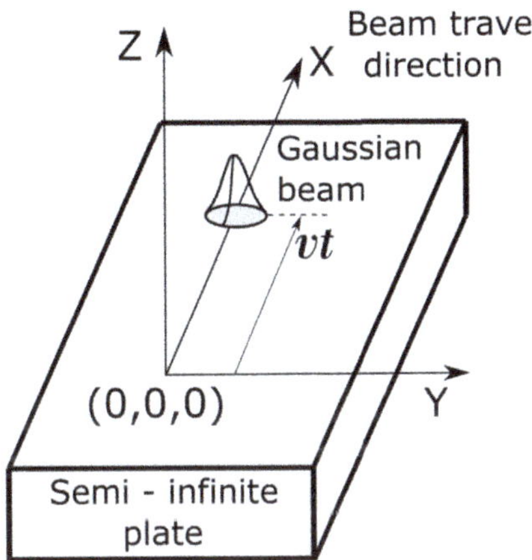

Figure 2. Schematic illustrating the coordinate system of the analytical model. The figure is reproduced from [37] under the terms of the Creative Commons Attribution 4.0 License from http://creativecommons.org/licenses/by/4.0/ (accessed on 9 April 2022).

A reasonable agreement between the theoretical and experimental data for steel, titanium and aluminum is demonstrated by Eager and Tsai [25]. This model has been widely used by researchers and has been incorporated with experimental studies for evaluating the melt pool evolution in L-DED processes to develop process mapping strategies.

2.1.2. Autodesk Netfabb® Model

Netfabb® model by Autodesk, a non-linear decoupled 3D transient FEM solver, is used as the HF model [36,38]. The underlying methodology of Netfabb® rests on the decoupled or weakly coupled modeling assumption that the relationship between the thermal and mechanical behaviors are unidirectional so that the thermal history affects the mechanical behavior, but the vice-versa does not hold. The Netfabb® model includes Marangoni convection, convection and radiation heat losses, and the temperature-dependent thermo-physical properties omitted by the Eagar-Tsai model.

The model domain is determined based on the scan velocity range so that the domain is large enough in the scan direction such that the melt pool reaches a steady-state. For thermal investigations, the energy balance is the governing equation, which is converted to a weak formulation using the Galerkin approach [39]. The distribution of heat through the part is described by the Fourier's conduction equation. The model uses a 3D volumetric heat source. The thermal boundary losses include thermal radiation, free convection, and forced convection. The total heat loss flux from the model is thus given by,

$$q = h_{\text{eff}}(T_s - T_\infty) \tag{2}$$

$$h_{\text{eff}} = h_{\text{free}} + h_{\text{forced}} + h_{\text{radiation}} \tag{3}$$

Here, T_s is the surface temperature, T_∞ is the ambient temperature, h_{eff} is the effective heat transfer coefficient which is a summation of the free convection (h_{free}), forced convection (h_{forced}), and radiation ($h_{\text{radiation}}$). Free convection arises due to the thermal gradients developed during the L-DED process while forced convection arises from the shielding gas and powder flowing over the melt pool. This model has been experimentally validated for L-DED processes for a popular nickel-base superalloy, IN625 [39] showing excellent agreement between the simulation and experimental data.

2.2. *Surrogate Development*

This subsection briefly presents the mathematical fundamentals of the GP and MF surrogate that integrates the LF and HF thermal models [40,41].

2.2.1. Gaussian Process (GP) Surrogate

The surrogate development strategy is based on a class of stochastic processes called 'GPs' that assume any finite collection of random variables to follow a multivariate jointly Gaussian distribution. For a finite collection of inputs, $\mathbf{x}$, the corresponding function outputs, $\mathbf{y}$ are assumed to have a multivariate jointly Gaussian distribution,

$$\mathbf{y} \sim \mathcal{N}(\mu(\mathbf{x})\,, k(\mathbf{x}, \mathbf{x}')) \tag{4}$$

Here, $\mathcal{N}$ implies a Gaussian distribution. The underlying GP is completely characterized by a mean function: $\mu(\mathbf{x}) \triangleq E[\mathbf{y}]$, and a covariance function: $k(\mathbf{x}, \mathbf{x}') \triangleq E[\mathbf{y} - \mu(\mathbf{x}))(\mathbf{y}' - \mu(\mathbf{x}'))]$ [27]. Here, $E[\mathbf{y}]$ denotes the expectation of $\mathbf{y}$. $\mathbf{x}'$ and $\mathbf{y}'$ denote an input other than $\mathbf{x}$ and the corresponding functional output of it, respectively. In the context of the melt pool prediction, each input point comprises of different (P, v) combinations. The melt pool depths (d) corresponding to the (P, v) combinations are jointly Gaussian. The training data-set comprises of $\mathbf{x}^{trn} \triangleq (P^{trn}, v^{trn})$ as the inputs and the known melt pool depths, $y^{trn} \triangleq d^{trn}$ as the outputs. The test data-set comprises of $\mathbf{x}^{tst} \triangleq (P^{tst}, v^{tst})$ as the inputs for which the melt pool depths are unknown. The conditional distribution of the outputs at the test locations is given by:

$$\mathbf{y}^{tst}|\mathbf{y}^{trn}, \mathbf{x}^{trn}, \mathbf{x}^{tst} \sim \mathcal{N}(\mu^{tst}, \Sigma^{tst}) \tag{5}$$

Here,

$$\mu^{tst} = \mathbf{K}(\mathbf{x}^{tst}, \mathbf{x}^{trn})[\mathbf{K}(\mathbf{x}^{trn}, \mathbf{x}^{trn}) + \sigma_\epsilon{}^2\mathbf{I}]^{-1}\mathbf{y}^{trn} \tag{6}$$

$$\Sigma^{tst} = \mathbf{K}(\mathbf{x}^{tst}, \mathbf{x}^{tst}) - \mathbf{K}(\mathbf{x}^{tst}, \mathbf{x}^{trn})[\mathbf{K}(\mathbf{x}^{trn}, \mathbf{x}^{trn}) + \sigma_\epsilon{}^2\mathbf{I}]^{-1}\mathbf{K}(\mathbf{x}^{trn}, \mathbf{x}^{tst}) \tag{7}$$

Here, $\mathbf{I}$ is the identity matrix and $\mathbf{K}$ is the covariance matrix. Thus, the predicted posterior distribution of the outputs at every test data point is also a Gaussian distribution, characterized by the mean, μ^{tst} and covariance, Σ^{tst}. A detailed mathematical account of GPs can be found in [27].

2.2.2. Multi-Fidelity (MF) GP Surrogate

Often computational models present a hierarchy of fidelities for a given process. HF models are generally more capable, but expensive. On the other hand, LF models are typically less capable, but cheaper. To develop process maps, an extensive use of HF models can be computationally infeasible. In such cases, it would be judicious to adopt intelligent strategies that leverage the computational inexpensiveness of the LF models by using them more frequently.

MF surrogate is a statistically-learned framework [12] that integrates the information present in all fidelities to develop a 'proxy' which can predict outputs with the accuracy of HF models, but with significantly inexpensive computational overhead. The general structure of MF information source is shown in Figure 3, which shows several levels of fidelities in the models. Such a framework relies on data-driven learning of the correlation among the different fidelities. Co-kriging approaches have been studied extensively for performing the joint estimation of the outputs from correlated variables [12–14]. The co-kriging approach employed in this work is based on the autoregressive formulation of Keneddy and O'Hagan [15]. MFGPs using co-kriging approaches rely on formulating separate surrogates for each fidelity which are coupled together through an appropriate covariance function in a GP setting.

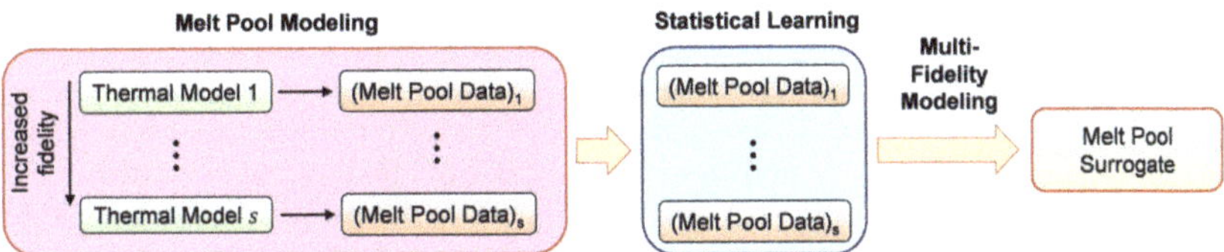

Figure 3. Presence of a hierarchy of fidelities in modeling platforms. Statistical learning integrates multi-fidelity (MF) information.

For an input set of process parameters, $\mathbf{x}$, the autoregressive formulation for 2 fidelities is given by:

$$\mathbf{y}_2 = \rho \mathbf{y}_1 + \delta(\mathbf{x}) \tag{8}$$

Here, $\mathbf{y}_1$ is the low-fidelity output and $\mathbf{y}_2$ is the high-fidelity output. ρ quantifies the correlation between $\mathbf{y}_2$ and $\mathbf{y}_1$. The Gaussian process, $\delta(\mathbf{x})$ represents the discrepancy between $\mathbf{y}_1$ and $\mathbf{y}_2$. These outputs, $\mathbf{y}_1$ and $\mathbf{y}_2$, take the jointly Gaussian distribution of the form:

$$\mathbf{Y} = \begin{bmatrix} \mathbf{y}_1 \\ \mathbf{y}_2 \end{bmatrix} \sim \mathcal{N}\left(\mathbf{0}, \begin{bmatrix} k_1(\mathbf{x}_1, \mathbf{x}_1'; \boldsymbol{\theta}_1) + \sigma_{\epsilon_1}^2 \mathbf{I} & \rho k_1(\mathbf{x}_1, \mathbf{x}_2'; \boldsymbol{\theta}_1) \\ \rho k_1(\mathbf{x}_2, \mathbf{x}_1'; \boldsymbol{\theta}_1) & \rho^2 k_1(\mathbf{x}_2, \mathbf{x}_2'; \boldsymbol{\theta}_1) + \\ & k_2(\mathbf{x}_2, \mathbf{x}_2'; \boldsymbol{\theta}_2) + \sigma_{\epsilon_2}^2 \mathbf{I} \end{bmatrix}\right) \tag{9}$$

Here, k_1 and k_2 are the kernel functions, $\sigma_{\epsilon_1}^2$ and $\sigma_{\epsilon_2}^2$ are the noise levels, θ_1 and θ_2 are the hyperparameters, where the subscripts 1 and 2 correspond to LF and HF models, respectively. The Negative Log Marginal Likelihood (NLML) in a two-fidelity setting is given by:

$$-\log p(\mathbf{Y}|\mathbf{X}, \boldsymbol{\theta}_1, \boldsymbol{\theta}_2, \rho, \sigma_{\epsilon_1}^2, \sigma_{\epsilon_2}^2) = \frac{1}{2}\log|\mathbf{K}| + \frac{1}{2}\mathbf{Y}^T\mathbf{K}^{-1}\mathbf{Y} - \frac{N_L + N_H}{2}\log 2\pi \tag{10}$$

Here, $\mathbf{X}$ and $\mathbf{Y}$ are the combined observed inputs and their outputs from the LF and HF models, respectively. N_L and N_H correspond to the number of observed input-output data from LF and HF models, respectively. The MFGP methodology is outlined in Algorithm 1.

Algorithm 1 Multi Fidelity Gaussian Process

Require: Low-fidelity input, $\mathbf{x_1}$; Low-fidelity output, $\mathbf{y_1}$; Hyperparameter of the low fidelity kernel, θ_1; High-fidelity input, $\mathbf{x_2}$; High-fidelity output, $\mathbf{y_2}$; Hyperparameter of the high fidelity kernel, θ_2; Kernel function, k (for simplicity, here $k_1 = k_2$); Test input, $\mathbf{x}^{tst}$; Noise-level, σ_ϵ^2 (for simplicity, here $\sigma_{\epsilon_1}^2 = \sigma_{\epsilon_2}^2$)

$\mathbf{L}$ = cholesky($\mathbf{K} + \sigma_\epsilon^2 \mathbf{I}$) ▷ $\mathbf{K}$ as calculated from Equation (9)
$\mathbf{Y} = [\mathbf{y_1}\ \mathbf{y_2}]$
$\alpha = \mathbf{L}^T \setminus (\mathbf{L} \setminus \mathbf{Y})$
$\psi_1 = \rho k(\mathbf{x}^{tst}, \mathbf{x}_1, \theta_1)$
$\psi_2 = \rho^2 k(\mathbf{x}^{tst}, \mathbf{x}_2; \theta_1) + k(\mathbf{x}^{tst}, \mathbf{x}_2; \theta_2)$
$\mathbf{\Psi} = [\psi_1\ \psi_2]$
$\hat{f}_{x^{tst}} = \mathbf{\Psi}.\alpha$ ▷ predictive mean
$\beta = \mathbf{L}^T \setminus (\mathbf{L} \setminus \Psi^T)$
$\mathbb{V}[f_{x^{tst}}] = \rho^2 k(\mathbf{x}^{tst}, \mathbf{x}^{tst}, \theta_1) + k(\mathbf{x}^{tst}, \mathbf{x}^{tst}; \theta_2) - \mathbf{\Psi}\beta$ ▷ predictive variance

The kernel function at every fidelity level has its own hyperparameter. The choice of the kernel function is a critical element for the success of a GP algorithm since it encodes the correlation between the points in the feature space. Typically, squared exponential kernels are best suited for interpolating smooth functional relationships. However, in this paper, Matérn kernels, with shape parameters of 5/2, are used since their length-scales are less prone to be affected by non-smooth regions, thereby improving performance in these regions. [40,42]. The MFGP is learnt by optimizing the hyperparameters through minimizing NLML using the Limited-memory Broyden–Fletcher–Goldfarb–Shanno (L-BFGS) algorithm [43]. While the current formulation focuses on a two-fidelity setup, this can be extended to higher fidelities by making appropriate modifications in the covariance matrix which correlates different levels of fidelity [15].

2.3. *Bayesian Optimization (BO)*

This subsection briefly presents the mathematical fundamentals of the Bayesian Optimization for single-fidelity (SF) and multi-fidelity (MF) surrogates [40,41].

2.3.1. Single-Fidelity Gaussian Process with Bayesian Optimization (SFGP-BO)

In this description, the term *optimization* is used to denote maximization of an objective function. A minimization problem can be posed similarly by taking the negative of the objective function. In a single-fidelity setting, there is a single objective function f. To optimize f over its domain, the solver needs to find:

$$\hat{\mathbf{x}} = \underset{\mathbf{x} \in X^*}{\text{argmax}}\, f(\mathbf{x}) \tag{11}$$

Here, 'argmax' finds the argument that gives the maximum value from an objective function, f. The functional form of f is typically unknown and, hence, a gradient-free or *black-box* optimization is often utilized. BO is one such *black-box* optimization technique [44] that leverages the predictions through a surrogate for sequential *active learning* to find the global optima of the objective function. The AL strategies find a trade-off between *exploration* and *exploitation* in possibly noisy settings [45–47], which facilitates a balance between the global search and local optimization through *acquisition functions*. One commonly used acquisition function in BO is Expected Improvement (*EI*). A detailed formulation of EI can be found in the work of Mockus et al. [48] and Jones et al. [45].

The objective function, $f(x)$ is often represented as a GP, which yields a posterior predictive Gaussian distribution characterized by the mean $\mu(\mathbf{x})$ and standard deviation $\sigma(\mathbf{x})$ for $\mathbf{x} \in X^*$, where X^* is the search space of the optimization challenge. When the optimization framework involves a single-fidelity GP, it is referred to as single fidelity GP (SFGP)-based optimization. The optimization algorithm proceeds sequentially by sampling

$\hat{\mathbf{x}} = argmax_{\mathbf{x}} EI(\mathbf{x})$ at every step of the iteration process to add on to the dataset, after which the GP surrogate is retrained with the new data set to predict the acquisition potential for the next iterative step. This process continues until an optimum is reached, or the computational budget is extinguished. Since the acquisition potential is predicted over the entire search space by the surrogate, BO can achieve fast predictions without a lot of function calls in the search space (i.e., without having to run the simulations to obtain the objectives at all the search locations). This process otherwise, might be computationally infeasible when the search space is high-dimensional and the simulations are expensive.

2.3.2. Multi-Fidelity Gaussian Process with Bayesian Optimization (MFGP-BO)

In the presence of multiple levels of fidelities, the recent work of Sarkar et al. [40] presents a demonstration of choosing appropriate acquisition functions for the HF and LF models to guide the search for optimum using MF surrogates. The HF predictions are chosen using the constrained EI acquisition function, while the LF predictions are selected using the GP-based Mutual Information Acquisition function (GP-MI) [49]. The rationale behind employing GP-MI for LF selections is in its mathematical formulation: GP-MI would preferentially promote *exploration* at the initial stages of optimization, and gradually drive *exploitation* as the global prediction becomes progressively more accurate in the subsequent iterations. Since a majority of engineering applications involves expensive and possibly limited HF evaluations, GP-MI fits the requirements of being exploratory at the initial stages where inexpensive LF evaluations can be leveraged to learn the process parameter space [40]. The algorithm implemented for MFGP-BO methodology is outlined in Algorithm 2.

Algorithm 2 MFGP—Bayesian Optimization

Require: Search space for optimization, X^*; MFGP prior with mean function, $\mu(\mathbf{x})$ and kernel function, $k(\mathbf{x}, \mathbf{x}')$; objective function, $J(t, P, v)$; number of optimization steps allowed, N_{opt}; N_H, N_L, accepted tolerance for the melt pool depth, ϵ

for $i \leftarrow 1$ to N_{opt} **do**
  if $|d(t, P, v) - d^*| < \epsilon$ **then**
    Perform HF simulations at $\hat{\mathbf{x}} = argmax_{\mathbf{x}} EI(\mathbf{x})$
    Augment data with HF predictions, update MFGP
    Perform LF simulations at $\hat{\mathbf{x}} = argmax_{\mathbf{x}} GP - MI(\mathbf{x})$
    Augment data with LF predictions, update MFGP
  else
    break
  end if
end for

3. Results and Discussion

3.1. Melt Pool Predictions from Thermal Models

Single crystal nickel-based superalloys have been increasingly deployed in gas turbines owing to their excellent high-temperature properties. CMSX-4® is one such popular second-generation ultra high-strength superalloy that is selected as the candidate alloy in this paper [50]. The life-cycle of these expensive parts comprising CMSX-4® is often limited by blade tip wear and crack, thereby requiring a feasible method of repair that ensures directional solidification in the repaired zones. L-DED has proved to be successful in achieving this objective [51]. A major determinant of the growth of SX epitaxial layers during L-DED is the laser process parameters. A lower laser power can increase the thermal gradient and promote epitaxial growth. A higher laser scanning velocity yields a shallow melt pool and stimulates epitaxial growth [52]. This outlines a critical need to propose processing windows for the manufacture and/or repair by L-DED. This section summarizes the results obtained from the low- and high-fidelity thermal models used to simulated the single scan deposits of CMSX-4®.

3.1.1. Predictions from LF Eagar-Tsai Model

The thermo-physical properties of CMSX-4® chosen for the LF model are those reported by Gäumann et al. [51]: k_p = 22 W/(m·K), ρ_p = 8700 kg/m^3, c_p = 690 J/(kg·K), and the liquidus temperature T_L = 1660 K. Other modeling parameters and their selected values are: T_∞ = 25 °C and α_L = 0.35. The laser beam radius is maintained at 0.39 mm.

Figure 4 shows the two-dimensional melt pool at different time instants illustrating the temporal nature of the melt pool evolution for laser power, P = 1000 W and scan velocity, v = 1 mm/s. The results are also reported in Table 1. The melt pool reaches a steady-state after t = 20 s. Similar behavior is observed for other P and v combinations as well.

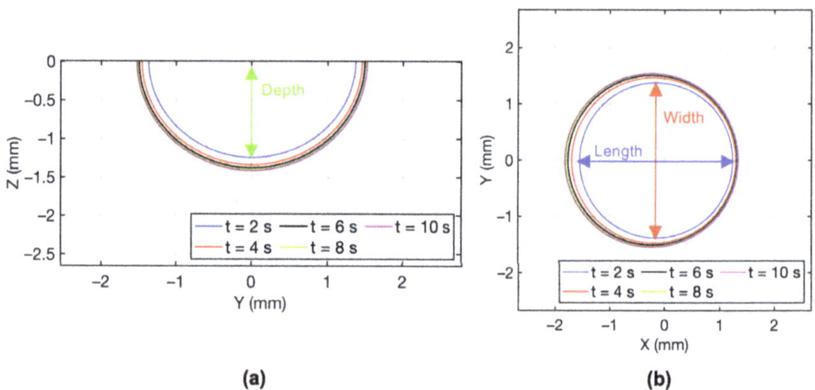

Figure 4. Temporal variation of the (**a**) melt pool depth and (**b**) melt pool length and width, for P = 1000 W and v = 1 mm/s.

Table 1. Melt pool dimensions calculated using Eagar-Tsai for P = 1000 W and v = 1 mm/s.

Time (s)	Melt Pool Depth (mm)	Melt Pool Width (mm)	Melt Pool Length (mm)
2	1.238	2.751	2.792
4	1.334	2.934	2.994
6	1.375	3.012	3.083
8	1.397	3.056	3.133
10	1.410	3.083	3.164

3.1.2. Predictions from HF Netfabb® Model

Single-track single-layer simulations are performed using the Netfabb® model. The simulation domain is shown in Figure 5. The laser parameters, i.e., laser radius and absorptivity, and the ambient temperature are kept identical to the Eagar-Tsai model. The effective heat transfer coefficient is set as h_{eff} = 25 W/m^{2}C [53]. The temperature of the substrate plate is kept at 25 °C. The laser vector file containing the laser power, laser vector, start and end positions of the laser, laser radius, scan speed, and start time, is fed into the Netfabb® model before creating the simulation domain. The temperature-dependent properties for CMSX-4® are obtained from JMatPro® [54] and shown in Table 2.

The L-DED-specific mesh features are varied to implement adaptive meshing which reduces the number of mesh elements [55] away from the melt pool as shown in Figure 6a. These features are (i) the number of elements per heat source radius, (ii) the number of fine layers below the heat source, and (iii) the number of mesh adaptivity levels. A mesh convergence test is conducted by varying the quality of the mesh for the simulation domain described in Figure 5. The difference between the melt pool depths calculated from the model having the finest mesh (340,400 elements) and the selected model (119,392 elements) is 0.54%, which is lesser than the minimum of 5% specified by Netfabb® for a good mesh convergence [53]. Figure 6b shows the results obtained from the mesh convergence study.

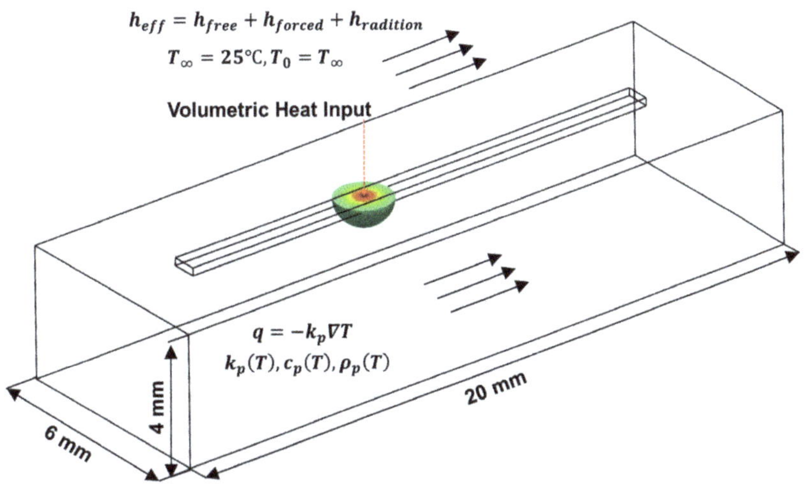

Figure 5. Simulation domain with boundary conditions.

Table 2. Temperature dependent properties for CMSX-4®.

T °C	Density (g/cm^3)	Thermal Conductivity (W/(m.K))	Specific Heat (J/(g.K))
27	8.592	11.902	0.426
127	8.552	12.655	0.454
327	8.49	14.794	0.53
527	8.41	16.875	0.542
727	8.318	19.274	0.589
927	8.211	22.226	0.652
1127	8.079	26.922	0.795
1327	7.921	35.331	0.62
1527	7.396	35.395	0.693
1727	7.208	38.915	0.698

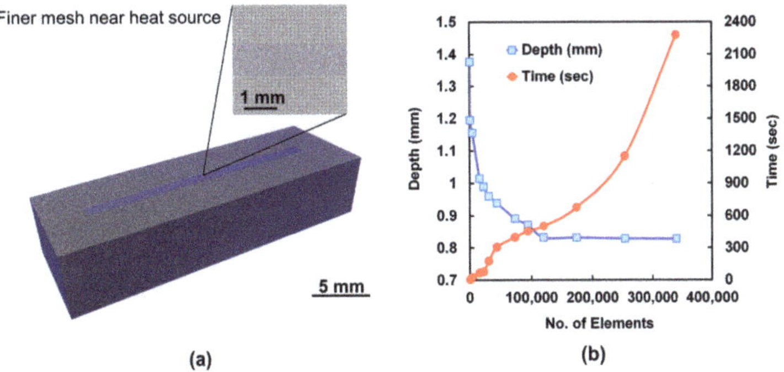

Figure 6. (a) Adaptive meshing applied to the simulation domain. (b) The variation of the melt pool depth and the simulation time with the number of mesh elements.

The melt pool properties are obtained from the Netfabb® model using ParaView [56], an open-source, multi-platform data analysis and visualization application. The thermal models are post processed to extract an isovolume corresponding to the melt pool that is defined by the liquidus temperature of the material i.e., T_L = 1660 K for CMSX-4®. Figure 7 shows the evolution of melt pool at four different time instants for P = 1000 W and v = 1 mm/s with the corresponding three-dimensional melt pool volume extracted. The results obtained from the Netfabb® model also reinstate the temporal nature of the melt pool necessitating the requirement of temporal process maps.

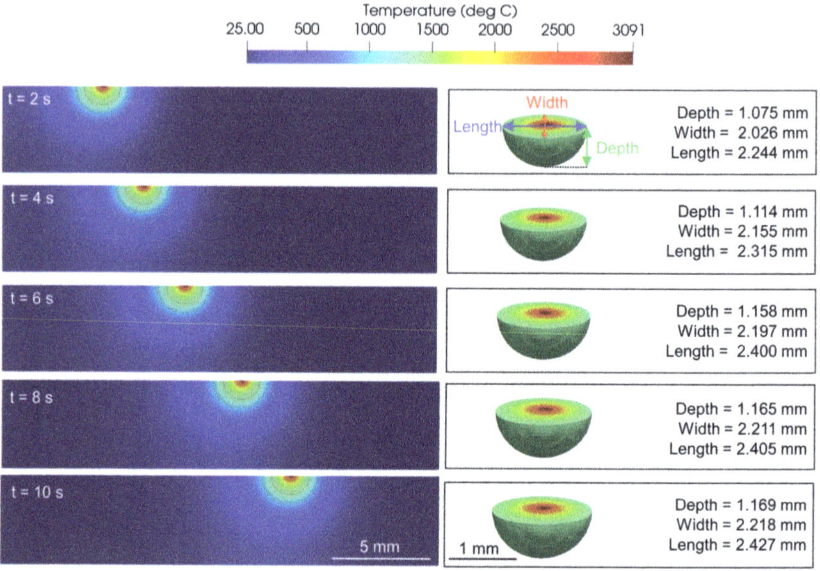

Figure 7. ZX cross section of the model simulations and the corresponding extracted isovolume to its right, showing the temporal variation of melt pool dimensions P of 1000 W and scan velocity v of 1 mm/s.

3.1.3. Comparison of Melt Pool Properties—Eagar-Tsai vs. Netfabb®

While in practice, the ranges of P and v depend on the experimental apparatus, a higher scan velocity will require a larger domain in the scan direction to ensure that the melt pool has reached a steady-state, which, consequently, will increase the cost of simulation of the Netfabb® model. The relationships between the the melt pool geometry (e.g., length, depth, and length to depth ratio) and the scan velocity over a wide range of power is shown in Figure 8a. The steady state melt pool length and depth are calculated using Netfabb® for CMSX-4®. This near-linear behavior indicates that the conclusions obtained from the current study will be applicable to other domain sizes as well. It is found that the simulation domain size of 22 mm × 6 mm × 4 mm accommodates the P range of (300 W–1300 W) and v range of (0.1 mm/s–1.5 mm/s) selected for the process parameter space, X_{space}.

The LF Eager-Tsai model rapidly conducts simulations over the process parameter space compared to the HF Netfabb® model. Figure 8b shows the time taken by each LF and HF model, and the ratio of time taken by the HF model to the LF model for each discrete time instants of the melt pool simulation. In this study, the cost of the Eagar-Tsai model is the time taken to solve the temperature over a three-dimensional domain that has been divided into 100 equal divisions in each direction. The cost of the Netfabb® model pertains to the total simulation time for the final mesh selected after the mesh convergence study.

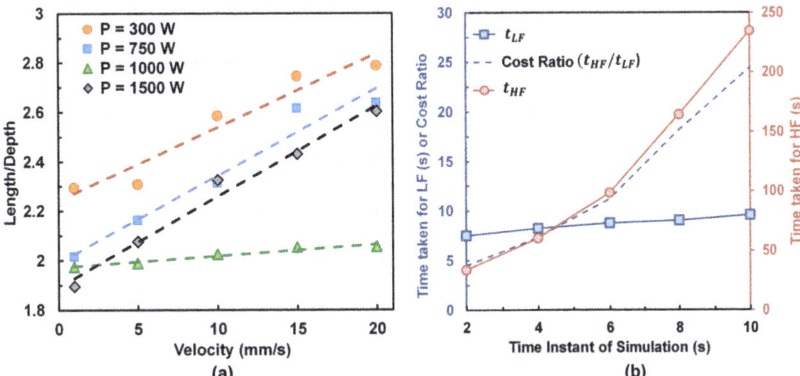

Figure 8. (**a**) The length to depth ratio of the melt pool for different combinations of laser power and velocity. Similar relationships are also observed for the melt pool length and depth and, hence, not shown here for brevity. (**b**) Variation of the cost ratio of HF to LF model at several discrete time instants of the melt pool simulation. Here t_{LF} is the time taken by LF simulations and t_{HF} is the time taken by HF simulations.

3.2. Design of Forward Process Maps

The overall methodology for developing the forward process maps is shown in Figure 9. The first step toward developing a process map is to perform an Latin Hypercube Sampling (LHS) [57] in the input data space, X_{space}. Each input data point is a combination of (P, v) values. Each output data point is the corresponding melt pool depths, d_H (depth obtained from HF model) and d_L (depth obtained from LF model).

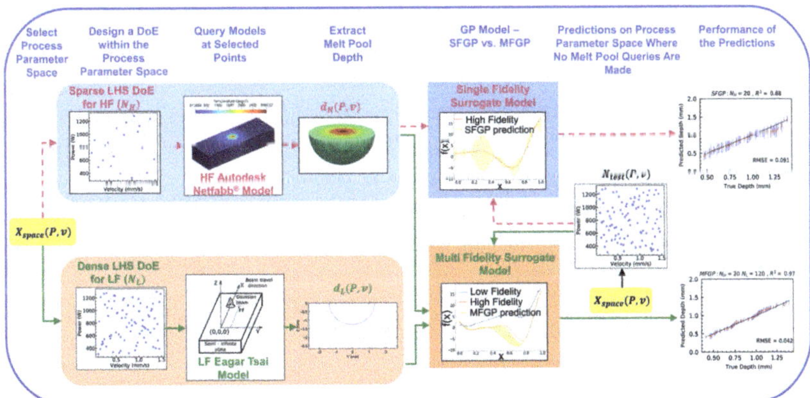

Figure 9. Framework followed for developing the forward process maps at discrete time instants using SFGP/MFGP. The solid green line indicates the path of the MFGP regression while the dashed magenta line indicates the path of the SFGP regression.

The MFGP surrogate is, thereafter, formulated as follows:

- LHS is employed to select N_L input data points in X_{space} for which the LF model is used to obtain the steady-state melt pool depths.
- Similarly, N_H input data points are selected in X_{space} through a separate LHS, for which the melt pool depths are obtained from the HF model.
- Using N_L LF input-output data points and N_H HF input-output data points, the MFGP surrogate is trained using the maximum likelihood estimation [12] at each discrete time instants of simulation.

LHS is adopted in this work because it is one of the most commonly used statistical methods for DoE [57], which, by virtue of its high sampling efficiency, is capable of generating a good spread of the initial input data points over the process parameter space within a limited number of iterations. The prediction performance of the trained MFGP surrogate is carried out on a set of N test input points (X_{test}) in the process parameter space (i.e., $X_{test} \in X_{space}$). The predicted values are, then, compared with the true melt pool depths (d_{true}) which refer to the outputs from the HF Autodesk Netfabb® model at X_{test}. When the number of LF simulations is 0, the MFGP surrogate essentially becomes an SFGP.

3.2.1. Performance Metrics

To investigate the performance of GP surrogates, the following evaluation metrics are used:

- Root Mean Square Error, $RMSE \triangleq \sqrt{\frac{1}{N}\sum_{i=1}^{N}|d_{pred_\mu}(i) - d_{true}(i)|^2}$. Here, $d_{pred_\mu}(i)$ denotes the mean of the posterior predictive Gaussian distribution at a test input indexed i in X_{test}. $d_{true}(i)$ denotes the true melt pool depth at the same test input and N is the number of test inputs. Additionally, $RMSE_{avg} \triangleq RMSE/50$, when 50 different DoE initializations are used.
- Total σ of prediction = $\sum_{i=1}^{N}\sigma_{pred}(i)/50$. Here, $\sigma_{pred}(i)$ is the standard deviation of the posterior predictive Gaussian distribution at a test input indexed with i in X_{test}.

3.2.2. Effect of Adding LF Predictions

Each initialization of the MFGP and SFGP surrogates involves a unique choice of the HF and LF input-output data points for the surrogate formulation. This is achieved through the generation of unique input data points for each fidelity model using LHS.

Since HF models are typically computationally expensive, it is of critical interest to investigate if adding more LF simulations to an MFGP surrogate with a limited number of HF simulations can improve the prediction performance. Figure 10 shows the effect of adding LF simulations to an MFGP surrogate for which the number of HF simulations (N_H) is constant. Since, in practice, the predictions are often dependent on the training data choice [58], the regression performance of the MFGP and SFGP surrogates reported in Figure 10 are based on the average performance over multiple initializations of the initial DoE of the respective surrogates. An intentional choice of low N_H values is adopted since it is challenging to fit a reasonably good MFGP surrogate when N_H is low [40].

Adding data from the LF model results in the reduction of both $RMSE_{avg}$ and total σ of the prediction. A reduction in $RMSE_{avg}$ corresponds to more accurate predictions of the melt pool depth, while a reduction in σ indicates higher confidence in the MFGP surrogate predictions [40]. For $N_H = 20$, Figure 10c shows a 55% reduction in $RMSE_{avg}$, and 61% reduction in total σ for $N_L/N_H = 6$, as compared to the SFGP surrogate ($N_L/N_H = 0$).

A similar trend is also observed for the other two cases of fixed N_H data. The highest reduction in $RMSE_{avg}$ and σ occurs for $N_H = 20$ as N_L is increased. This is due to the presence of a larger amount of HF data in the corresponding MFGP surrogate which is expected to reduce $RMSE_{avg}$. But, even for the case with $N_H = 5$, the reductions in $RMSE_{avg}$ and total σ are found to be 35% and 71%, respectively. This demonstrates the capability of the MFGP surrogate to efficiently incorporate MF information to improve the predictive performance as well as to achieve higher confidence in the predictions in a Bayesian setting, without increasing the total simulation cost significantly.

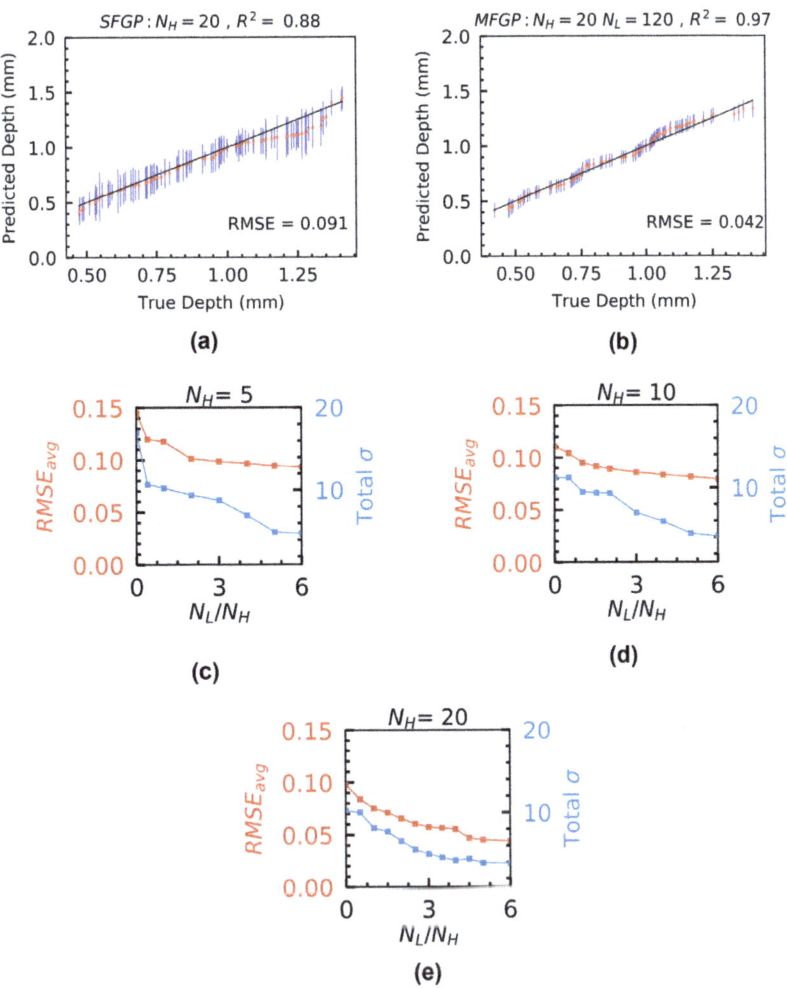

Figure 10. Parity plots for (**a**) SFGP with $N_H = 20$ and (**b**) MFGP with $N_H = 20, N_L = 120$. Variation of $RMSE_{avg}$ and Total σ as a function of N_L by keeping N_H fixed at (**c**) 5 (**d**) 10, and (**e**) 20. N_L / N_H has been varied from 0 to 6. The performance of MFGP in predicting the melt pool depth is evaluated on a test set, X_{test} of 100 randomly sampled points from X_{space}. The results reported are the corresponding averages of $RMSE$ and σ over 50 different initializations of the MFGP.

3.2.3. Effect of Adding HF Predictions

The effect of adding HF data to MFGP surrogates is shown in Figure 11. N_L is kept fixed at 80 for all MFGPs. The error probability distribution indicates a significant increase in prediction accuracy as N_H is increased from 20 to 40, with almost 60% of the predictions falling in the bin corresponding to the smallest absolute prediction error for $N_H = 40$. This probability is calculated as $p[0\ mm \leq |d_{pred_\mu} - d_{true}| \leq 0.025\ mm]$. The probability shows a slight increase with N_H = 60, but the overall mean prediction performance appears to saturate after N_H = 40, when a coefficient of determination, R^2 = 0.99 and $RMSE$ of 0.029 are achieved. Adding more HF data shows little effect in the mean prediction, as evinced by the respective R^2 and $RMSE$ scores. The uncertainty in prediction, as expected, progressively decreases with the addition of HF data.

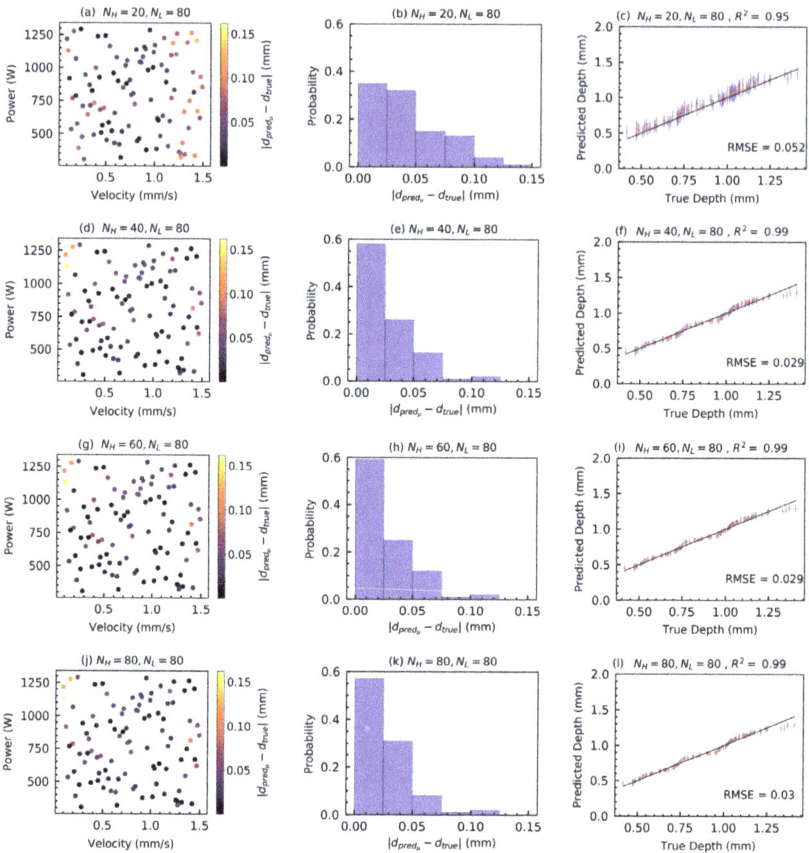

Figure 11. Effects of adding HF data to MFGP surrogates. (**a**,**d**,**g**,**j**) Variation of the absolute error $|d_{pred_\mu} - d_{true}|$ over X_{test} as a function of N_H. N_H is varied from 20 to 80, in steps of 20, while keeping N_L constant at 80. Each circle on the figure indicates the error at a test input data point in X_{test}. (**b**,**e**,**h**,**k**) Histograms approximating the probability distribution of the absolute errors in the prediction over X_{test}. The bin probabilities indicate the fraction of points in the test set for which the absolute values of the prediction error lie within the respective bin limits. (**c**,**f**,**i**,**l**) Parity plots comparing the true depths at X_{test} (sorted in an increasing order of magnitude) with the corresponding depths predicted by the respective MFGPs, along with the RMSEs of prediction. The predicted depth, being probabilistic, is represented by a filled circle indicating the mean (μ), and vertical bars indicating the associated uncertainty in prediction ($\mu \pm 1.0\sigma_{pred}$). The individual figure captions also include the coefficient of determination, R^2 in the predicted mean melt pool depths for the respective MFGP surrogates.

Figure 12 shows the predicted melt pool depths at X_{test} along with the prediction error % and the standard deviation associated with it obtained via an MFGP with N_H = 40 and N_L = 80 that results in the high R^2 and low $RMSE$ values reported for the same case in Figure 11f. The results show that a maximum number of 40 HF input-output data points and 80 LF input-output data points is enough to develop a robust MFGP surrogate to predict the melt pool properties (e.g., melt pool depth) for the case studied in this work.

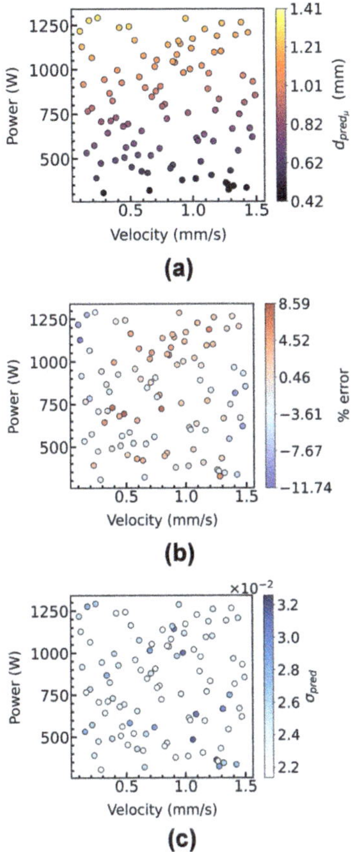

Figure 12. Comparison of (**a**) the mean predicted melt pool depth with (**b**) % error calculated as $\left(\frac{d_{pred_\mu} - d_{true}}{d_{true}}\right) \times 100\%$, and (**c**) standard deviation, σ_{pred} associated with each prediction, for MFGP with $N_H = 40$ and $N_L = 80$ on X_{test} of 100 randomly sampled input data points from X_{Space}.

NLML is minimized using the L-BFGS optimization scheme. Figure 13 shows the NLML convergence curves for SFGP and MFGP for each case of N_H where the MFGP surrogate is augmented with 80 LF points. The SFGP surrogate is found to be optimized with a lower number of iterations compared to the MFGP surrogate, for all cases of N_H. However, the MFGP surrogate converges to a better optimum for cases with a lower number of HF predictions, i.e, $N_H = 20, 40$, and 60. A lower value of NLML corresponds to a better optimum and, hence, an improved prediction by the surrogate. The MFGP convergence curves are less susceptible to variations in the number of HF predictions, unlike SFGP where large disparities in the optimum values are observed with a change in N_H. Naturally, SFGP converges to a better optimum for $N_H = 80$ since it is being trained with more HF predictions, that are closer to the true values, without any influence from the less accurate LF predictions. The results bolster the robustness and reliability of MFGP especially in the data-scarce regime.

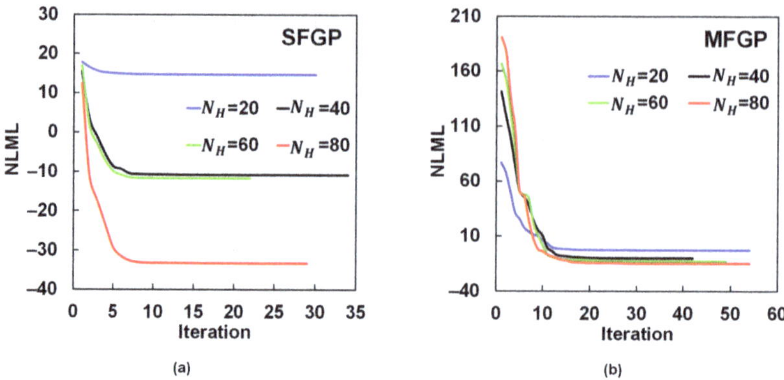

Figure 13. Comparison of convergence of NLML: (**a**) SFGP for 20, 40, 60, and 80 HF points and (**b**) MFGP for 20, 40, 60, and 80 HF points integrated with 80 LF points.

3.3. Design of Inverse Process Maps

While the forward process maps including UQ are extremely critical in identifying the effects of process parameters on the melt pool depths, it is not sufficient for intelligent process planning. AM is a spatio-temporally evolving process because of which the thermal conditions continuously change as a part is built-in layers [37]. This can cause significant disparities in the microstructures and properties of the first and the last layers [59]. The melt pool depths vary across different layers and tracks due to the progressive addition of thermal energy to the part during the build process. Non-uniform melt pool depths in AM are widely reported in the open literature involving experimental [60,61] as well as numerical [37,62] investigations. The development of inverse process maps is, therefore, a critical requirement for achieving a consistent melt pool depth throughout the build process, even when the thermal conditions change continuously [63]. The inverse process maps are developed by solving an optimization problem that searches for process parameters to maintain the melt pool depth at the desired value during a single-layer single-track L-DED process. Since an integration of the HF model with an optimization tool will be computationally expensive, this paper proposes to develop the inverse process maps in an MF setting, whereby the MFGP surrogates are used to solve an optimization problem under a limited (pre-defined) computational budget.

The MFGP approach described in the previous subsection is extended in the setting of BO to minimize an objective function: $J(t,P,v) \triangleq \frac{|d(t,P,v)-d^*|}{|d^*|}$, where $d(t,P,v)$ is the melt pool depth obtained at discrete time instants during a representative build process for a given choice of process parameters (P,v), and d^* is the desired melt pool depth. The total duration of the build process is discretized into time intervals Δt, during which the process parameters are kept constant at the optimized values of the previous time step. Making the Δt finer would potentially allow for a smoother variation of process parameters during the build process. It is easily understood that $J(t,P,v)$ is minimized when $d(t,P,v) = d^*$ at all time instants under consideration, and hence, solving the optimization problem amounts to finding process parameters (P,v) that maintain the melt pool depth close to d^*.

The MFGP-BO optimization algorithm is schematically shown in Figure 14 and formulated as follows:

1. First, an initial MFGP surrogate is learned with N_H number of HF input-output data points and N_L number of LF input-output data points. LHS is employed to select N_L input data points in the process parameter space, X_{space}, for which the LF model is used to obtain the melt pool depths. Similarly, N_H input data points are selected in the process parameter space, $X^* \in X_{space}$, through a separate LHS, for which the melt pool depths are obtained from the HF model. It is to be noted that $N_H << N_L$.

2. A prospective HF input data point is selected from the search space X_{space}, using the EI acquisition function [48], and the corresponding output data point is obtained from the HF model. This input-output data is added to the initial MFGP surrogate.
3. The MFGP surrogate is then retrained, and a prospective LF input data point is selected from X^* using the GP-MI acquisition function [49], and the corresponding output data point obtained from the LF model. Thereafter, this input-output data is added to the MFGP surrogate, followed by another step of surrogate retraining.
4. This sequential selection of new HF and LF data, followed by surrogate retraining (Steps 2 and 3) is performed until the optimization budget expires. The optimization budget limitation is manifested by a maximum allowable N_{opt}, denoting the number of sequential optimization steps that are allowed to be performed, and is a user-defined parameter.

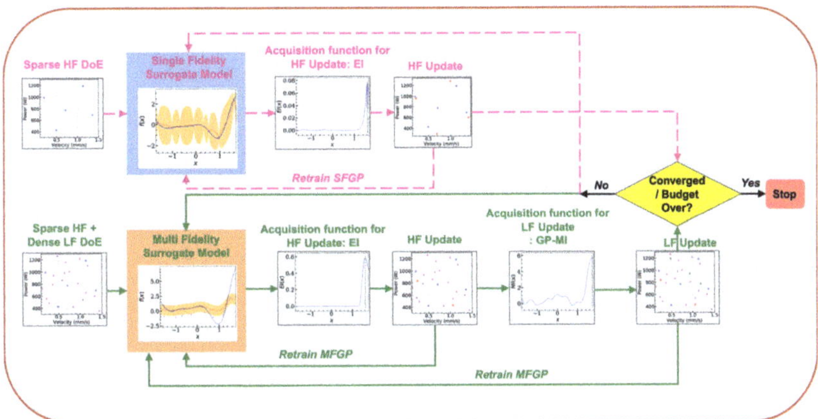

Figure 14. Framework of the MFGP and the SFGP Bayesian Optimization routines. The solid green line indicates the path of MFGP-BO and the dashed magenta line indicates the path of SFGP-BO.

The optimization framework involving the SFGP surrogate is similar, albeit it only involves the HF model as the sole fidelity level. Hence, the SFGP optimization algorithm starts with N_H number of HF data for training the surrogate. LHS is employed to select N_H input data points in the process parameter space, X_{space}, for which the HF model is used to obtain the melt pool depths. The selection of new HF input data in $X^* \in X_{space}$ is performed using the EI acquisition function in BO. The output data is obtained from the HF model. Thereafter, the input-output data is added back to the initial SFGP surrogate for retraining it, and the sequence continues till the optimization budget expires. Similar to the regression studies, the optimizations are also carried out over 50 initializations, and the results reported are the average over these initializations. Since the MFGP surrogate is used for optimization, analyzing its performance over multiple initializations ensures that the performance metrics do not reflect a bias inadvertently introduced by the choice of the initial MFGP surrogate. For example, if an initial MFGP caters to a local cluster in the search space of process parameters, the surrogate would likely be poor in other areas of the input domain. One way to counter that problem is to have an initial input-output training data set that is well spread out, e.g., using separate LHSs for selecting N_H and N_L input points from X_{space}. However, if the initial MFGP surrogate contains input points that are close to the true optimal point, there may be a tendency for the optimization routine to converge quickly to the global optima.

3.3.1. Performance Metrics

The MFGP optimization framework is compared with its SFGP counterpart with respect to computational savings and the quality of the optimized process parameters. The optimization performance of both are based on the average performance over multiple initializations of the initial DoE to avoid any bias arising from the choice of training data.The optimization routine is executed only if the initial MFGP/ SFGP surrogate has no (P, v) input for which the melt pool depth is within $(d^* \pm \epsilon)$ for a pre-defined $\epsilon > 0$. This ensures that no optimization routine gets accidentally *fortunate* with an initial LHS input data point yielding close to the desired melt pool depth. The performance metrics are: (1) Fraction of optimization budget consumption (χ_{budget}) and (2) Quality Improvement (QI). The metrics are defined as follows:

1. $\chi_{\text{budget}} = \frac{N^*/N_T}{N_{\text{opt}}}$, where $N^* = \sum_{i=1}^{N_T} N_{d_i \sim d^*}$. Here, $N_{d_i \sim d^*}$ indicates the optimization iteration number at which the obtained depth d_i is closest to d^* in absolute norm, N_T = number of initializations out of 50 for which the optimization routine is executed, according to the ϵ criterion described above. Hence, N^*/N_T indicates the average number of optimization steps required to obtain the process parameters for which the melt pool depth is closest to the desired value. Normalizing N^*/N_T with N_{opt}, thus, reflects the fraction of the optimization budget that is consumed.
2. $QI = \frac{RMSE_{\text{SFGP}} - RMSE_{\text{MFGP}}}{RMSE_{\text{SFGP}}} \times 100\%$ where, $RMSE_{\text{MFGP/SFGP}} = \sqrt{\frac{\Sigma(d_i - d^*)^2}{N_T}}$ where d_i is the melt pool depth (mm) closest to d^* in absolute norm obtained within N_{opt} iterations for each initialization, d^* is the desired depth (mm). The subscripted MFGP/SFGP denote the $RMSE$ obtained from the MFGP/SFGP surrogates respectively. QI is a measure of comparing the process parameter combinations obtained from MFGP-BO and SFGP-BO with respect to the closeness of the respective melt pool depths to the desired depth.

3.3.2. Optimization Performance—SFGP-BO vs. MFGP-BO

The comparison of QIs among different surrogates (Figure 15) shows that the quality of the optimized design points obtained from MFGP tends to be better than those obtained from SFGP. QI is the highest with $N_H = 5$ and the lowest with $N_H = 20$ for all time steps. This is expected since adding HF data points in the initial DoE allows the SFGP surrogate to learn the response surface better in the input-output space, which results in better predictions from the SFGP optimization. This analysis shows that MFGP surrogates can significantly improve the optimization performance particularly in the scarce-data regime.

Figure 16 shows the comparison of the optimization performance between MFGP and SFGP. Low values of initial N_H ($N_H < 40$) are selected for this comparison, since it is previously observed from Figure 11 that $N_H = 40$ points result in a highly accurate fit in the input-output space under consideration, and, hence, is expected to perform well in the optimization phase. The true potential of the optimization algorithm is, therefore, tested when the initial HF information is not significant enough to start with a good response surface of the objective function. Such a scarcity of data is a much closer representation of the real-world design optimization tasks when dealing with very expensive process models. From Figure 16a–c, it is observed that the MFGP surrogate results in lower χ_{budget} for all N_H selections at several discrete time instants. This indicates, on an average, around 12% reduction in the consumption of the optimization budget is observed for the cases investigated in this paper. This indicates the benefit of integrating LF information through MFGP surrogates to identify optimal points faster than SFGP surrogates that solely use the HF information.

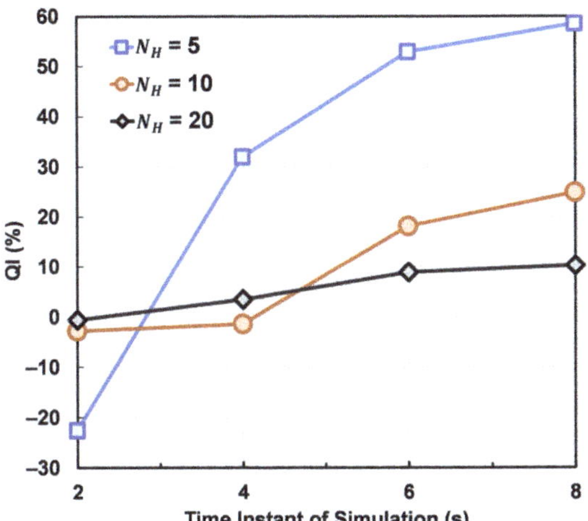

Figure 15. Quality improvement (QI) compared for the three N_H values. $N_{\text{opt}} = 5$ has been chosen for all optimization exercises.

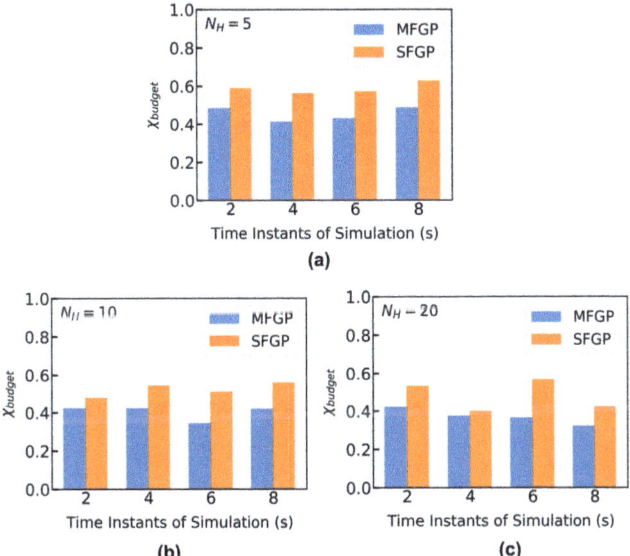

Figure 16. Optimization performance metrics for maintaining the melt pool depth at $d^* = 1$ mm. $d^* = 1$ mm is selected based on the work by Toyserkani and Khajepour [61]. The search space, X^* consists of 1000 randomly sampled input data points from X_{space}. For all MFGP surrogates, $N_L/N_H = 4$ is maintained. The optimization is performed starting from $t = 2$ s till $t = 8$ s at every 2 s time interval. The maximum time is 8 s since the melt pool depth reaches a steady-state value within 8 s for the selected geometry and the process parameter range. The optimization is performed when there are no initial melt pool depth values in the initial MFGP surrogate within 1 ± 0.03 mm ($\epsilon = 30$ μm). A comparison of χ_{budget} for MFGP and SFGP with (**a**) $N_H = 5$ (**b**) $N_H = 10$ and (**c**) $N_H = 20$. $N_{\text{opt}} = 5$ is chosen for all optimization exercises.

4. Conclusions and Future Work

This paper has developed a methodology using MFGP and MFGP-BO for designing forward and inverse temporal process maps in L-DED. The continuous changes in the melt pool geometry are predicted by a low-cost MFGP surrogate developed by integrating two melt pool simulation models. The LF model is based on the analytical Eagar-Tsai's model while the HF model is based on Autodesk Netfabb®'s FEM model. The uncertainties associated with the predictions of the melt pool depths are quantified using GPs. It is demonstrated that MFGP predictions are more accurate and have higher confidence than SFGP. Once the temporal forward process maps are developed, MFGP and SFGP are coupled with BO for developing the inverse process maps. These maps are used to estimate the process parameters required to achieve the desired melt pool depth. The BO algorithm minimizes an appropriate objective function that quantifies the deviation of the melt pool depth from the desired value under computational budget constraints to yield the optimal process parameters under varying thermal conditions.

The reliability of the optimization algorithm, however, depends on the fundamental physics addressed by the models. While the HF models are more capable in resolving the fundamental physics, the computational cost involved in the optimization process can be significantly high. For example, the cost of running a Netfabb® L-DED model for a single-track and single-layer simulation as described in this paper is $\sim$10 times higher than the Eagar-Tsai's model at t = 2 s, and $\sim$150 times higher at t = 10 s. The MFGP-BO, where the HF model is integrated with the LF model, thus fairs better than the SFGP-BO by reducing the computational overhead by 12% percent without compromising on the quality of the optimized process parameters. Such a benefit will continue to increase for larger domain sizes having multi-layer multi-track depositions. Hence, this algorithm is particularly conducive for process planning purposes in data-scarce regimes.

The demonstrations of MFGP and MFGP-BO are presented, in this paper, for designing the forward and inverse temporal process maps, respectively, incorporating UQ for the melt pool depth. However, the formulation is flexible to accommodate for other properties such as the residual stress or the mean grain size as long as multiple process models having different fidelities exist. The formulation can also incorporate more than two levels of fidelity by appropriately modifying the covariance matrix which determines the correlations among the different levels of fidelity [15]. The realization of such a formulation has the potential to reduce the requirement for extensive computational investigation toward the development of sophisticated model-based feedforward and feedback control strategies [64] in and beyond L-DED AM. Additional investigations are also planned in the future as summarized below:

1. Developing multi-dimensional process maps that include other process parameters such as scan spacing, powder feed rate, and build plate temperature.
2. Augmenting the present two-fidelity surrogate by incorporating experimental data that would serve as the highest-fidelity level (true values).
3. Using the MF surrogate for constrained optimization e.g., estimating the optimal parameters for controlling the melt pool depth while being constrained to maintain the desired microstructure (e.g., % of equiaxed or columnar grain morphology).
4. Formulating an MF framework that allows for the inclusion of heterogeneous input spaces across different fidelities. For example, the HF model can take multiple process parameters e.g., scan pattern, hatch spacing, etc. while the LF model can incorporate only the primary process parameters, P and v. Optimization with such different input parameter space needs special methods such as heterogeneous transfer learning [65] to learn from a common subspace of the inputs.

Author Contributions: Conceptualization, A.B. and S.M.; methodology, N.M., S.M. and A.B.; software, N.M. and S.M.; validation, N.M. and S.M.; formal analysis, N.M. and S.M.; investigation, N.M. and S.M.; resources, A.B.; data curation, N.M.; writing—original draft preparation, N.M., S.M. and A.B.; writing—review and editing, A.B.; visualization, N.M. and S.M.; supervision, A.B.; project administration, A.B.; funding acquisition, A.B. All authors have read and agreed to the published version of the manuscript.

Funding: The work reported in this paper is funded in part by the Pennsylvania State University, PA 16802, USA and in part by the U.S. Army Engineer Research and Development Center through Contract Number W912HZ21C0001. Any opinions, findings, and conclusions in this paper are those of the authors and do not necessarily reflect the views of the supporting institution.

Institutional Review Board Statement: Not applicable.

Informed Consent Statement: Not applicable.

Data Availability Statement: The datasets generated during and/or analyzed during the current study are available from the corresponding author on reasonable request.

Acknowledgments: The authors would like to thank Michael Gouge, Research Engineer, Autodesk for his help with Netfabb®. The authors are indebted to Chris Hirsh, IT Consultant, Department of Mechanical Engineering for his help with Netfabb® installation and remote operation.

Conflicts of Interest: The authors declare no conflict of interest.

Abbreviations

The following abbreviations are used in this manuscript:

AL	Active Learning
AM	Additive Manufacturing
BO	Bayesian Optimization
DoE	Design of Experiments
EI	Expected Improvement
FEM	Finite Element Modeling
GP	Gaussian Process
HF	High Fidelity
L-BFGS	Limited Memory Broyden-Fletcher-Goldfarb-Shanno
L-DED	Laser-Directed Energy Deposition
LF	Low Fidelity
LHS	Latin Hypercube Sampling
MF	Multi Fidelity
MFGP	Multi Fidelity Gaussian Process
MFGP-BO	Multi Fidelity Gaussian Process—Bayesian Optimization
MI	Mutual Information
NLML	Negative Log Marginal Likelihood
SF	Single Fidelity
SFGP	Single Fidelity Gaussian Process
SFGP-BO	Single Fidelity Gaussian Process—Bayesian Optimization
UQ	Uncertainty Quantification
Nomenclature	
α_L	Absorptivity of laser beam
$\delta(\mathbf{x})$	Discrepancy function of $\mathbf{x}$
ρ	Correlation function
$\boldsymbol{\theta_1}, \boldsymbol{\theta_2}$	Hyperparameters of the covariance
q	Heat flux
$\mu(\mathbf{x})$	Mean function of $\mathbf{x}$
χ_{budget}	Fraction of the optimization budget that is consumed
cholesky(A)	Cholesky decomposition: L is a lower triangular matrix such that $LL^T = A$
ρ_p	Material density
ϵ	Pre-defined tolerance in optimized value
$\sigma(\mathbf{x})$	Variance function of $\mathbf{x}$
σ_L	Distribution parameter

σ_{pred}	Standard deviation of the posterior predictive Gaussian distribution of the melt pool depth
${\sigma_\epsilon}^2, {\sigma_{\epsilon_1}}^2, {\sigma_{\epsilon_2}}^2$	Noise variance
a_p	Thermal diffusivity of material
c_p	Specific heat of material
d	Melt pool depth
d_i	Melt pool depth closest to desired depth at ith initialization for which optimization routine is executed
d^*	Desired melt pool depth
d_L, d_H	Predicted melt pool depth from LF and HF models, respectively
d_{true}	True melt pool depth
$d_{pred\,\mu}$	Mean of the posterior predictive Gaussian distribution of the melt pool depth
$E(\mathbf{y})$	Expectation of $\mathbf{y}$
$f(x)$	Gaussian process function values, $f = (f(x_1), \ldots, f(x_n))$
f_{x^*}	Gaussian process (posterior) prediction (random variable)
$\hat{f}_{x^*}$	Gaussian process posterior mean
h_{eff}	Effective heat transfer coefficient
h_{forced}	Forced convection heat transfer coefficient
h_{free}	Free convection heat transfer coefficient
$h_{\text{radiation}}$	Radiation convection heat transfer coefficient
J	Objective function for BO
$k, k(\mathbf{x}, \mathbf{x}')$	Kernel functions of GPs
k_2, k_1	Kernel functions of HF and LF GPs
$\mathbf{K}$	Covariance matrix
k_p	Thermal conductivity of material
N_L, N_H	Number of LF points, Number of HF points
N_{opt}	Maximum allowable optimization iterations
N_T	Number of initializations for which the optimization routine is executed
N^*	Sum of optimization iteration numbers at which the obtained depth is closest to d^* in absolute norm
$N_{d_i \sim d^*}$	Optimization iteration number at which the obtained depth d_i is closest to d^* in absolute norm
P	Laser power
P^{trn}, v^{trn}	Training input data: Laser power, velocity
P^{tst}, v^{tst}	Test input data: Laser power, velocity
QI	Quality Improvement
$RMSE$	Root mean square error
$RMSE_{avg}$	RMSE averaged over all initializations
$RMSE_{MFGP}, RMSE_{MFGP}$	RMSE calculated for MFGP-BO,RMSE calculated for MFGP-BO
R^2	Coefficient of Determination
t	Time
t'	Dummy integration variable
t_{LF}, t_{HF}	Time taken by LF model (s), Time taken by HF model (s)
T_0	Initial temperature
Δt	Time step
T_s	Surface temperature
$T(x_c, y_c, z_c, t)$	Temperature as a function of coordinates (x_c, y_c, z_c) and time (t)
T_∞	Ambient temperature
T_L	Liquidus temperature
v	Laser scan velocity
Σ^{tst}	Covariance of the posterior Gaussian distribution at test input
μ^{tst}	Mean of the posterior Gaussian distribution at test input
$\mathbf{x}, \mathbf{x}'$	Input variables

$\hat{\mathbf{x}}$	Value of $\mathbf{x}$ that maximizes an objective function
$\mathbf{x}^{trn}, \mathbf{x}^{tst}$	Training input, Test input
$\mathbf{x}_2, \mathbf{x}_1$	Inputs to HF and LF models
$\mathbf{X}$	Combined input to a GP consisting of $\mathbf{x}_2, \mathbf{x}_1$
X_{Space}	Process parameter space
X_{Test}	Test space
X^*	Search space for optimization
$\mathbf{y}, \mathbf{y}'$	Output of a GP
$\mathbf{Y}$	Combined output of a GP consisting of $\mathbf{y}_2, \mathbf{y}_1$
$\mathbf{y}_2, \mathbf{y}_1$	Outputs from HF and LF models
$\mathbf{y}^{trn}, \mathbf{y}^{tst}$	Training output, Test output
$\mathbb{V}$	Variance
$\mathcal{N}$	Gaussian distribution
$\alpha, \mathbf{\Upsilon}, \psi_1\, \psi_2, \beta$	Intermediate parameters in Algorithm 1

References

1. Basak, A.; Das, S. Epitaxy and microstructure evolution in metal additive manufacturing. *Annu. Rev. Mater. Res.* **2016**, *46*, 125–149. [CrossRef]
2. Sun, S.; Liu, Q.; Brandt, M.; Janardhana, M.; Clark, G. Microstructure and Mechanical Properties of Laser Cladding repair of AISI 4340 Steel. In Proceedings of the 28th Int Congress of the Aeronautical Sciences, Brisbane, Australia, 23–28 September 2012; Volume 6.
3. Kumar, L.J.; Nair, C.K. Laser metal deposition repair applications for Inconel 718 alloy. *Mater. Today Proc.* **2017**, *4*, 11068–11077. [CrossRef]
4. Kistler, N.A.; Corbin, D.J.; Nassar, A.R.; Reutzel, E.W.; Beese, A.M. Effect of processing conditions on the microstructure, porosity, and mechanical properties of Ti-6Al-4V repair fabricated by directed energy deposition. *J. Mater. Process. Technol.* **2019**, *264*, 172–181. [CrossRef]
5. Zheng, B.; Haley, J.; Yang, N.; Yee, J.; Terrassa, K.; Zhou, Y.; Lavernia, E.; Schoenung, J. On the evolution of microstructure and defect control in 316L SS components fabricated via directed energy deposition. *Mater. Sci. Eng. A* **2019**, *764*, 138243. [CrossRef]
6. Sampson, R.; Lancaster, R.; Sutcliffe, M.; Carswell, D.; Hauser, C.; Barras, J. The influence of key process parameters on melt pool geometry in direct energy deposition additive manufacturing systems. *Opt. Laser Technol.* **2021**, *134*, 106609. [CrossRef]
7. Zuback, J.; Knapp, G.; Palmer, T.; DebRoy, T. Deposit geometry and oxygen concentration spatial variations due to composition change in printed functionally graded components. *Int. J. Heat Mass Transf.* **2021**, *164*, 120526. [CrossRef]
8. Jones, K.; Yang, Z.; Yeung, H.; Witherell, P.; Lu, Y. Hybrid Modeling of Melt Pool Geometry in Additive Manufacturing Using Neural Networks. In Proceedings of the International Design Engineering Technical Conferences and Computers and Information in Engineering Conference, Virtual, 17–19 August 2021; Volume 85376, p. V002T02A031.
9. Wei, H.; Mukherjee, T.; Zhang, W.; Zuback, J.; Knapp, G.; De, A.; DebRoy, T. Mechanistic models for additive manufacturing of metallic components. *Prog. Mater. Sci.* **2020**, *116*, 100703. [CrossRef]
10. Johnson, L.; Mahmoudi, M.; Zhang, B.; Seede, R.; Huang, X.; Maier, J.T.; Maier, H.J.; Karaman, I.; Elwany, A.; Arróyave, R. Assessing printability maps in additive manufacturing of metal alloys. *Acta Mater.* **2019**, *176*, 199–210. [CrossRef]
11. Beuth, J.; Fox, J.; Gockel, J.; Montgomery, C.; Yang, R.; Qiao, H.; Soylemez, E.; Reeseewatt, P.; Anvari, A.; Narra, S.P.; et al. Process Mapping for Qualification Across Multiple Direct Metal Additive Manufacturing Processes. In Proceedings of the 2013 International Solid Freeform Fabrication Symposium, Austin, TX, USA, 12–14 August 2013.
12. Perdikaris, P.; Venturi, D.; Royset, J.O.; Karniadakis, G.E. Multi-fidelity modelling via recursive co-kriging and Gaussian-Markov random fields. *Proc. R. Soc. A Math. Phys. Eng. Sci.* **2015**, *471*, 20150018. [CrossRef]
13. Wang, X.; Liu, Y.; Sun, W.; Song, X.; Zhang, J. Multidisciplinary and Multifidelity Design Optimization of Electric Vehicle Battery Thermal Management System. *J. Mech. Des.* **2018**, *140*, 094501. [CrossRef]
14. Wang, G.G.; Shan, S. Review of Metamodeling Techniques in Support of Engineering Design Optimization. *J. Mech. Des.* **2006**, *129*, 370–380. [CrossRef]
15. Kennedy, M.; O'Hagan, A. Predicting the Output from a Complex Computer Code When Fast Approximations Are Available. *Biometrika* **2000**, *87*, 1–13. [CrossRef]
16. Han, Z.H.; Görtz, S. Hierarchical Kriging Model for Variable-Fidelity Surrogate Modeling. *AIAA J.* **2012**, *50*, 1885–1896. [CrossRef]
17. Le Gratiet, L. Multi-Fidelity Gaussian Process Regression for Computer Experiments. Ph.D Thesis, Université Paris-Diderot—Paris VII, Paris, France, 2013.
18. Gratiet, L.L.; Garnier, J. Recursive co-kriging model for Design of Computer experiments with multiple levels of fidelity. *arXiv* **2014**, arXiv:1210.0686.
19. Perdikaris, P.; Venturi, D.; Karniadakis, G. Multifidelity Information Fusion Algorithms for High-Dimensional Systems and Massive Data sets. *SIAM J. Sci. Comput.* **2016**, *38*, B521–B538. [CrossRef]

20. Sen, R.; Kandasamy, K.; Shakkottai, S. Multi-Fidelity Black-Box Optimization with Hierarchical Partitions. In Proceedings of the 35th International Conference on Machine Learning, Stockholm, Sweden, 10–15 July 2018; Dy, J., Krause, A., Eds.; PMLR: Stockholmsmässan: Stockholm Sweden, 2018; Volume 80, pp. 4538–4547.
21. Kandasamy, K.; Dasarathy, G.; Schneider, J.; Póczos, B. Multi-fidelity Bayesian Optimisation with Continuous Approximations. In Proceedings of the 34th International Conference on Machine Learning, Sydney, Australia, 6–11 August 2017; Precup, D., Teh, Y.W., Eds.; PMLR: International Convention Centre: Sydney, Australia, 2017; Volume 70, pp. 1799–1808.
22. Perdikaris, P.; Karniadakis, G.E. Model inversion via multi-fidelity Bayesian Optimization: A new paradigm for parameter estimation in haemodynamics, and beyond. *J. R. Soc. Interface* **2016**, *13*, 20151107. [CrossRef]
23. Peherstorfer, B.; Willcox, K.; Gunzburger, M. Survey of Multifidelity Methods in Uncertainty Propagation, Inference, and Optimization. *SIAM Rev.* **2018**, *60*, 550–591. [CrossRef]
24. Netfabb®—Software for 3D Printing. Available online: http://www.netfabb.com/ (accessed on 31 May 2021).
25. Eagar, T.; Tsai, N. Temperature fields produced by traveling distributed heat sources. *Weld. J.* **1983**, *62*, 346–355.
26. Shahriari, B.; Swersky, K.; Wang, Z.; Adams, R.P.; de Freitas, N. Taking the Human Out of the Loop: A Review of Bayesian Optimization. *Proc. IEEE* **2016**, *104*, 148–175. [CrossRef]
27. Rasmussen, C.E.; Williams, C.K. *Gaussian Processes for Machine Learning (Adaptive Computation and Machine Learning)*; The MIT Press: Cambridge, MA, USA, 2005.
28. Rosenthal, D. Mathematical theory of heat distribution during welding and cutting. *Weld. J.* **1941**, *20*, 220–234.
29. Gan, Z.; Yu, G.; He, X.; Li, S. Numerical simulation of thermal behavior and multicomponent mass transfer in direct laser deposition of Co-base alloy on steel. *Int. J. Heat Mass Transf.* **2017**, *104*, 28–38. [CrossRef]
30. Tsai, M.; Kou, S. Marangoni convection in weld pools with a free surface. *Int. J. Numer. Methods Fluids* **1989**, *9*, 1503–1516. [CrossRef]
31. Wolff, S.J.; Gan, Z.; Lin, S.; Bennett, J.L.; Yan, W.; Hyatt, G.; Ehmann, K.F.; Wagner, G.J.; Liu, W.K.; Cao, J. Experimentally validated predictions of thermal history and microhardness in laser-deposited Inconel 718 on carbon steel. *Addit. Manuf.* **2019**, *27*, 540–551. [CrossRef]
32. Huang, Y.; Khamesee, M.B.; Toyserkani, E. A comprehensive analytical model for laser powder-fed additive manufacturing. *Addit. Manuf.* **2016**, *12*, 90–99. [CrossRef]
33. Anderson, T.; DuPont, J.; DebRoy, T. Origin of stray grain formation in single-crystal superalloy weld pools from heat transfer and fluid flow modeling. *Acta Mater.* **2010**, *58*, 1441–1454. [CrossRef]
34. Mundra, K.; DebRoy, T.; Kelkar, K. Numerical prediction of fluid flow and heat transfer in welding with a moving heat source. *Numer. Heat Transf. A Part Appl.* **1996**, *29*, 115–129. [CrossRef]
35. Kamara, A.; Wang, W.; Marimuthu, S.; Li, L. Modelling of the melt pool geometry in the laser deposition of nickel alloys using the anisotropic enhanced thermal conductivity approach. *Proc. Inst. Mech. Eng. Part B J. Eng. Manuf.* **2011**, *225*, 87–99. [CrossRef]
36. Michaleris, P. Modeling metal deposition in heat transfer analyses of additive manufacturing processes. *Finite Elem. Anal. Des.* **2014**, *86*, 51–60. [CrossRef]
37. Mondal, S.; Gwynn, D.; Ray, A.; Basak, A. Investigation of Melt Pool Geometry Control in Additive Manufacturing Using Hybrid Modeling. *Metals* **2020**, *10*, 683. [CrossRef]
38. Gouge, M.; Michaleris, P.; Denlinger, E.; Irwin, J. The finite element method for the thermo-mechanical modeling of additive manufacturing processes. In *Thermo-Mechanical Modeling of Additive Manufacturing*; Elsevier: Oxford, UK, 2018; pp. 19–38. [CrossRef]
39. Gouge, M.F.; Heigel, J.C.; Michaleris, P.; Palmer, T.A. Modeling forced convection in the thermal simulation of laser cladding processes. *Int. J. Adv. Manuf. Technol.* **2015**, *79*, 307–320. [CrossRef]
40. Sarkar, S.; Mondal, S.; Joly, M.; Lynch, M.E.; Bopardikar, S.D.; Acharya, R.; Perdikaris, P. Multifidelity and Multiscale Bayesian Framework for High-Dimensional Engineering Design and Calibration. *J. Mech. Des.* **2019**, *141*, 121001. [CrossRef]
41. Mondal, S.; Joly, M.; Sarkar, S. Multi-fidelity Global-Local Optimization of a Transonic Compressor Rotor. In Proceedings of the ASME Turbo Expo 2019: Turbomachinery Technical Conference and Exposition, American Society of Mechanical Engineers, Phoenix, AZ, USA, 17–21 June 2019. [CrossRef]
42. Duvenaud, D. Automatic Model Construction with Gaussian Processes. Ph.D. Thesis, Computational and Biological Learning Laboratory, University of Cambridge, Cambridge, UK, 2014.
43. Liu, D.; Nocedal, J. On the limited memory BFGS method for large scale optimization. *Math. Program.* **1989**, *45*, 503–528. [CrossRef]
44. Brochu, E.M.; Cora, V.; De Freitas, N. *A Tutorial on Bayesian Optimization of Expensive Cost Functions, with Application to Active User Modeling and Hierarchical Reinforcement Learning*; Technical Report TR-2009-23; Department of Computer Science, University of British Columbia: Vancouver, BC, Canada, 2010.
45. Jones, D.; Schonlau, M.; Welch, W. Efficient Global Optimization of Expensive Black-Box Functions. *J. Glob. Optim.* **1998**, *13*, 455–492. [CrossRef]
46. Streltsov, S.; Vakili, P. A Non-myopic Utility Function for Statistical Global Optimization Algorithms. *J. Glob. Optim.* **1999**, *14*, 283–298. [CrossRef]
47. Mockus, J. Application of Bayesian approach to numerical methods of global and stochastic optimization. *J. Glob. Optim.* **1994**, *4*, 347–365. [CrossRef]

48. Mockus, J. On Bayesian Methods for Seeking the Extremum. In Proceedings of the IFIP Technical Conference, Novosibirsk, Russia, 1–7 July 1974; Springer: London, UK, 1974; pp. 400–404.
49. Contal, E.; Perchet, V.; Vayatis, N. Gaussian Process Optimization with Mutual Information. In Proceedings of the 31st International Conference on Machine Learning, Beijing, China, 21–26 June 2014; Xing, E.P., Jebara, T., Eds.; PMLR: Bejing, China, 2014; Volume 32, pp. 253–261.
50. CMSX-4 Cannon Muskegon. Available online: https://cannonmuskegon.com/cmsx-4/ (accessed on 9 June 2021).
51. Gäumann, M.; Henry, S.; Cléton, F.; Wagnière, J.D.; Kurz, W. Epitaxial laser metal forming: Analysis of microstructure formation. *Mater. Sci. Eng. A* **1999**, *271*, 232–241. [CrossRef]
52. Chen, H.; Lu, Y.; Luo, D.; Lai, J.; Liu, D. Epitaxial laser deposition of single crystal Ni-based superalloys: Repair of complex geometry. *J. Mater. Process. Technol.* **2020**, *285*, 116782. [CrossRef]
53. Mesh Convergence using Autodesk Netfabb Simulation—Netfabb. Available online: https://blogs.autodesk.com/netfabb/2018/08/30/mesh-convergence-using-autodesk-netfabb-simulation/ (accessed on 9 June 2021).
54. Sente Software—JMatPro®. Available online: https://www.sentesoftware.co.uk/jmatpro (accessed on 9 June 2021).
55. Denlinger, E.R. Chapter 12—Development and Numerical Verification of a Dynamic Adaptive Mesh Coarsening Strategy for Simulating Laser Power Bed Fusion Processes. In *Thermo-Mechanical Modeling of Additive Manufacturing*; Gouge, M., Michaleris, P., Eds.; Butterworth-Heinemann: Oxford, UK, 2018; pp. 199–213. [CrossRef]
56. Ahrens, J.; Geveci, B.; Law, C. *ParaView: An End-User Tool for Large Data Visualization*; Visualization Handbook; Elsevier: Amsterdam, The Netherlands, 2005; ISBN: 978-0123875822.
57. McKay, M.D.; Beckman, R.J.; Conover, W.J. A Comparison of Three Methods for Selecting Values of Input Variables in the Analysis of Output from a Computer Code. *Technometrics* **1979**, *21*, 239–245. [CrossRef]
58. Parr, J.M.; Keane, A.J.; Forrester, A.I.; Holden, C.M. Infill sampling criteria for surrogate-based optimization with constraint handling. *Eng. Optim.* **2012**, *44*, 1147–1166. [CrossRef]
59. Menon, N.; Mahdi, T.H.; Basak, A. Microstructure of IN738LC Fabricated Using Laser Powder Bed Fusion Additive Manufacturing. *J. Turbomach.* **2022**, *144*, 031011. [CrossRef]
60. Angel, N.M.; Basak, A. On the Fabrication of Metallic Single Crystal Turbine Blades with a Commentary on Repair via Additive Manufacturing. *J. Manuf. Mater. Process.* **2020**, *4*, 101. [CrossRef]
61. Toyserkani, E.; Khajepour, A. A mechatronics approach to laser powder deposition process. *Mechatronics* **2006**, *16*, 631–641. [CrossRef]
62. Liu, P.; Wang, Z.; Xiao, Y.; Horstemeyer, M.F.; Cui, X.; Chen, L. Insight into the mechanisms of columnar to equiaxed grain transition during metallic additive manufacturing. *Addit. Manuf.* **2019**, *26*, 22–29. [CrossRef]
63. Wang, L.; Felicelli, S.; Gooroochurn, Y.; Wang, P.; Horstemeyer, M. Optimization of the LENS® process for steady molten pool size. *Mater. Sci. Eng. A* **2008**, *474*, 148–156. [CrossRef]
64. Sammons, P.M.; Bristow, D.A.; Landers, R.G. A Model Predictive Repetitive Process Control Formulation for Additive Manufacturing Processes. In Proceedings of the ASME 2015 Dynamic Systems and Control Conference, Columbus, OH, USA, 28–30 October 2015; p. V002T32A003.
65. Day, O.; Khoshgoftaar, T.M. A survey on heterogeneous transfer learning. *J. Big Data* **2017**, *4*, 29. [CrossRef]

 materials

Article

Heat Treatments of Metastable β Titanium Alloy Ti-24Nb-4Zr-8Sn Processed by Laser Powder Bed Fusion

Maxwell Hein [1,2,*], Nelson Filipe Lopes Dias [3], Sudipta Pramanik [1], Dominic Stangier [3], Kay-Peter Hoyer [1,2], Wolfgang Tillmann [3] and Mirko Schaper [1,2]

1 Chair of Materials Science (LWK), Paderborn University, Warburger Str. 100, 33098 Paderborn, Germany; pramanik@lwk.upb.de (S.P.); hoyer@lwk.upb.de (K.-P.H.); schaper@lwk.upb.de (M.S.)
2 DMRC—Direct Manufacturing Research Center, Paderborn University, Mersinweg 3, 33100 Paderborn, Germany
3 Institute of Materials Engineering (LWT), TU Dortmund University, Leonhard-Euler-Str. 2, 44227 Dortmund, Germany; filipe.dias@tu-dortmund.de (N.F.L.D.); dominic.stangier@tu-dortmund.de (D.S.); wolfgang.tillmann@udo.edu (W.T.)
* Correspondence: hein@lwk.upb.de; Tel.: +49-5251-60-5447

Abstract: Titanium alloys, especially β alloys, are favorable as implant materials due to their promising combination of low Young's modulus, high strength, corrosion resistance, and biocompatibility. In particular, the low Young's moduli reduce the risk of stress shielding and implant loosening. The processing of Ti-24Nb-4Zr-8Sn through laser powder bed fusion is presented. The specimens were heat-treated, and the microstructure was investigated using X-ray diffraction, scanning electron microscopy, and transmission electron microscopy. The mechanical properties were determined by hardness and tensile tests. The microstructures reveal a mainly β microstructure with α″ formation for high cooling rates and α precipitates after moderate cooling rates or aging. The as-built and α″ phase containing conditions exhibit a hardness around 225 HV5, yield strengths (YS) from 340 to 490 MPa, ultimate tensile strengths (UTS) around 706 MPa, fracture elongations around 20%, and Young's moduli about 50 GPa. The α precipitates containing conditions reveal a hardness around 297 HV5, YS around 812 MPa, UTS from 871 to 931 MPa, fracture elongations around 12%, and Young's moduli about 75 GPa. Ti-24Nb-4Zr-8Sn exhibits, depending on the heat treatment, promising properties regarding the material behavior and the opportunity to tailor the mechanical performance as a low modulus, high strength implant material.

Keywords: laser powder bed fusion; Ti-24Nb-4Zr-8Sn; titanium alloy; biomedical engineering; mechanical properties; microstructure; X-ray diffraction

Citation: Hein, M.; Lopes Dias, N.F.; Pramanik, S.; Stangier, D.; Hoyer, K.-P.; Tillmann, W.; Schaper, M. Heat Treatments of Metastable β Titanium Alloy Ti-24Nb-4Zr-8Sn Processed by Laser Powder Bed Fusion. *Materials* **2022**, *15*, 3774. https://doi.org/10.3390/ma15113774

Academic Editors: Wai Yee Yeong and Swee Leong Sing

Received: 19 April 2022
Accepted: 22 May 2022
Published: 25 May 2022

1. Introduction

Nowadays, there is still a lack of individualized implants, facing safety and risk concerns [1]. Individually adapted implants may reduce revision surgeries and rehabilitation times. They can also restore joint kinetics and improve implant fixation during healing [2]. Patient-specific, customized implants can significantly increase the success rate of the medical treatment of a patient [3,4]. Additive manufacturing (AM) and, specifically, laser powder bed fusion (LPBF) provide the possibility to manufacture these customized implants [5]. LPBF is a powder-based AM technology in which parts are built layer by layer using a laser beam melting the specified area. This layer-wise fabrication of parts leads to geometrical freedom, with few design constraints, and the possibility to manufacture, economically, down to a batch size of one [6,7].

Titanium and titanium alloys are receiving increasing attention in different industries such as aerospace and especially medical industries, due to their superior mechanical performance, as well as their excellent corrosion resistance and biocompatibility [8–10]. For most applications and biomedical use cases, (α + β) phase titanium alloys, such as Ti-6Al-4V or Ti-6Al-7Nb, are used with the main disadvantage of a relatively high Young's modulus around 110 GPa, which is over 3 to 4 times higher compared to the cortical human bone [11]. Titanium alloys are often used for load-bearing implants, where this mismatch between the implant and surrounding bone leads to stress shielding [12]. The difference in stiffness leads to a reduced loading of the bone. Stress shielding is an adaption of bone cells to varying stress states in the bone and, therefore, may result in bone density reduction and implant loosening [13]. Recent studies reflect the development of titanium alloys towards low elastic moduli accompanied by high strengths, and non-toxic and non-allergic elements [14–17]. Trying to increase the biocompatibility by substituting the toxic elements aluminum and vanadium, and still obtaining sufficient mechanical properties, the β titanium alloy Ti-24Nb-4Zr-8Sn (Ti2448) was developed. The Ti2448 alloy exhibits a low Young's modulus of 42 GPa to 55 GPa and a high strength of 800 MPa to 1200 MPa, depending on processing and heat treatment routes [18–21]. The biocompatibility and corrosion behavior are improved through the substitution of the hazardous alloying elements with non-toxic elements such as niobium, zirconium, and tin. Ti2448 is a promising alloy, containing non-toxic, biocompatible elements, but is also compromised in terms of microstructural features and mechanical properties, due to the high concentration of β phase stabilizing elements. Analogous to pure titanium, the α phase has a hexagonal closed packed (hcp) structure at ambient temperature. Above 882 °C, titanium consists of body-centered cubic (bcc) structures, which is the β phase. The β phase becomes stable and can be maintained in the metastable state below 882 °C by adding β stabilizers. The amount of β stabilizer to obtain purely β phases at ambient temperature depends on the molybdenum equivalency Mo_{eq}, which is an empirical rule derived from an analysis of binary titanium alloys, and can be described as follows [22–26]:

$$Mo_{eq} = 1.0\,Mo + 0.67\,V + 0.44\,W + 0.28\,Nb + 0.22\,Ta + 1.6\,Cr + 2.5\,Fe\ldots - 1.0\,Al\ [wt.\ \%]. \quad (1)$$

In general, the addition of 10 wt. % molybdenum suppresses the transformation from β to α upon quenching. Below a Mo_{eq} of 10, the alloy is metastable [22,26]. The effect of molybdenum is, on the one hand, a reduction in the critical cooling rate to retain the β phase and, on the other hand, a significant reduction in the martensitic start temperature $M_{S\alpha'}$ [25]. Bania's equation for the Mo_{eq} includes a factor for aluminum. The aluminum equivalent Al_{eq} is added to reflect the tendency to support the formation of the α phase concerning the α stabilizers aluminum, tin, oxygen, and nitrogen as follows [26,27]:

$$Al_{eq} = 1.0\,Al + 0.17\,Zr + 0.33\,Sn + 10\,O + 10\,N\ [wt.\ \%]. \quad (2)$$

In the condition of energy disturbance, such as heat treatment and deformation, the metastable bcc β phase decomposes into the hcp α phase, the hcp α′ martensite phase, the orthorhombic α″ martensite, or/and the ω phase [27–32]. The α′ martensite formation in titanium alloys, in general, results in increasing tensile strength and hardness accompanied by decreasing plasticity [33,34]. The formation of α″ martensite under rapid quenching tends to reduce hardness, tensile, and fatigue strength [18,35–38]. Additionally, the ω phase is strengthening the titanium alloy in a certain volume fraction range, but also can lead to embrittlement [39,40]. Heat treatments of metastable β titanium alloys show promising features to improve and enhance the mechanical properties and are considered for different fields of applications, including aerospace or medical applications [22,26,41,42].

Various titanium alloys have already been manufactured by LPBF, such as Ti-6Al-4V [43–45] and Ti-6Al-7Nb [46–50]. As described, these alloys obtain high elastic moduli compared to human bones, which could lead to stress shielding, and contain, in part, hazardous elements such as aluminum and vanadium, which are referred to allergic reactions, neurotoxic effects, and probably Alzheimer's disease [51–53]. Therefore, in this study, the low-modulus metastable β titanium alloy Ti2448 is manufactured by LPBF and investigated. The complex phase transformation in combination with the LPBF is unknown and these investigations are to contribute to understanding the mechanisms. Subsequent heat treatments are applied to modify and adjust the mechanical properties. To tailor the material behavior and to achieve sufficient mechanical properties, different heat treatments are conducted. Conclusively, the effects of different microstructures in the as-built and heat-treated conditions on the mechanical behavior is determined.

2. Materials and Methods

The Ti2448 powder was gas atomized by GfE Metalle und Materialien GmbH (Nürnberg, Germany) and examined concerning particle size distribution with a Mastersizer 2000 (Malvern Panalytical GmbH, Kassel, Germany) using laser diffraction. The powder is mainly spherical, see Figure 1a. It has a nominal particle size distribution between 28.5 µm (D_{10}) and 73.9 µm (D_{90}) with a log-transformed normal distribution centered at 45.9 µm (D_{50}). The chemical composition of the initial Ti2448 powder was determined by the Revierlabor (Chemische Laboratorien für Industrie und Umwelt GmbH, Essen, Germany) by X-ray fluorescence analysis, combustion analysis with infrared detection, and standard carrier gas hot extraction. The measured composition is compared to the target values and the LPBF manufactured specimens. The chemical composition of the as-built specimens was determined using the Bruker Q4 Tasman optical emission spectrometer (OES, Bruker AXS GmbH, Karlsruhe, Germany).

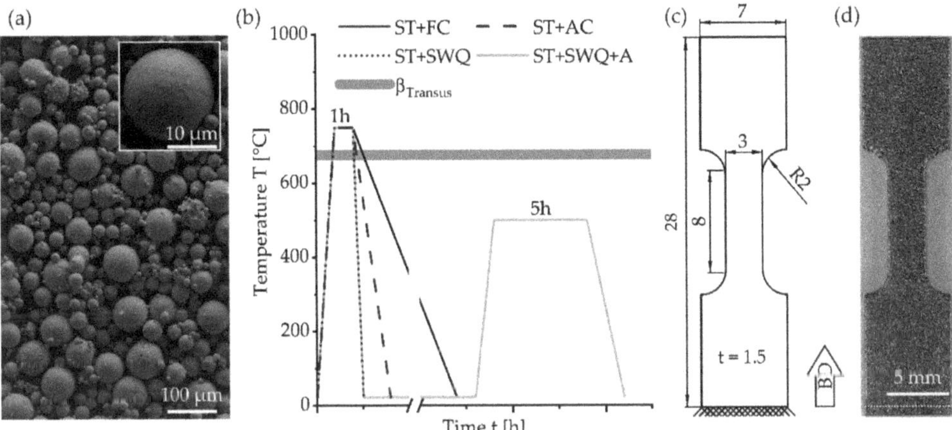

Figure 1. (**a**) SEM image of the morphology of initial Ti2448 powder, inset showing a spherical powder particle in higher magnification; (**b**) schematic overview of the heat treatments (ST = solution treatment, FC = furnace cooling, AC = air cooling, SWQ = slow water quenching in a glass ampule, A = aging) as well as the $\beta_{Transus}$ temperature; (**c**) geometry in reference to the building direction (BD) of the tensile specimens; (**d**) image of an as-built sample, support structure below the white dashed line.

The specimens were manufactured using an LT12 SLM machine (DMG MORI AG, Bielefeld, Germany) in an inert argon atmosphere containing less than 1000 ppm oxygen to minimize oxidation of the molten pool and the risk of introducing contaminants. The machine is equipped with a 400 W fiber laser with a spot size of 35 µm. Specimens were manufactured using a laser power P = 225 W, with a laser scanning speed v = 1.5 m s^{-1} and hatch distance h = 0.1 mm to obtain a relative density of more than 99.5%. As the contour parameters, a laser power P_c = 150 W and a scanning speed v_c = 0.4 m s^{-1} were applied. The layer thickness was kept constant at 50 µm, while the layer-wise rotation of the scanning vectors of 67°, and 5 mm stripes were applied. The building platform was preheated to 200 °C.

The as-built specimens were studied and compared to heat-treated specimens to analyze the differences in microstructure and mechanical properties. For identification of the effects of heat treatments, four different treatments were conducted in evacuated glass ampules in a Nabertherm furnace N41/13 (Nabertherm GmbH, Lilienthal, Germany). Solution treatment (ST) was performed at 750 °C for 1 h for all heat treatments followed by different cooling rates. Furnace cooling (FC), air cooling (AC), and slow water quenching (SWQ) of the glass ampules in water were applied. Additional specimens were aged at 500 °C for 4 h after SWQ (SWQ+A) with subsequent AC, see Figure 1b [54].

The crystalline phases were analyzed by X-ray diffraction (XRD) using a D8 Advance diffractometer (Bruker AXS GmbH, Karlsruhe, Germany) equipped with a polycapillary parallel X-ray lens of 2 mm and a LynxEye silicon strip detector (Bruker AXS GmbH, Karlsruhe, Germany). The specimens were ground with sandpaper (grain size 2500) and then vibration polished for 24 h on a VibroMet (Buehler, ITW Test & Measurement GmbH, Düsseldorf, Germany). The XRD measurements were performed with Cu-Kα1 radiation (λ = 1.5406 Å) at an acceleration voltage of 40 kV and a current of 40 mA. The diffractograms were obtained in Bragg-Brentano geometry over a scanning range of 2θ within 30° and 90° with a scan step of $\Delta 2\theta = 0.034°$ and an exposure time of 1 s. In addition, the lattice parameters of the hcp α, orthorhombic α'', and bcc β phases were determined.

For microstructural and hardness investigations, the specimens were ground and vibration polished. KOH-solution (32% H_2O; 8% H_2O_2; 60% KOH (40%)) was used for etching the specimens from 2.5 min to 5 min. The etched specimens were investigated with a light microscope (LiMi) Zeiss Axiophot (Carl Zeiss AG, Oberkochen, Germany). Powder morphology and microstructure were examined with a scanning electron microscope (SEM) Zeiss Ultra Plus (Carl Zeiss AG, Oberkochen, Germany).

Further examination of the microstructure was performed using transmission electron microscopy (TEM). Thin foils were prepared. Slices of ≈400 μm thickness were cut employing the Struers Sectom-5 (Struers GmbH, Willich, Germany). The slices were ground using SiC abrasive papers to a thickness of ≈100 μm. Then, 3 mm diameter discs were punched out of the slices. The discs were further thinned with Struers Tenupol-5 (Struers GmbH, Willich, Germany) using an electrolyte containing 5% perchloric acid-methanol solution at a voltage of 21 V, a current of 16 mA, and at a solution temperature of −23 °C. TEM was performed using a JEOL JEM-ARM 200F (JOEL Ltd., Tokyo, Japan). TEM, high-resolution TEM (HRTEM), high angle annular dark-field scanning TEM (HAADF-STEM), and energy dispersive spectroscopy (EDS) was performed. EDS maps were measured on selected regions of the HAADF-STEM images with a 5 nm step size and 2 s dwell time per step.

The Vickers hardness was measured on surfaces perpendicular to the build direction with a hardness tester KB 30 FA (KB Prüftechnik GmbH, Hochdorf-Assenheim, Germany) according to Vickers HV5. A minimum of two specimens per condition were tested with 30 hardness indentions across the surface of each specimen.

The monotonic tensile tests were performed utilizing a servo-hydraulic test-rig MTS 858 Tabletop System (MTS Systems Corporation, Eden Prairie, MN, USA) equipped with a 20 kN load cell and an extensometer 632.29F-30 (MTS Systems Corporation, Eden Prairie, MN, USA). The design of the miniature dogbone specimens is according to DIN EN ISO 6892-1, but does not fulfill the proportional specimen criterium [55]. The loading direction was parallel to the building direction (BD). The geometry and the BD are depicted in Figure 1c. An as-built sample is shown in Figure 1d. The tensile test procedure corresponded to a displacement-controlled execution with a crosshead speed of 1.5 mm min^{-1} according to DIN EN ISO 6892-1 [55].

3. Results and Discussion

3.1. Chemical Composition

The chemical composition of the initial Ti2448 powder was determined and compared to the target values and the LPBF manufactured specimens, see Table 1. The initial powder niobium content is slightly higher than prescribed, whereas the tin content is slightly lower.

Due to the different melting temperatures of the alloying elements, it is challenging to obtain the desired chemical configuration of the initial powder. The contents of zirconium and tin in the as-built specimens are in the prescribed range, the content of niobium is above 9 wt. %. The chemical composition could be affected by the vaporization of the alloying elements during the fabrication process [56–60]. Nevertheless, the oxygen content could be increased due to residual oxygen in the process chamber and the oxygen affinity of titanium at elevated temperatures [61,62]. Based on Equations (1) and (2), the Mo_{eq} is calculated for the initial powder state to be 2.94. Based on the chemical composition of the alloy, one may assume that the alloy is a metastable β to β rich titanium alloy [22,26].

Table 1. Chemical composition of initial powder and LPBF manufactured Ti2448 alloy in wt. %.

Condition	Ti	Nb	Zr	Sn	O	Mo_{eq}
Target Value	Bal.	23.5–24.5	3.5–4.5	7.5–8.5	<0.12	2.32
Initial Powder	Bal.	25.2	4.2	7.1	0.11	2.94
as-built	Bal.	>9.0	3.5	8.2	0.37	-

3.2. Phase Constituents and Microstructure

The microstructure of Ti2448 highly depends on the processing route and the heat treatment. The XRD diffractograms of the LPBF processed Ti2448 in the initial powder, as-built, and heat-treated conditions are shown in Figure 2. It can be observed that Ti2448 mainly consists of the cubic β phase (space group $Im\bar{3}m$) and, depending on the condition, also contains low amounts of the hexagonal α phase (space group $P6_3/mmc$), orthorhombic α″ martensite (space group *Cmcm*), or presumably hexagonal ω phase (space group $P6/mmm$). Compared to the β phase, the Bragg reflections of the remaining phases are broader, indicating a high degree of microstrain (deformed crystal lattice) and/or small grain sizes [63,64]. The Ti2248 powder contains the α″ phase next to the β phase. For Ti2248 in the as-built state, the Bragg reflections of the β (200) and β (211) planes show a shoulder formation towards higher angles, indicating a superposition with another Bragg reflection. At these angles, the Bragg reflections may originate from the ω phase. A clear assignment is not possible due to the overlapping and broad reflections. Contrarily, the heat-treated Ti2248 alloys possess either the α″ or α phase. The microstructure and subsequent transformation behavior are highly sensitive to the chemical composition and various cooling rates from the β phase field. In general, metastable β decomposes in the α (hcp), the α′ martensite (hcp), the α″ martensite (orthorhombic), and the ω phases [27–30]. The transformation can be tailored by employing heat treatments, thereby controlling the content of β stabilizers in the β phase through the cooling conditions [64–66]. For niobium concentrations less than 13 wt. % in Ti-Nb alloys, the α′ martensite is usually present [37]. Therefore, the Ti2248 alloy with its higher niobium content is expected to contain the ω or α″ phase inside a β matrix. With decreasing temperature, the instability of the bcc lattice in β phase alloys increases. As described by Moffat and Larbalestier for Ti-Nb alloys, the α″ and ω phases compete to evolve in an unstable lattice [67]. The β phase decomposition is determined by the quench rate, while the two decomposition modes to α″ or ω phase are mutually exclusive, due to the different transformation variants in metastable titanium alloys, as observed by Duering et al. [68]. For higher cooling rates, the formation of α″ martensite is favorable, while material cooled with lower rates comprise β and ω phase. The lower the cooling rate, the more complete is the collapse of the $\{222\}_\beta$ planes and the larger the ω precipitates should be [67]. It is assumed that the cooling condition in the LPBF process fosters the formation of the ω phase. As already proposed by Qi et al., a continuous diffusional-displacive β → α″ → α transformation may be present [69]. High cooling rates, such as in SWQ, AC, and during the powder fabrication process, promote the α″ phase formation, while for slow cooling rates such as in FC or for aging treatments, the α″ phase was not detected, see Figure 2. The α″ phase can be regarded as the intermediate stage in between the transformation from bcc to hcp. In the as-built condition, only minor reflections of the α″ phase are observable and, in addition, some of the peaks of other phases are broadened, probably due to the overlay with the α″ phase peaks, see black arrows in Figure 2. It is assumed, that due to insufficient undercooling during the process and/or decomposition through in situ heat treatments, the formation of α″ martensite is inhibited, suppressed, or reversed [34,70,71]. Heat treatments, including slow cooling or aging, enable the decomposition of α″ and the precipitation of the α phase [72]. In summary, slow cooling from the β phase-field, as well as aging, results in the formation of α precipitates. Higher cooling rates result in the martensitic formation of the α″ phase, while in the as-built state, at intermediate cooling rates and/or in-process heat treatment, the formation of the ω and α″ phases seems to be supported, which is mutually competitive.

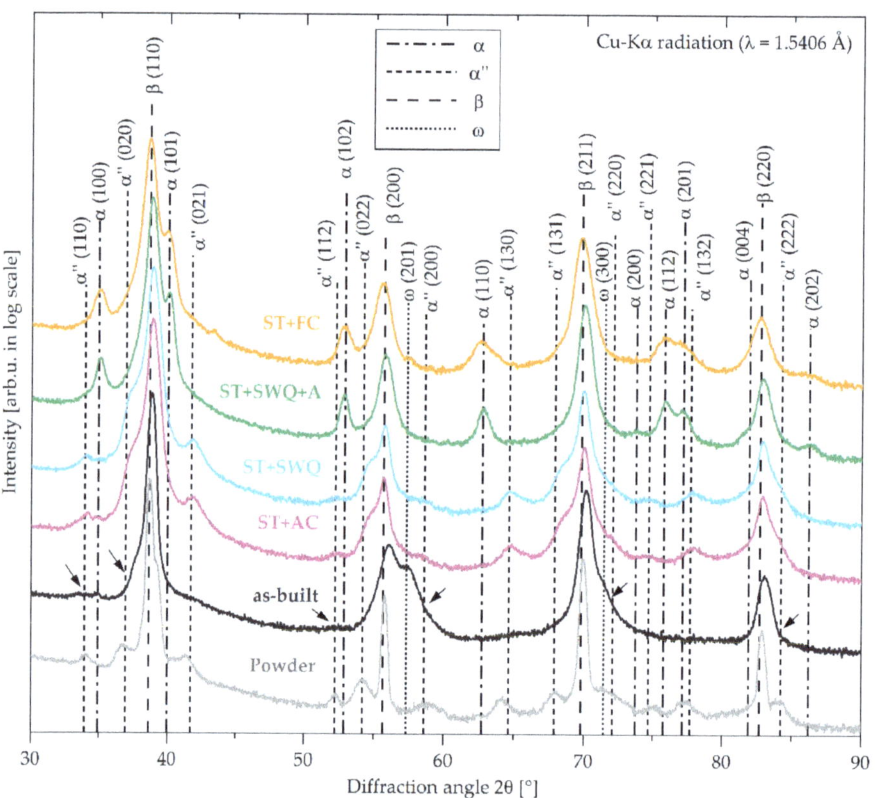

Figure 2. X-ray diffractograms of Ti2448 in as-built, ST+FC, ST+AC, ST+SWQ, and ST+SWQ+A conditions as well as the initial powder; black arrows indicating overlay of α″ phase peaks in the as-built condition; not depicted in the diagram: α″ orthorhombic (111) (superposition with β (110)), α″ orthorhombic (113) (superposition with β (211)); α hexagonal (002) (superposition with β (110)), and α hexagonal (103) (superposition with β (211)).

The lattice parameters of the hcp α phase, orthorhombic α″ martensite, and bcc β phase of Ti2448 in different conditions were determined from the XRD diffractograms and are summarized in Table 2. For the hexagonal ω phase, the calculation of the lattice parameters was not possible due to the overlapping of the Bragg reflections. The lattice parameter of the β phase is in the range between 3.287 Å and 3.303 Å. Deviations can be caused by distortions of the lattices, interference between phases, different chemical compositions, and therefore various atom radii. As calculated by XRD, the α phase exhibits lattice parameters of a = 2.959 Å and c = 4.758 Å for the furnace cooled condition and a = 2.959 Å and c = 4.726 Å for the aged condition. The obtained values for the α and the β phase follow the results of previous studies [62,73]. The α phase lattice parameters of the furnace-cooled determined by TEM are a = 3.191 Å and c = 5.045 Å and deviate from the other values. The orthorhombic α″ phase shows lattice parameters of approximately a = 3.1 Å, b = 4.9 Å, and c = 4.7 Å for the XRD calculation and about a = 3.28 Å, b = 4.88 Å, and c = 4.617 Å for the TEM analysis, which slightly differ from reported values of deformation-induced α″ martensitic transformation [74,75]. It is assumed that the thermal formation of the α″ phase differs from mechanically induced transformations, leading to slightly different lattice parameters.

Table 2. Lattice parameters, determined by XRD, for hcp α, orthorhombic α″, and bcc β of Ti2448 in the as-built, ST+FC, ST+AC, ST+SWQ, and ST+SWQ+A condition as well as the initial powder; lattice parameters determined by TEM (marked in brackets).

Condition	Hcp α		Orthorhombic α″			Bcc β
	a = b [Å]	c [Å]	a [Å]	b [Å]	c [Å]	a = b = c [Å]
Initial Powder	-	-	3.107	4.924	4.716	3.298
as-built	-	-	-	-	-	3.287
as-built (TEM)	-	-	3.28	4.88	4.617	3.297
ST+FC	2.959	4.758	-	-	-	3.303
ST+FC (TEM)	3.191	5.045	-	-	-	3.294
ST+AC	-	-	3.110	4.860	4.716	3.294
ST+SWQ	-	-	3.121	4.871	4.679	3.295
ST+SWQ+A	2.959	4.726	-	-	-	3.294

The KOH-etched surfaces of the different conditions are depicted in Figure 3. Figure 3a,b demonstrate the as-built condition. β grain boundaries are observable in the LiMi and SEM images. Laths are visible in the grains and in the inset in (a), which is presumably referred to as the α″ martensitic phase. The ω phase was not detected. The air-cooled and water quenched conditions have a similar microstructure, see Figure 3c–f, respectively. Coarse grains are observed for both conditions. In contrast to the as-built condition, no α″ phase laths are observable in the microscopic images for the AC and SWQ condition, although the XRD measurements suggest this. Figure 3g,h show the LiMi and SEM images of the aged specimen. The prior β grains are not as clearly visible as in the other conditions. A very fine, diffuse microstructure is visible in the LiMi, which turned out to be α precipitates at a higher magnification in the SEM. Primary β grain boundaries are visible in the SEM image of the aged condition, see Figure 3h. The initial microstructure, such as in the as built condition, is visible in the furnace cooled specimen, see Figure 3i. Yet, the SEM close-up in Figure 3j indicates α precipitates similar to those in the aged specimens, but not as homogeneous distributed. As observed by Ren et al., aging has a significant effect on the precipitation behavior of the secondary α phase in titanium alloys [76]. The ST+FC specimens exhibit an irregular distribution and heterogeneous size of intra-granular α precipitates (white arrows) with grain boundary α (black arrows), see Figure 3j. An adapted heat treatment (ST+SWQ+A) can modify the microstructure to achieve a uniform size and orientation (approximately 60°, white triangle) of the acicular α precipitations, see Figure 3h. Varying the aging time and temperature can tailor the morphology of the intra-granular α precipitates and the grain boundary α phase can be coarsened and, therefore, adjust the mechanical behavior [76,77].

A uniform microstructure with grain sizes less than 50 nm was observed by Hao et al. and Li et al. in cold-rolled sheets of Ti2448 [18,20]. The microstructure of hot-forged and cold-rolled Ti2448 was investigated and compared by Li et al. [19]. The hot-forged material consists of equiaxed β grains with a grain size of around 5 μm, divided into subgrains with sizes less than 1 μm. Aging of the hot-forged alloy resulted in α phase precipitations with a needle-like shape. The cold-rolled Ti2448 has a β + α″ microstructure with coarse grains but also nanostructured regions. Yang et al. cold-rolled previous hot-forged Ti2448 cylinder and performed subsequent solution treatment at 900 °C with water quenching (ST) or flash treatment at 700 °C with air cooling (FT). The resulting microstructures were homogenous single β phase with grain sizes about 50 μm and 7 μm, respectively [78]. The microstructure of Ti2448 processed by warm swaging and warm rolling is presented by Hao et al. [79]. The hot-forged alloy showed an equiaxed β phase microstructure with grain sizes of 100 μm, consisting of equiaxed subgrains with sizes of around 1 μm. After warm swaging, the microstructure became swirled marble-like, with differences from the surface to the core. Subsequent warm rolling resulted in a homogeneous microstructure comprising β phase with grain sizes less than 200 nm and nanosized α phase. Zhang et al. observed a microstructure for as-hot-rolled Ti2448 with grains around tens of microns, containing subgrains of the sizes of hundreds of nanometers, and consisting of single β phase without the formation of ω phase or α″ martensite [21].

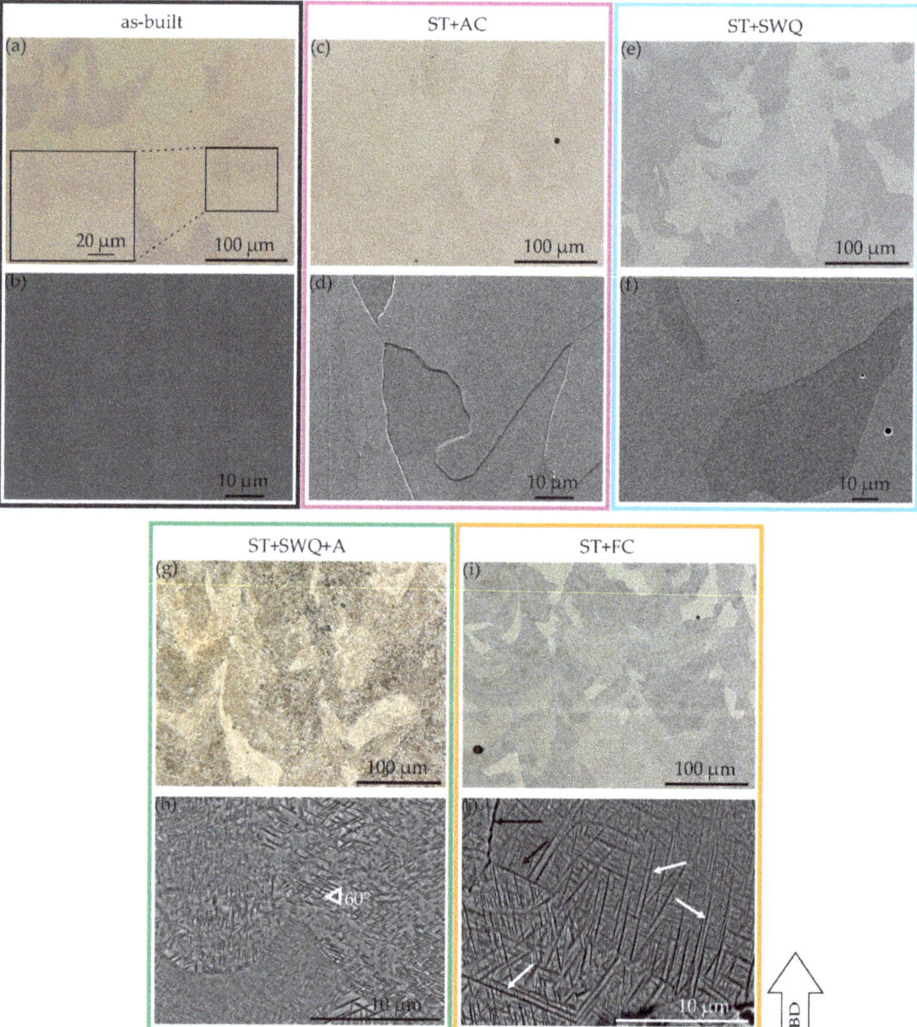

Figure 3. LiMi images of KOH-etched as-built (**a**) inset showing the black box in higher magnification), ST+AC (**c**), ST+SWQ (**e**), ST+SWQ+A (**g**), and ST+FC (**i**) conditions; and SEM images of KOH-etched as-built (**b**), ST+AC (**d**), ST+SWQ (**f**), ST+SWQ+A (**h**), and ST+FC (**j**) conditions; building direction (BD) for all conditions is indicated by the arrow.

Figure 4 shows the HAADF-STEM images and EDS maps of the as-built specimen. Figure 4a,b illustrate the nanostructure consisting of alternately α″ martensite and β titanium laths. Figure 4b is the magnification of the white square region in (a). The images prove the presence of parallel plates. The thickness of the plates varies between 42 and 85 nm. The plates are orthorhombic α″ martensite phase (bright laths) and the surrounding matrix is β phase (dark laths). The EDS maps for titanium, niobium, zirconium, and tin are depicted in Figure 4c–f and are taken from the white, dashed square in (b). The composition of the alloying elements is uniform in the α″ plates and β matrix. This indicates the diffusion-free nature of the orthorhombic α″ martensite phase formation.

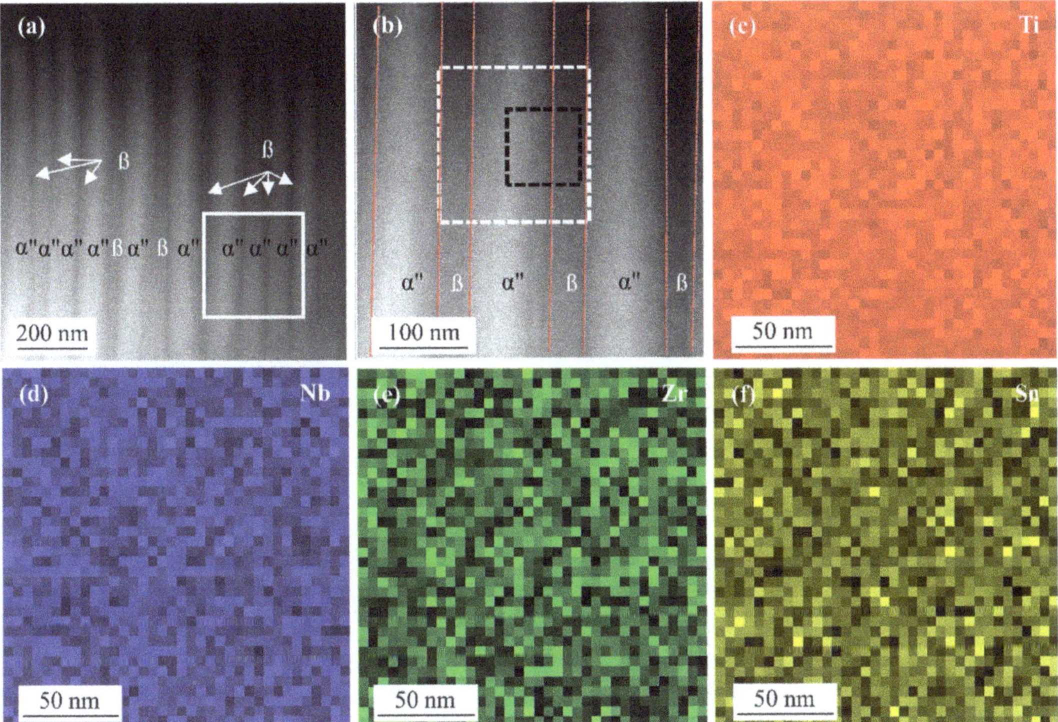

Figure 4. (**a**) HAADF STEM image of the as-built specimen, (**b**) HAADF STEM image of the white square region in (**a**), brighter laths are α″ martensite phase, while the surrounding area is the β matrix, red, dashed lines demarcate the boundary between α″ martensite and β phase; EDS maps from the white, dashed square in (**b**) and chemical distribution of (**c**) titanium, (**d**) niobium, (**e**) zirconium, and (**f**) tin.

HRTEM, FFT, and Fourier filtered HRTEM images of the as-built specimen from the black, dashed square region in Figure 4b are shown in Figure 5. Figure 5a–c are from the β phase, whereas Figure 5d–f are from the α″/β interface region. The HRTEM image in Figure 5a highlights the lattice of the β phase. The corresponding FFT image in (b) is captured from the [011]β zone axis and shows the $0\bar{1}1$β, $\bar{2}00$β and $\bar{2}11$β spots. Figure 5c is the zoomed-in image of (a), indicating the (110)β and (200)β planes. Figure 5d shows the HRTEM image from the lattice fringes of the orthorhombic α″ martensite phase and β phase lattice planes. Figure 5e is taken from the [011]β and [001]α″ zone axis. The FFT image in (e) presents the $0\bar{1}1$β, $\bar{2}00$β, and $\bar{2}11$β spots as well as the $0\bar{2}0$α″, 200α″, $2\bar{2}0$α″, and $1\bar{1}0$α″ spots. Multiple spots are observed to overlap with each other: (i) $0\bar{2}0$α″ and $0\bar{1}1$β, (ii) 200α″ and $\bar{2}00$β and (iii) $2\bar{2}0$α″ and $\bar{2}11$β. From the FFT image, the orientation relationship between the β and α″ martensite phase is [001]α″||[011]β, [0$\bar{2}$0]α″||[0$\bar{1}\bar{1}$]β, [2$\bar{2}$0]α″||[$\bar{2}\bar{1}$1]β and [200]α″||[$\bar{2}$00]β. Figure 5f illustrates a magnified HRTEM marking the fringes from the (200)β, (200)α″, (110)β, (020)α″, and (110)α″ lattice planes.

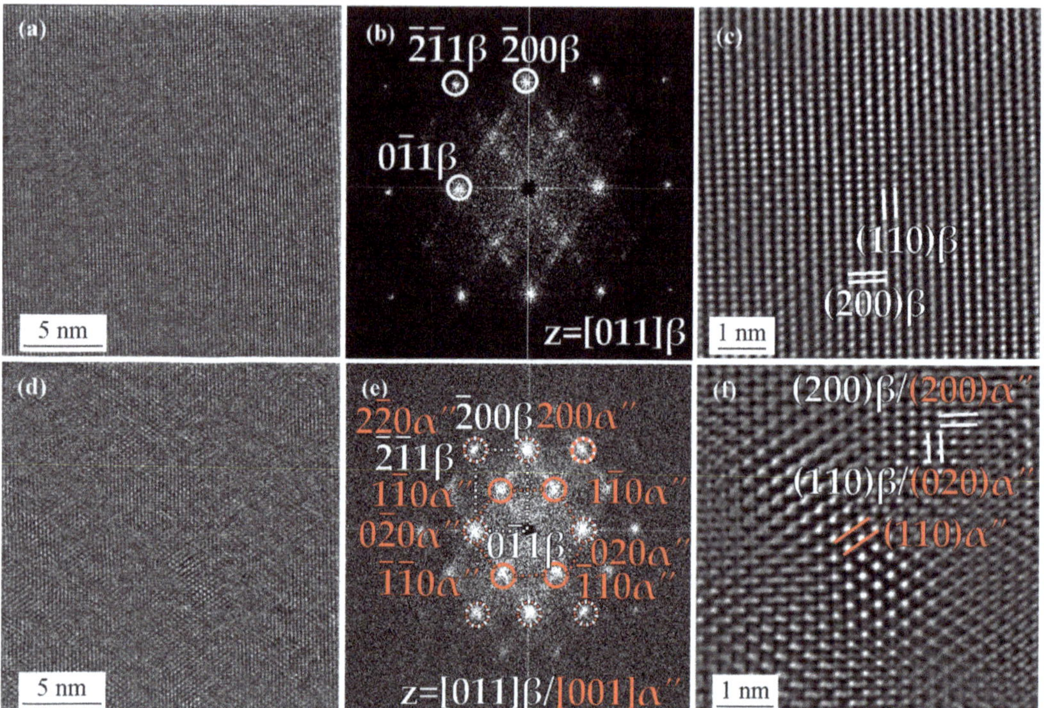

Figure 5. HRTEM, FFT, and Fourier filtered HRTEM images from the black, dashed square region in Figure 4b; (**a**) HRTEM and (**b**) FFT images of the β phase in (**a**), zone axis is [011]β; (**c**) higher magnification Fourier filtered HRTEM image of the β matrix in (**a**); (**d**) HRTEM and (**e**) FFT images of the α″ and β phase in Figure 4b, zone axis in is [011]β/[001]α″; (**f**) higher magnification Fourier filtered HRTEM image of the α″ and β phase in (**d**).

HAADF-STEM images of the ST+FC specimens are depicted in Figure 6a,b, showing the α and β phases. The α phase is present in the form of plates (white arrows) inside the β matrix. The EDS maps in Figure 6c–f are taken from the region in (b), showing the distribution of titanium, niobium, zirconium, and tin in the α and β phases. A depletion (lower color intensity) of niobium and an enrichment (higher color intensity) of titanium is observed in the α phase, indicating a diffusional mechanism of α phase formation due to the diffusion of mainly niobium. Zirconium and tin are stabilizing or neutral to the β phase-formation of titanium [80,81]. Minor differences in the distribution of zirconium and tin in terms of depletion can be observed between the α and β phases.

The TEM image of the ST+FC specimen depicted in Figure 7a shows the α phase plates and the β phase matrix. Figure 7b illustrates the HRTEM image of the α plate taken from the white square region in (a). Figure 7c is a magnified view of the top β/α interface in (b). The FFT image in Figure 7d is calculated from (c) with the zone axis of [$\bar{1}$13]β. The 0002α spot overlaps with the 110β spot, indicating that (0002)α||(110)β. Figure 7e is a close-up HRTEM image of the β/α interface of (c), marking the fringes from the (0002)α, (110)β, (2$\bar{2}$00)α, (1$\bar{2}$1)β, and (2$\bar{1}$1)β lattice planes. At the β/α interface, the lattice fringe of (0002)α is parallel to (110)β. A magnified Fourier filtered image of the β/α interface is presented in Figure 7f, illustrating the (0002)α||(110)β relationship.

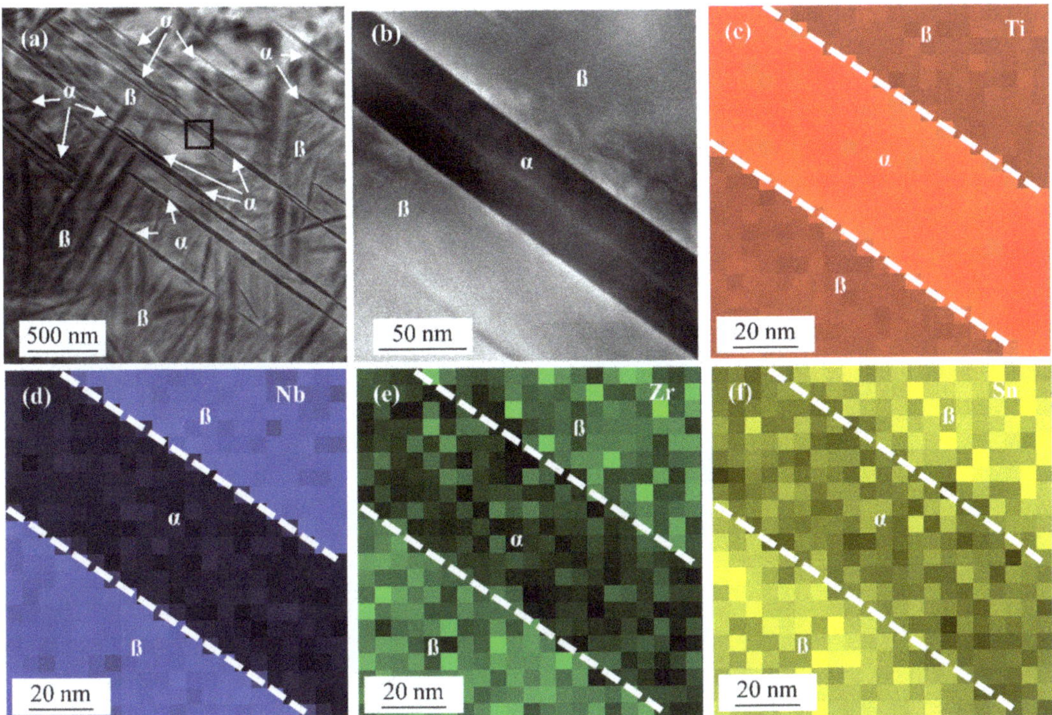

Figure 6. (**a**) HAADF STEM image of the ST+FC specimen with α phase laths inside the β phase matrix; (**b**) HAADF STEM image of the black square in (**a**), showing a higher magnification of an α phase lath surrounded by the β phase; EDS maps and chemical distribution of (**c**) titanium, (**d**) niobium, (**e**) zirconium, and (**f**) tin, demonstrating titanium enrichment inside the α phase (higher color intensity) and titanium depletion inside β phase, and enrichment of niobium, zirconium, and tin inside the β phase (higher color intensity) and depletion inside α phase, respectively.

Metastable β titanium alloys tend to decompose to the ω, α, and α″ phases. The ω phase probably occurs at lower temperatures, as described by Ohmori et al. [82]. In the as-built condition, the ω phase was not observed during TEM investigations, but, due to the XRD results, is assumed to be present in this condition. For the furnace cooled conditions, nevertheless, the α phase laths probably nucleate at the ω/β interfaces and grow in the β matrix, consuming the ω phase particles, as described by Ohmori et al. during quenching from the β phase region. The α″ martensite plates nucleated preferentially at the β grain boundaries [82].

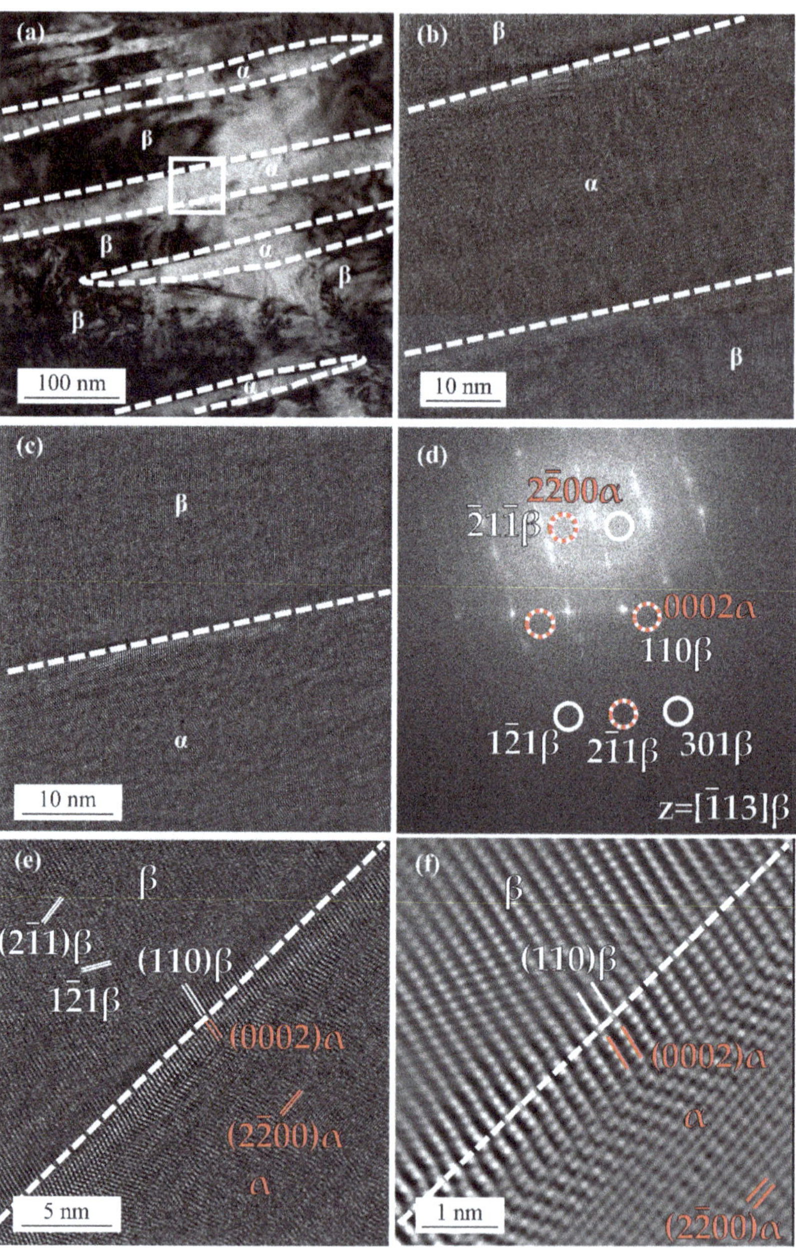

Figure 7. (**a**) TEM image of α phase and β phase in the ST+FC specimen; (**b**) HRTEM image of the α lath and β phase; (**c**) HRTEM image of the upper α/β interface in (**b**); (**d**) FFT image of the area in (**c**) with [113]β zone axis; (**e**) magnified HRTEM image of the α/β interface in (**c**) with the (2$\bar{1}$1)β, (1$\bar{2}$1)β, (110)β, (0002)α, and (2$\bar{2}$00)α planes marked; (**f**) Fourier filtered HRTEM image showing the α/β interface in (**e**) with the (110)β, (0002)α, and (2$\bar{2}$00)α planes marked.

3.3. Mechanical Properties

The hardness of the additively manufactured Ti2448 in as-built and heat-treated conditions is summarized in Figure 8a. In the as-built condition, the hardness is 219 ± 8 HV5 and slightly lower but in good agreement with other research on LPBF-fabricated Ti2448 with near full dense (230 HV0.5 [83], 240 HV1 [54], 230 HV0.5 [84]), and lower than for EBM-fabricated Ti2448 (255 HV [85]). For conventional conditions, the hardness varies between 215 HV (hot-forged), 230 HV (warm swaged), and 265 HV (warm rolled) [79]. For ST+AC and ST+SWQ conditions, the hardness is slightly higher, 232 ± 5 HV5 and 230 ± 5 HV5, respectively, compared to the as-built condition, probably due to the presence of α″ martensite [18,35–38]. The ST+SWQ+A heat treatment leads to a hardness of 290 ± 4 HV5. The highest hardness results in ST+FC condition with 305 ± 3 HV5. The increased hardness of the aged and furnace cooled condition is assumed to be precipitation hardening due to α precipitates formation in the β matrix [69,86]. The hardness values are summarized in Table 3.

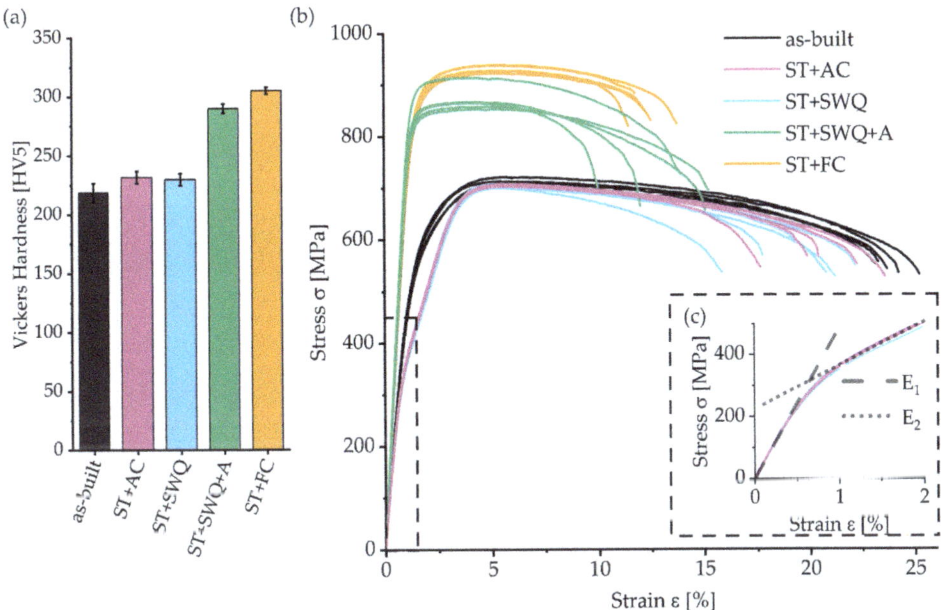

Figure 8. (**a**) Vickers hardness for different conditions; (**b**) stress–strain curves of different conditions; (**c**) inset of the dashed boxed in (**b**) presenting the nonlinear elastic behavior of the ST+AC and ST+SWQ conditions.

The stress–strain curves of the as-built and heat-treated specimens of Ti2448 are depicted in Figure 8b. The lowest ultimate tensile strength (UTS) of 700 ± 6 MPa and the highest fracture elongation A of 22 ± 1% were measured for the as-built condition. AC or SWQ after ST result in similar UTS of approximately 705 MPa and elongations of around 20%. The Young's moduli E_1 of the as-built, ST+AC, and ST+SWQ conditions are similar at the beginning of the elastic deformation, until a strain of 0.5%. The tensile tests of the ST+AC and ST+SWQ conditions show significant and continuous elastic softening with increasing strain and, therefore, stress. The slope changes for strain values higher than 1% to E_2 of approximately 14 ± 0.11 GPa, see Figure 8c. The α″ martensite phase is present in these two conditions. Kolli et al. propose deformation mechanisms by conventional slip and stress-induced transformation due to α′ martensite, α″ martensite, ω phase, and deformation twinning, whereas the mechanisms depend on the β phase stability of the titanium alloy [42,87]. With increasing Mo_{eq}, the stress-induced deformation mechanisms follow the sequence $\alpha' \rightarrow \alpha'' \rightarrow \omega$ + twinning $\rightarrow$ twinning $\rightarrow$ twinning + slip $\rightarrow$ slip. With a Mo_{eq} of around 2.94 and a niobium content higher than 13 wt. %, the alloys are expected to contain the α'' phase and the deformation mechanism is likely to be based on α'' formation [37]. The presence of metastable α'' martensite or ω phase can promote stress-induced transformation and serve as nucleation sites during deformation [42,87]. As described by Furuta et al., both conditions show pseudo-elastic deformation. This behavior is based on stress-induced α'' martensite, which is determined through in situ XRD measurements during tensile tests. Along with increasing tensile strain and stress, the α'' phase retransforms to β and can be explained by reversible martensitic transformation, also known as "pseudo-elastic deformation" [88]. Hao et al. assume stress-induced phase transformation and/or incipient kink bands as the origin of the peculiar elastic behavior [89]. Although the LPFB process leads to α'' martensite microstructure, shown by the TEM investigations, this exceptional pseudo-elastic deformation behavior was not detectable for the as-built condition. The thermal history of each layer during the process may result in a characteristic microstructure of the as-built specimen and, therefore, a unique material behavior. The material undergoes a series of thermal cycles, where the previously fabricated material can be partially re-melted and re-solidified [90,91]. Non-linear elastic behavior of the as-built condition may be attributed to the microstructure, probably characterized by localized distorted region with the elastic strain being located hierarchically in the alloy, as assumed by Furuta et al. [88]. Subsequent aging or FC after ST led to higher strengths, 871 ± 22 MPa and 931 ± 7 MPa, a significant reduction of elongations, 12 ± 2% and 11 ± 1 MPa, and increased Young's moduli E_1, 76 ± 3 GPa and 73 ± 1 GPa, respectively. These treatments produce fine dispersion of α precipitates as the principal strengthening mechanism. The aging temperature and time have a significant effect on the size and spacing of the α phase formed and, therefore, on the mechanical performance [92].

Table 3. Comparison of average mechanical properties, including the hardness and monotonic tensile properties, of the as-built and heat-treated Ti2448, as well as literature values including conventional processing methods: yield strength YS, ultimate tensile strength UTS, fracture elongation A, and Young's Moduli E_1 and E_2.

Condition	Hardness [HV5]	YS [MPa]	UTS [MPa]	A [%]	E_1/E_2 [GPa]
as-built	219 ± 8	490 ± 16	700 ± 6	22 ± 1	49 ± 1/-
ST+AC	232 ± 5	362 ± 7	707 ± 2	20 ± 2	51 ± 2/14 ± 0.1
ST+SWQ	230 ± 5	339 ± 10	705 ± 3	19 ± 2	50 ± 2/14 ± 0.1
ST+SWQ+A	290 ± 4	819 ± 27	871 ± 22	12 ± 2	76 ± 3/-
ST+FC	305 ± 3	805 ± 11	931 ± 7	11 ± 1	73 ± 1/-
LPBF [83]	220–230	563 ± 38	665 ± 18	14 ± 4	53 ± 1/-
Hot rolled [21]	-	700	830	15	46
Hot-forged [19]	230–370	570	750	13	55
Hot-forged [79]	215	-	800	18	52
Warm swaged [79]	230	-	850	14	55
Warm rolled [79]	265	-	1150	8	56

The mechanical properties are summarized in Table 3. Rapid cooling during fabrication or heat treatment results in lower strength properties. High cooling rates from above the $\beta_{Transus}$ temperature result in the formation of the α'' martensite and, consequently, lower strength and elongation, compared to the as-built condition. Previous studies have shown that the α'' martensite phase reduces hardness, tensile, and fatigue properties [36,93]. Through slow cooling or aging treatments at temperatures below $\beta_{Transus}$, α phase laths can evolve and grow. Annealing treatments and slower cooling lead to strengthening due to enhanced nucleation of the α precipitates [20,94]. Compared to other research on additively manufactured Ti2448, the properties are in good agreement regarding the hardness and UTS, whereas the YS and fracture elongation slightly differ, due to varying proportionality factors concerning the specimen geometry [83]. In comparison to other processing routes, the different heat treatments, in particular the aging and FC, enhance the mechanical properties in terms of UTS and hardness. The properties of the warm swaging and rolling treatments are caused by the ultrafine-grained microstructures, not achievable by the applied treatments [79]. Further work shows similar properties, depending on the processing routes and heat treatment strategy [95], although, for the additively manufactured Ti2448, different treatments show huge potential in improving the mechanical behavior and properties, especially aging or FC. Besides improved UTS and YS remaining similar elongation compared to the conventional process routes, the Young's moduli increase for these conditions due to the presence of α precipitations. The hardness of conventional processed Ti2448 can be outperformed by the additively processed and heat-treated conditions [19]. Nevertheless, additively manufactured and heat-treated (FC or SWQ+A) Ti2448 achieves similar or better mechanical performances than conventional processed Ti2448 regarding UTS and YS, with slightly increased Young's moduli and still retained good ductility.

The possibility of in situ α″ martensitic transformations during deformation of the β phase for air-cooled and slow water quenched specimens, during hardness or tensile tests, could be present. The determination is challenging and would require the observation of the specimens at the nanoscopic level during the tests. Further investigations on the stress-induced deformation and formation of the α″ phase are necessary for a comprehensive understanding of the mechanical behavior, exclusively concerning the fatigue behavior and the utilization as an implant alloy. In addition, the individual tailoring of the mechanical properties by adapting the aging heat treatment parameters has to be considered for further research.

4. Conclusions

Micro- and nanostructure, as well as Vickers hardness and monotonic tensile properties of Ti2448 manufactured with LPBF, have been studied in the as-built state and heat-treated conditions. Solution heat treatment at 750 °C with various cooling rates (AC, FC, and SWQ) or ST with subsequent aging at 500 °C were performed and compared to the as-built additively manufactured condition. The effect of different heat treatments is determined as follows:

- The as-built condition shows mainly β phase containing acicular α″ martensite phase, as detected via TEM. Air-cooled and water-quenched states exhibit similar microstructures consisting of β grains, investigated by LiMi and SEM. The aged condition exhibits a diffuse microstructure with homogeneous and uniform α precipitates inside the β matrix. The furnace-cooled specimens have a microstructure comparable to the aged conditions but have heterogeneously distributed and coarser α precipitates within the β phase.
- The nanostructure, determined by TEM, of the as-built condition is compared to the furnace cooled condition, which exhibits the best tensile properties regarding hardness, UTS, and YS. The α″ martensite laths were detected inside the β matrix. Based on FFT images, the orientation relationship of the α″ martensite and β phase is determined. The homogeneous distribution of alloying elements in the EDS maps indicates a diffusion-free phase transformation during cooling. The furnace-cooled specimens consist of α precipitates inside a β phase matrix. Based on crystallographic relation, the precipitates are oriented approximately 60° to each other.
- X-ray diffractograms are sensitive to the various heat treatments and, in particular, the cooling rates. High cooling rates, e.g., AC, SWQ, and the powder fabrication process, lead to the formation of the martensitic α″ phase. Low cooling rates (FC) or aging after ST result in the formation of the α phase.
- FC or aging after solution treatment results in a microstructure containing acicular α precipitates in the β matrix, leading to high tensile strength with relatively low ductility. Phase transformation, such as stress-induced α″ phase transformation, probably leads to pseudo-elastic deformation behavior in the air-cooled and slow water-quenched conditions. As the α″ phase was not detected with XRD in the furnace cooled and

aged conditions, a linear stress–strain relationship was observable in the elastic range. The as-built conditions show elastic anomaly, which is attributed to the LPBF resulting microstructure. The furnace cooled condition exhibits the best mechanical properties regarding UTS, YS, and hardness with a slightly worse fracture elongation compared to the aged conditions.

Controlling the development and refinement of precipitated α phase in metastable β titanium alloys allows the achievement of an excellent combination of strength and ductility, including superb biocompatible properties. In this work, the furnace-cooled specimens obtain the best mechanical properties regarding hardness, UTS, and YS. Nevertheless, the aged condition has to be considered for further investigations regarding the possibility of tailoring the material properties. Adapted heat treatments enable adjusting the microstructural features in terms of the α precipitates. Furthermore, in situ investigations at the nanoscopic level should be performed to determine and understand the material behavior and possible α'' phase transformation behavior during the tensile tests of the air-cooled and water-quenched conditions. If fully understood, the two-part Young's modulus can be utilized. Regarding the overall performance, Ti2448 shows promising features regarding the tailoring of mechanical properties and applying an implant material with a low Young's modulus, as well as adequate and adaptable strength properties.

Author Contributions: Conceptualization, M.H., K.-P.H. and M.S.; methodology, M.H., K.-P.H. and M.S.; software, M.H. and N.F.L.D.; validation, M.H., N.F.L.D., K.-P.H. and M.S.; formal analysis, M.H., K.-P.H. and M.S.; investigation, M.H., N.F.L.D. and S.P.; resources, W.T. and M.S.; data curation, M.H.; writing—original draft preparation, M.H.; writing—review and editing, M.H., N.F.L.D., S.P., D.S., K.-P.H., W.T. and M.S.; visualization, M.H.; supervision, K.-P.H., D.S., W.T. and M.S.; project administration, M.H.; funding acquisition, K.-P.H., W.T. and M.S. All authors have read and agreed to the published version of the manuscript.

Funding: This research was funded by the Deutsche Forschungsgemeinschaft (DFG), grant numbers SCHA 1484/45-1, and TI 343/167-1.

Data Availability Statement: The data that support the findings of this study are available from the corresponding author upon reasonable request.

Acknowledgments: The research was performed with the equipment and base of the LWK and DMRC research infrastructure. The authors are grateful to the staff members of the LWK and DMRC.

Conflicts of Interest: The authors declare no conflict of interest. The funders had no role in the design of the study; in the collection, analyses, or interpretation of data; in the writing of the manuscript, or in the decision to publish the results.

References

1. Gubin, A.V.; Kuznetsov, V.P.; Borzunov, D.Y.; Koryukov, A.A.; Reznik, A.V.; Chevardin, A.Y. Challenges and Perspectives in the Use of Additive Technologies for Making Customized Implants for Traumatology and Orthopedics. *Biomed. Eng.* **2016**, *50*, 285–289. [CrossRef]
2. Bechtold, J.E. Application of Computer Graphics in the Design of Custom Orthopedic Implants. *Orthop. Clin. N. Am.* **1986**, *17*, 605–612. [CrossRef]
3. Heller, M.; Bauer, H.-K.; Goetze, E.; Gielisch, M.; Roth, K.E.; Drees, P.; Maier, G.S.; Dorweiler, B.; Ghazy, A.; Neufurth, M.; et al. Applications of patient-specific 3D printing in medicine. *Int. J. Comput. Dent.* **2016**, *19*, 323–339. [PubMed]
4. Munsch, M. Laser additive manufacturing of customized prosthetics and implants for biomedical applications. In *Laser Additive Manufacturing*; Elsevier: Amsterdam, The Netherlands, 2017; pp. 399–420, ISBN 9780081004333.
5. Narra, S.P.; Mittwede, P.N.; DeVincent Wolf, S.; Urish, K.L. Additive Manufacturing in Total Joint Arthroplasty. *Orthop. Clin. N. Am.* **2019**, *50*, 13–20. [CrossRef] [PubMed]
6. Gebhardt, A.; Hötter, J.-S. *Additive Manufacturing: 3D Printing for Prototyping and Manufacturing*; Hanser Publishers: Cincinnati, OH, USA; Munich, Germany, 2016; ISBN 978-1-56990-582-1.
7. AlMangour, B. (Ed.) *Additive Manufacturing of Emerging Materials*; Springer: Cham, Switzerland, 2019; ISBN 978-3-319-91713-9.
8. Long, M.; Rack, H. Titanium alloys in total joint replacement—A materials science perspective. *Biomaterials* **1998**, *19*, 1621–1639. [CrossRef]
9. Niinomi, M. Recent metallic materials for biomedical applications. *Metall. Mater. Trans.* **2002**, *33*, 477–486. [CrossRef]

10. Blakey-Milner, B.; Gradl, P.; Snedden, G.; Brooks, M.; Pitot, J.; Lopez, E.; Leary, M.; Berto, F.; Du Plessis, A. Metal additive manufacturing in aerospace: A review. *Mater. Des.* **2021**, *209*, 110008. [CrossRef]
11. Niinomi, M. Mechanical properties of biomedical titanium alloys. *Mater. Sci. Eng.* **1998**, *243*, 231–236. [CrossRef]
12. Sumitomo, N.; Noritake, K.; Hattori, T.; Morikawa, K.; Niwa, S.; Sato, K.; Niinomi, M. Experiment study on fracture fixation with low rigidity titanium alloy: Plate fixation of tibia fracture model in rabbit. *J. Mater. Sci. Mater. Med.* **2008**, *19*, 1581–1586. [CrossRef] [PubMed]
13. Cheruvu, B.; Venkatarayappa, I.; Goswami, T. Stress Shielding in Cemented Hip Implants Assessed from Computed Tomography. *Biomed. J. Sci. Tech. Res.* **2019**, *18*, 13637–13641. [CrossRef]
14. Liu, Y.J.; Li, S.J.; Wang, H.L.; Hou, W.T.; Hao, Y.L.; Yang, R.; Sercombe, T.B.; Zhang, L.C. Microstructure, defects and mechanical behavior of beta-type titanium porous structures manufactured by electron beam melting and selective laser melting. *Acta Mater.* **2016**, *113*, 56–67. [CrossRef]
15. Bahl, S.; Suwas, S.; Chatterjee, K. Comprehensive review on alloy design, processing, and performance of β Titanium alloys as biomedical materials. *Int. Mater. Rev.* **2021**, *66*, 114–139. [CrossRef]
16. Geetha, M.; Singh, A.K.; Asokamani, R.; Gogia, A.K. Ti based biomaterials, the ultimate choice for orthopaedic implants—A review. *Prog. Mater. Sci.* **2009**, *54*, 397–425. [CrossRef]
17. Niinomi, M. Mechanical biocompatibilities of titanium alloys for biomedical applications. *J. Mech. Behav. Biomed. Mater.* **2008**, *1*, 30–42. [CrossRef]
18. Hao, Y.L.; Li, S.J.; Sun, S.Y.; Zheng, C.Y.; Hu, Q.M.; Yang, R. Super-elastic titanium alloy with unstable plastic deformation. *Appl. Phys. Lett.* **2005**, *87*, 91906. [CrossRef]
19. Li, S.J.; Cui, T.C.; Hao, Y.L.; Yang, R. Fatigue properties of a metastable beta-type titanium alloy with reversible phase transformation. *Acta Biomater.* **2008**, *4*, 305–317. [CrossRef]
20. Li, S.J.; Zhang, Y.W.; Sun, B.B.; Hao, Y.L.; Yang, R. Thermal stability and mechanical properties of nanostructured Ti–24Nb–4Zr–7.9Sn alloy. *Mater. Sci. Eng.* **2008**, *480*, 101–108. [CrossRef]
21. Zhang, S.Q.; Li, S.J.; Jia, M.T.; Hao, Y.L.; Yang, R. Fatigue properties of a multifunctional titanium alloy exhibiting nonlinear elastic deformation behavior. *Scr. Mater.* **2009**, *60*, 733–736. [CrossRef]
22. Bania, P.J. Beta titanium alloys and their role in the titanium industry. *JOM* **1994**, *46*, 16–19. [CrossRef]
23. Collings, E.W. *The Physical Metallurgy of Titanium Alloys*; American Society for Metals: Metals Park, OH, USA, 1984; ISBN 9780871701817.
24. Sing, S.L.; Yeong, W.Y.; Wiria, F.E. Selective laser melting of titanium alloy with 50 wt% tantalum: Microstructure and mechanical properties. *J. Alloys Compd.* **2016**, *660*, 461–470. [CrossRef]
25. Vrancken, B.; Thijs, L.; Kruth, J.-P.; van Humbeeck, J. Microstructure and mechanical properties of a novel β titanium metallic composite by selective laser melting. *Acta Mater.* **2014**, *68*, 150–158. [CrossRef]
26. Chen, L.-Y.; Cui, Y.-W.; Zhang, L.-C. Recent Development in Beta Titanium Alloys for Biomedical Applications. *Metals* **2020**, *10*, 1139. [CrossRef]
27. Jaffee, R.I.; Promisel, N.E. *The Science, Technology and Application of Titanium*; Elsevier: Amsterdam, The Netherlands, 1970; ISBN 9780080065649.
28. Burgers, W.G. On the process of transition of the cubic-body-centered modification into the hexagonal-close-packed modification of zirconium. *Physica* **1934**, *1*, 561–586. [CrossRef]
29. Newkirk, J.B.; Geisler, A.H. Crystallographic aspects of the beta to alpha transformation in titanium. *Acta Metall.* **1953**, *1*, 370–374. [CrossRef]
30. Schmidt, F.F.; Wood, R.A. Heat Treatment of Titanium and Titanium Alloys: Heat Treatments and Tensile Properties of Titanium and Titanium Alloys for Structures. 1966. Available online: https://ntrs.nasa.gov/citations/19660015720 (accessed on 19 February 2022).
31. Ehtemam-Haghighi, S.; Liu, Y.; Cao, G.; Zhang, L.-C. Influence of Nb on the $\beta \rightarrow \alpha''$ martensitic phase transformation and properties of the newly designed Ti-Fe-Nb alloys. *Mater. Sci. Eng. Mater. Biol. Appl.* **2016**, *60*, 503–510. [CrossRef]
32. Haghighi, S.E.; Lu, H.B.; Jian, G.Y.; Cao, G.H.; Habibi, D.; Zhang, L.C. Effect of α'' martensite on the microstructure and mechanical properties of beta-type Ti–Fe–Ta alloys. *Mater. Des.* **2015**, *76*, 47–54. [CrossRef]
33. He, J.; Li, D.; Jiang, W.; Ke, L.; Qin, G.; Ye, Y.; Qin, Q.; Qiu, D. The Martensitic Transformation and Mechanical Properties of Ti6Al4V Prepared via Selective Laser Melting. *Materials* **2019**, *12*, 321. [CrossRef]
34. Motyka, M. Martensite Formation and Decomposition during Traditional and AM Processing of Two-Phase Titanium Alloys—An Overview. *Metals* **2021**, *11*, 481. [CrossRef]
35. Banumathy, S.; Mandal, R.K.; Singh, A.K. Structure of orthorhombic martensitic phase in binary Ti–Nb alloys. *J. Appl. Phys.* **2009**, *106*, 93518. [CrossRef]
36. Hao, Y.L.; Yang, R.; Niinomi, M.; Kuroda, D.; Zhou, Y.L.; Fukunaga, K.; Suzuki, A. Young's modulus and mechanical properties of Ti-29Nb-13Ta-4.6Zr in relation to α'' martensite. *Metall. Mater. Trans.* **2002**, *33*, 3137–3144. [CrossRef]
37. Tang, X.; Ahmed, T.; Rack, H.J. Phase transformations in Ti-Nb-Ta and Ti-Nb-Ta-Zr alloys. *J. Mater. Sci.* **2000**, *35*, 1805–1811. [CrossRef]
38. Li, S.J.; Niinomi, M.; Akahori, T.; Kasuga, T.; Yang, R.; Hao, Y.L. Fatigue characteristics of bioactive glass-ceramic-coated Ti-29Nb-13Ta-4.6Zr for biomedical application. *Biomaterials* **2004**, *25*, 3369–3378. [CrossRef] [PubMed]

39. Hickman, B.S. The formation of omega phase in titanium and zirconium alloys: A review. *J. Mater. Sci.* **1969**, *4*, 554–563. [CrossRef]
40. Williams, J.C.; Hickman, B.S.; Marcus, H.L. The effect of omega phase on the mechanical properties of titanium alloys. *Metall. Mater. Trans.* **1971**, *2*, 1913–1919. [CrossRef]
41. Colombo-Pulgarín, J.C.; Biffi, C.A.; Vedani, M.; Celentano, D.; Sánchez-Egea, A.; Boccardo, A.D.; Ponthot, J.-P. Beta Titanium Alloys Processed by Laser Powder Bed Fusion: A Review. *J. Mater. Eng. Perform.* **2021**, *30*, 6365–6388. [CrossRef]
42. Kolli, R.; Devaraj, A. A Review of Metastable Beta Titanium Alloys. *Metals* **2018**, *8*, 506. [CrossRef]
43. Murr, L.E.; Gaytan, S.M.; Ramirez, D.A.; Martinez, E.; Hernandez, J.; Amato, K.N.; Shindo, P.W.; Medina, F.R.; Wicker, R.B. Metal Fabrication by Additive Manufacturing Using Laser and Electron Beam Melting Technologies. *J. Mater. Sci. Technol.* **2012**, *28*, 1–14. [CrossRef]
44. Murr, L.E.; Quinones, S.A.; Gaytan, S.M.; Lopez, M.I.; Rodela, A.; Martinez, E.Y.; Hernandez, D.H.; Martinez, E.; Medina, F.; Wicker, R.B. Microstructure and mechanical behavior of Ti-6Al-4V produced by rapid-layer manufacturing, for biomedical applications. *J. Mech. Behav. Biomed. Mater.* **2009**, *2*, 20–32. [CrossRef]
45. Thijs, L.; Verhaeghe, F.; Craeghs, T.; van Humbeeck, J.; Kruth, J.P. A study of the microstructural evolution during selective laser melting of Ti–6Al–4V. *Acta Mater.* **2010**, *58*, 3303–3312. [CrossRef]
46. Polozov, I.; Sufiiarov, V.; Popovich, A.; Masaylo, D.; Grigoriev, A. Synthesis of Ti-5Al, Ti-6Al-7Nb, and Ti-22Al-25Nb alloys from elemental powders using powder-bed fusion additive manufacturing. *J. Alloys Compd.* **2018**, *763*, 436–445. [CrossRef]
47. Xu, C.; Sikan, F.; Atabay, S.E.; Muñiz-Lerma, J.A.; Sanchez-Mata, O.; Wang, X.; Brochu, M. Microstructure and mechanical behavior of as-built and heat-treated Ti–6Al–7Nb produced by laser powder bed fusion. *Mater. Sci. Eng. Struct. Mater.* **2020**, *793*, 139978. [CrossRef]
48. Sercombe, T.; Jones, N.; Day, R.; Kop, A. Heat treatment of Ti-6Al-7Nb components produced by selective laser melting. *Rapid Prototyp. J.* **2008**, *14*, 300–304. [CrossRef]
49. Hein, M.; Hoyer, K.-P.; Schaper, M. Additively processed TiAl6Nb7 alloy for biomedical applications. *Mater. Und Werkst.* **2021**, *52*, 703–716. [CrossRef]
50. Hein, M.; Kokalj, D.; Lopes Dias, N.F.; Stangier, D.; Oltmanns, H.; Pramanik, S.; Kietzmann, M.; Hoyer, K.-P.; Meißner, J.; Tillmann, W.; et al. Low Cycle Fatigue Performance of Additively Processed and Heat-Treated Ti-6Al-7Nb Alloy for Biomedical Applications. *Metals* **2022**, *12*, 122. [CrossRef]
51. Mold, M.; Linhart, C.; Gómez-Ramírez, J.; Villegas-Lanau, A.; Exley, C. Aluminum and Amyloid-β in Familial Alzheimer's Disease. *J. Alzheimers Dis.* **2020**, *73*, 1627–1635. [CrossRef]
52. Engelhart, S.; Segal, R.J. Allergic reaction to vanadium causes a diffuse eczematous eruption and titanium alloy orthopedic implant failure. *Cutis* **2017**, *99*, 245–249.
53. Klotz, K.; Weistenhöfer, W.; Neff, F.; Hartwig, A.; van Thriel, C.; Drexler, H. The Health Effects of Aluminum Exposure. *Dtsch. Arztebl. Int.* **2017**, *114*, 653–659. [CrossRef]
54. Yang, C.L.; Zhang, Z.J.; Li, S.J.; Liu, Y.J.; Sercombe, T.B.; Hou, W.T.; Zhang, P.; Zhu, Y.K.; Hao, Y.L.; Zhang, Z.F.; et al. Simultaneous improvement in strength and plasticity of Ti-24Nb-4Zr-8Sn manufactured by selective laser melting. *Mater. Des.* **2018**, *157*, 52–59. [CrossRef]
55. *DIN EN ISO 6892-1:2020-06*; Metallische Werkstoffe_-Zugversuch_-Teil_1: Prüfverfahren bei Raumtemperatur (ISO_6892-1:2019), Deutsche Fassung EN_ISO_6892-1:2019. Beuth Verlag GmbH: Berlin, Germany, 2020.
56. Bidare, P.; Bitharas, I.; Ward, R.M.; Attallah, M.M.; Moore, A.J. Laser powder bed fusion at sub-atmospheric pressures. *Int. J. Mach. Tools Manuf.* **2018**, *130*, 65–72. [CrossRef]
57. Collur, M.M.; Paul, A.; DebRoy, T. Mechanism of alloying element vaporization during laser welding. *MTB* **1987**, *18*, 733–740. [CrossRef]
58. Dai, D.; Gu, D. Effect of metal vaporization behavior on keyhole-mode surface morphology of selective laser melted composites using different protective atmospheres. *Appl. Surf. Sci.* **2015**, *355*, 310–319. [CrossRef]
59. Santecchia, E.; Spigarelli, S.; Cabibbo, M. Material Reuse in Laser Powder Bed Fusion: Side Effects of the Laser—Metal Powder Interaction. *Metals* **2020**, *10*, 341. [CrossRef]
60. Yin, J.; Zhang, W.; Ke, L.; Wei, H.; Wang, D.; Yang, L.; Zhu, H.; Dong, P.; Wang, G.; Zeng, X. Vaporization of alloying elements and explosion behavior during laser powder bed fusion of Cu–10Zn alloy. *Int. J. Mach. Tools Manuf.* **2021**, *161*, 103686. [CrossRef]
61. Hryha, E.; Shvab, R.; Bram, M.; Bitzer, M.; Nyborg, L. Surface chemical state of Ti powders and its alloys: Effect of storage conditions and alloy composition. *Appl. Surf. Sci.* **2016**, *388*, 294–303. [CrossRef]
62. Lütjering, G.; Williams, J.C. *Titanium*, 2nd ed.; Springer: Berlin/Heidelberg, Germany, 2007; ISBN 9783540713975.
63. Li, C.; Chen, J.H.; Wu, X.; van der Zwaag, S. Effect of strain rate on stress-induced martensitic formation and the compressive properties of Ti–V–(Cr,Fe)–Al alloys. *Mater. Sci. Eng. Struct. Mater.* **2013**, *573*, 111–118. [CrossRef]
64. Neelakantan, S. Tailoring the Mechanical Properties of Titanium Alloys via Plasticity Induced Transformations. Ph.D. Dissertation, Delft University of Technology, Delft, The Netherlands, 2010.
65. Meng, F.; Yan, J.-Y.; Olson, G.B. TRIP Titanium Alloy Design. *MATEC Web Conf.* **2020**, *321*, 11070. [CrossRef]
66. Yan, J.-Y. TRIP Titanium Alloy Design. Ph.D. Dissertation, Northwestern University, Evanston, IL, USA, 2014.
67. Moffat, D.L.; Larbalestier, D.C. The compctition between martensite and omega in quenched Ti-Nb alloys. *Metall. Mater. Trans.* **1988**, *19*, 1677–1686. [CrossRef]

68. Duerig, T.W.; Albrecht, J.; Richter, D.; Fischer, P. Formation and reversion of stress induced martensite in Ti-10V-2Fe-3Al. *Acta Metall.* **1982**, *30*, 2161–2172. [CrossRef]
69. Qi, L.; He, S.; Chen, C.; Jiang, B.; Hao, Y.; Ye, H.; Yang, R.; Du, K. Diffusional-displacive transformation in a metastable β titanium alloy and its strengthening effect. *Acta Mater.* **2020**, *195*, 151–162. [CrossRef]
70. Dutta, B.; Froes, F.H. The Additive Manufacturing (AM) of titanium alloys. *Metall. Powder Rep.* **2017**, *72*, 96–106. [CrossRef]
71. Barriobero-Vila, P.; Gussone, J.; Haubrich, J.; Sandlöbes, S.; Da Silva, J.C.; Cloetens, P.; Schell, N.; Requena, G. Inducing Stable α + β Microstructures during Selective Laser Melting of Ti-6Al-4V Using Intensified Intrinsic Heat Treatments. *Materials* **2017**, *10*, 268. [CrossRef] [PubMed]
72. Mantani, Y.; Tajima, M. Phase transformation of quenched α″ martensite by aging in Ti–Nb alloys. *Mater. Sci. Eng. Struct. Mater.* **2006**, *438*, 315–319. [CrossRef]
73. Zwicker, U. *Titan und Titanlegierungen*; Springer: Berlin/Heidelberg, Germany, 1974; ISBN 978-3-642-80588-2.
74. Liu, J.P.; Wang, Y.D.; Hao, Y.L.; Wang, H.L.; Wang, Y.; Nie, Z.H.; Su, R.; Wang, D.; Ren, Y.; Lu, Z.P.; et al. High-energy X-ray diffuse scattering studies on deformation-induced spatially confined martensitic transformations in multifunctional Ti–24Nb–4Zr–8Sn alloy. *Acta Mater.* **2014**, *81*, 476–486. [CrossRef]
75. Yang, Y.; Castany, P.; Bertrand, E.; Cornen, M.; Lin, J.X.; Gloriant, T. Stress release-induced interfacial twin boundary ω phase formation in a β type Ti-based single crystal displaying stress-induced α″ martensitic transformation. *Acta Mater.* **2018**, *149*, 97–107. [CrossRef]
76. Ren, L.; Xiao, W.; Han, W.; Ma, C.; Zhou, L. Influence of duplex ageing on secondary α precipitates and mechanical properties of the near β-Ti alloy Ti-55531. *Mater. Charact.* **2018**, *144*, 1–8. [CrossRef]
77. Mantri, S.A.; Choudhuri, D.; Alam, T.; Viswanathan, G.B.; Sosa, J.M.; Fraser, H.L.; Banerjee, R. Tuning the scale of α precipitates in β-titanium alloys for achieving high strength. *Scr. Mater.* **2018**, *154*, 139–144. [CrossRef]
78. Yang, Y.; Castany, P.; Cornen, M.; Prima, F.; Li, S.J.; Hao, Y.L.; Gloriant, T. Characterization of the martensitic transformation in the superelastic Ti–24Nb–4Zr–8Sn alloy by in situ synchrotron X-ray diffraction and dynamic mechanical analysis. *Acta Mater.* **2015**, *88*, 25–33. [CrossRef]
79. Hao, Y.L.; Zhang, Z.B.; Li, S.J.; Yang, R. Microstructure and mechanical behavior of a Ti–24Nb–4Zr–8Sn alloy processed by warm swaging and warm rolling. *Acta Mater.* **2012**, *60*, 2169–2177. [CrossRef]
80. Peters, M.; Leyens, C. *Titan und Titanlegierungen*, 3rd ed.; Wiley-VCH: Weinheim, Germany, 2002; ISBN 9783527611089.
81. Sibum, H. Titanium and Titanium Alloys—From Raw Material to Semi-finished Products. *Adv. Eng. Mater.* **2003**, *5*, 393–398. [CrossRef]
82. Ohmori, Y.; Ogo, T.; Nakai, K.; Kobayashi, S. Effects of ω-phase precipitation on β→α, α″ transformations in a metastable β titanium alloy. *Mater. Sci. Eng. Struct. Mater.* **2001**, *312*, 182–188. [CrossRef]
83. Zhang, L.C.; Klemm, D.; Eckert, J.; Hao, Y.L.; Sercombe, T.B. Manufacture by selective laser melting and mechanical behavior of a biomedical Ti–24Nb–4Zr–8Sn alloy. *Scr. Mater.* **2011**, *65*, 21–24. [CrossRef]
84. Zhang, L.C.; Sercombe, T.B. Selective Laser Melting of Low-Modulus Biomedical Ti-24Nb-4Zr-8Sn Alloy: Effect of Laser Point Distance. *KEM* **2012**, *520*, 226–233. [CrossRef]
85. Hernandez, J.; Li, S.J.; Martinez, E.; Murr, L.E.; Pan, X.M.; Amato, K.N.; Cheng, X.Y.; Yang, F.; Terrazas, C.A.; Gaytan, S.M.; et al. Microstructures and Hardness Properties for β-Phase Ti–24Nb–4Zr–7.9Sn Alloy Fabricated by Electron Beam Melting. *J. Mater. Sci. Technol.* **2013**, *29*, 1011–1017. [CrossRef]
86. Ardell, A.J. Precipitation hardening. *Metall. Mater. Trans.* **1985**, *16*, 2131–2165. [CrossRef]
87. Kolli, R.P.; Joost, W.J.; Ankem, S. Phase Stability and Stress-Induced Transformations in Beta Titanium Alloys. *JOM* **2015**, *67*, 1273–1280. [CrossRef]
88. Furuta, T.; Kuramoto, S.; Hwang, J.; Nishino, K.; Saito, T. Elastic Deformation Behavior of Multi-Functional Ti–Nb–Ta–Zr–O Alloys. *Mater. Trans.* **2005**, *46*, 3001–3007. [CrossRef]
89. Hao, Y.L.; Li, S.J.; Sun, S.Y.; Zheng, C.Y.; Yang, R. Elastic deformation behaviour of Ti-24Nb-4Zr-7.9Sn for biomedical applications. *Acta Biomater.* **2007**, *3*, 277–286. [CrossRef]
90. Pantawane, M.V.; Dasari, S.; Mantri, S.A.; Banerjee, R.; Dahotre, N.B. Rapid thermokinetics driven nanoscale vanadium clustering within martensite laths in laser powder bed fused additively manufactured Ti6Al4V. *Mater. Res. Lett.* **2020**, *8*, 383–389. [CrossRef]
91. Cao, S.; Zou, Y.; Lim, C.V.S.; Wu, X. Review of laser powder bed fusion (LPBF) fabricated Ti-6Al-4V: Process, post-process treatment, microstructure, and property. *Gxjzz* **2021**, *2*, 1. [CrossRef]
92. Froes, F.H.; Bomberger, H.B. The Beta Titanium Alloys. *JOM* **1985**, *37*, 28–37. [CrossRef]
93. Lin, D.J.; Chern Lin, J.H.; Ju, C.P. Structure and properties of Ti–7.5Mo–xFe alloys. *Biomaterials* **2002**, *23*, 1723–1730. [CrossRef]
94. Rajan Soundararajan, S.; Vishnu, J.; Manivasagam, G.; Rao Muktinutalapati, N. Heat Treatment of Metastable Beta Titanium Alloys. In *Welding—Modern Topics*; Crisóstomo Absi Alfaro, S., Borek, W., Tomiczek, B., Eds.; IntechOpen: Rijeka, Croatia, 2021; ISBN 978-1-83881-895-1.
95. Coakley, J.; Rahman, K.M.; Vorontsov, V.A.; Ohnuma, M.; Dye, D. Effect of precipitation on mechanical properties in the β-Ti alloy Ti–24Nb–4Zr–8Sn. *Mater. Sci. Eng. Struct. Mater.* **2016**, *655*, 399–407. [CrossRef]

Article

Metallization of Recycled Glass Fiber-Reinforced Polymers Processed by UV-Assisted 3D Printing

Alessia Romani [1,2,*], Paolo Tralli [3], Marinella Levi [1], Stefano Turri [1] and Raffaella Suriano [1]

[1] Department of Chemistry, Materials and Chemical Engineering "Giulio Natta", Politecnico di Milano, Piazza Leonardo da Vinci 32, 20133 Milano, Italy
[2] Design Department, Politecnico di Milano, Via Durando 10, 20158 Milano, Italy
[3] Divisione Green Coat, Novellini S.p.a., Strada Romana Nord, 1, 46027 San Benedetto Po, Italy
* Correspondence: alessia.romani@polimi.it; Tel.: +39-02-2399-4701

Abstract: An ever-growing amount of composite waste will be generated in the upcoming years. New circular strategies based on 3D printing technologies are emerging as potential solutions although 3D-printed products made of recycled composites may require post-processing. Metallization represents a viable way to foster their exploitation for new applications. This paper shows the use of physical vapor deposition sputtering for the metallization of recycled glass fiber-reinforced polymers processed by UV-assisted 3D printing. Different batches of 3D-printed samples were produced, post-processed, and coated with a chromium metallization layer to compare the results before and after the metallization process and to evaluate the quality of the finishing from a qualitative and quantitative point of view. The analysis was conducted by measuring the surface gloss and roughness, analyzing the coating morphology and thickness through the Scanning Electron Microscopy (SEM) micrographs of the cross-sections, and assessing its adhesion with cross-cut tests. The metallization was successfully performed on the different 3D-printed samples, achieving a good homogeneity of the coating surface. Despite the influence of the staircase effect, these results may foster the investigation of new fields of application, as well as the use of different polymer-based composites from end-of-life products, i.e., carbon fiber-reinforced polymers.

Keywords: additive manufacturing; 3D printing; surface finishing; physical vapor deposition; composites; recycling; polymer-matrix composites (PMCs); direct ink writing; liquid deposition modeling; glass fibers

Citation: Romani, A.; Tralli, P.; Levi, M.; Turri, S.; Suriano, R. Metallization of Recycled Glass Fiber-Reinforced Polymers Processed by UV-Assisted 3D Printing. *Materials* **2022**, *15*, 6242. https://doi.org/10.3390/ma15186242

Academic Editors: Swee Leong Sing and Wai Yee Yeong

Received: 29 July 2022
Accepted: 5 September 2022
Published: 8 September 2022

1. Introduction

Polymer-based composites are increasingly employed for a large number of applications ranging from aerospace and automotive to healthcare and construction [1]. Their extensive use in different sectors mainly derives from their versatility and technical properties, i.e., high strength-to-weight ratio. Although the COVID-19 pandemic strongly impacted the composites market, the production volumes of fiber-reinforced polymers are reaching the pre-crisis levels of 2019, especially considering short fiber-reinforced materials, i.e., glass fiber-reinforced polymers (GFRPs) [2]. Consequently, this ever-increasing production will result in higher volumes of GFRP waste to be correctly managed in the coming years [3,4]. For instance, the wind energy sector will generate a growing amount of GFRPs at their end-of-life (EoL) since the lifetime of wind turbine blades usually corresponds to approximately 20–25 years, leading to the disposal of an ever-growing number of windfarms [5]. Despite the current challenges in reusing and recycling composite materials [6,7], new solutions have been investigated to reintroduce GFRP waste in new closed- and open-loop systems, following the principles of circular economy [8,9].

Among the available technologies, 3D printing, or Additive Manufacturing (AM) represents a promising way to reprocess recycled GFRPs thanks to its flexibility in terms of material formulations and customizable complex geometries [10,11]. From the literature, different formulations of GFRPs and carbon fiber-reinforced polymers (CFRPs) were 3D-printed through FDM (fused deposition modeling, or FFF, fused filament fabrication) and DIW (direct ink writing, or LDM, liquid deposition modeling), showing their use for a wide range of applications [12–15]. Recently, some works have also demonstrated their capability to foster this sustainable transition within the composite market by reprocessing recycled GFRPs and CFRPs [16–21]. Nevertheless, the surface properties of these 3D-printed recycled composites may considerably differ from pristine materials, as well as their aesthetic appearance. Furthermore, some applications require different technical properties, which can be achieved only through a post-processing treatment.

Metallization represents an effective way to combine the advantages of polymer-based composites with those of metallic surfaces. Metallization coatings can be deposited through different chemical or physical techniques such as electroless plating or thermal spray deposition [22]. In particular, physical vapor deposition (PVD) sputtering stands out for its coating homogeneity, low processing temperatures, and the use of less dangerous chemicals during the metal deposition [23–26]. From the literature, different studies explored the deposition of metallic coatings onto 3D-printed thermoplastic substrates processed by FFF [27–34], whereas a limited number of works focused on the metallization of polymer-based composites [35], or parts obtained by DIW [36]. To the best of the authors' knowledge, no articles related to the metallization of 3D-printed recycled composites were found by using neither FFF nor DIW processes. Indeed, a hard chromium coating would significantly influence not only their mechanical, surface, and chemical properties but also the overall perception of the final part by end users [37,38].

This paper aims to show the successful use of PVD sputtering for the metallization of 3D-printed GFRPs from recycled composites. First, four batches of samples made of thermoset acrylic resin filled with mechanically-recycled GFRPs from EoL wind turbine blades were 3D-printed through UV-assisted three-dimensional (UV-3D) printing. These samples were post-processed and coated with a PVD chromium sputtering deposition process. The surfaces of these samples were then assessed from a qualitative and quantitative point of view, i.e., by evaluating the aesthetic appearance, surface properties, and morphology of the coating. Considering the promising results, this work may influence further research activities focused on the post-processing of additive manufactured parts for new potential applications, especially considering emerging 3D printing processes or new polymer-based materials from composite waste and parts at their EoL.

2. Materials and Methods

2.1. Materials and UV-3D Printing

The 3D printable ink consisted of a photo- and thermo-curable acrylic matrix mixed with mechanically recycled GFRPs for 20 min at 40 rpm through a Brabender mixer equipped with a roller blade (C.W. Brabender Instruments Inc., South Hackensack, NJ, US). The matrix is composed of 96.7 wt% of ethoxylate bisphenol A diacrylate resin named SR349 (Arkema, Colombes, France and locally distributed by Came S.r.l., Lainate, Italy), 3 wt% of ethyl phenyl (2,4,6-trimethyl benzoyl) phosphinate named TPO-L (Lambson Limited, Wetherby, UK), and 0.3 wt% of dicumyl peroxide (Sigma-Aldrich Corporation, St. Louis, MO, USA). The matrix was previously mixed with a magnetic stirrer at room temperature for 2 h. The ink formulation used in this work contains 55 wt% of shredded GFRPs from EoL wind turbine blades made of continuous epoxy-based GFRPs (Siemens Gamesa Renewable Energy S.A., Zamudio, Spain, and Consiglio Nazionale di Ricerca—Sistemi e Tecnologie Industriali Intelligenti per il Manifatturiero Avanzato—Stiima CNR, Milano, Italy) and 45 wt% of the acrylic matrix. The shredded recycled GFRPs have a glass fiber content of 70 wt% and a nominal granulometry of 80 μm (34 and 14 μm of average length and diameter) [17].

The 3D samples for the surface characterization and metallization assessment have a nominal overall dimension of 50 × 25 × 30 mm and were designed with Fusion 360 CAD software (Autodesk, San Rafael, CA, USA). The UV-3D printing was performed with a custom version of a 3Drag, a commercial FFF low-cost 3D printer (Futura Group S.r.l., Gallarate, Italy). The 3D printer was converted into a UV-assisted DIW apparatus by adding a UV LED source, a 20 mL syringe, and a stainless-steel UV-shielded conic nozzle (Techcon Systems Inc, Cypress, CA, USA) [17]. The G-code file of the 3D sample was prepared with the slicing software Prusa Slicer (Prusa Research, Prague, Czech Republic), and the main 3D printing parameters are shown in Table 1. The 3D-printed samples were then post-cured to completely crosslink the acrylic-based resin. The post-curing treatment was performed in a UV chamber Polymer 500 W with a UVA emittance mercury vapor lamp for 30 min (Helios Italquartz S.r.l., Cambiago, Italy) and in a non-controlled atmosphere oven for 2 h at 140 °C.

Table 1. Three-dimensional printing parameters of the samples for the surface characterization.

Parameters	Units	Values
Perimeters	-	1
Infill	%	0
Flow	%	100
Speed	mm/s	15
Layer height	mm	0.25
Nozzle diameter	mm	1
UV LED source	-	3 × 3 W (395 nm)

2.2. *PVD Sputtering and Surface Characterization*

Four different batches of minimum two samples were then produced to assess the influence of 3D printing, post-processing, and chromium metallization on the surfaces, as better explained in Section 3.1. The PVD sputtering deposition was carried out by Novellini S.p.a.—Divisione Green Coat with a proprietary process (San Benedetto Po, Italy), following the same procedure used for the metallization of FFF 3D-printed samples [35]. After cleaning the surfaces of the samples with a cloth moistened with isopropyl alcohol, a UV-curable acrylic-based primer named UNILAC UV BC 05 (Cromogenia Units S.A., Barcelona, Spain) was sprayed with an average thickness of 70–80 µm to improve the surface roughness of the 3D printing process. The primer deposition was performed with an anthropomorphic robot equipped with rotary atomizer (cup spinning at 25,000 rpm) once the samples were fixed on a rotational jig. The UV curable primer was then photo-crosslinked with 26 mercury vapor lamps with a peak irradiance of 500 mW/cm^2 and a 10″ bulb for 2 min. The jig was moved into the in-line metallization system for the surface activation with the oxygen plasma for 90 s before the chromium deposition (O_2 flow of 700 sccm, electrode power of 4 kW, rotation speed of 5 rpm). Afterward, the PVD magnetron sputtering was carried out to deposit a chromium layer with a nominal thickness of 200 nm. The process was performed for 16 min at a pressure of 1–5 × 10^{-3} mbar and a constant argon flow of 600 sccm. A maximum temperature of 60 °C was reached during the PVD sputtering process.

Gloss was evaluated with a Multi-gloss 268A gloss meter (Konica Minolta, Tokyo, Japan). The measurement results, expressed in Gloss Units (GU), were carried out on the planar surfaces of the samples according to ASTM D523-14 Standard "Standard Test Method for Specular Gloss" [39]. A 60° measuring geometry, or the angle of incidence of 60°, was used for the different batches to compare the gloss values and to determine when using the 20° and 85° geometries. The 20° geometry was used for gloss values higher than 70 GU, whereas the latter was for values lower than 70 GU. The measurements on each sample were repeated five times to calculate the average values and standard deviations.

The surface roughness of the 3D samples was measured with a laser profilometer (UBM) along five different lines for each surface with a point density of 150 points/mm. The values were then used to calculate the average experimental values and standard deviation of roughness average (Ra) and root mean square roughness (Rq).

The cross-section analysis of the average coating thicknesses and morphology was performed through Scanning Electron Microscopy (SEM) micrographs with an extended-pressure SEM Zeiss EVO 50 EP (Carl Zeiss S.p.A., Milano, Italy). Two different cross-sections were obtained from each sample. The surfaces of the samples were then prepared for observation through PVD of a thin gold layer. The samples were observed at 500 and 3k × magnification in high-vacuum conditions and an acceleration voltage of 10–20 kV.

Manual cross-cut tests were carried out on the planar surfaces of a batch of samples to evaluate the adhesion properties according to ISO 2409:2020 Standard "Paints and varnishes—Cross-cut test" [40]. The results were then expressed from 0 to 5, where 0 means no removals and 5 means that more than 65% of the cross-cut area is affected by detachments.

3. Results and Discussion

3.1. Three-Dimensional Printing and Metallization of the Samples

A specific 3D sample was designed to compare the influence of post-processing and PVD sputtering on the 3D-printed surfaces of the recycled GFRPs. Since UV-DIW may be considered an emerging 3D printing process, a simple geometry was defined to reduce the presence of defects due to DIW, such as missing layers for nozzle clogging [17,41]. The shape of the 3D sample was defined according to the PVD sputtering setup, whereas the overall dimensions were modified to control the printing times of each sample. In particular, the sample had a wider planar surface on the bottom to perform the main surface characterizations, i.e., gloss measurements, roughness evaluations, and cross-cut tests. The smaller planar surface, the curved profile, and the internal cavity can be used to evaluate the homogeneity of the chromium deposition on complex profiles by simulating complex features to be reached with the PVD sputtering.

To this end, four different batches were 3D printed. Each batch was composed of two samples at least and was obtained starting from the same G-code file. The batches are different in terms of post-processing and/or presence of the chromium metallization, as listed in Table 2. Different surface characterization tests were performed on each batch. Gloss measurements and roughness evaluations were carried out for each batch, whereas adhesion tests and SEM micrographs were performed on the most representative one, batch n. 2. This batch was not post-processed before the PVD sputtering; hence it should represent the most influenced batch from the UV-3D printing process and the recycled GFRP since the chromium deposition was performed on a surface with the typical layer-by-layer appearance, also known as the staircase effect.

Table 2. Sample batches with the corresponding specifics (post-processing, metallization) and characterization tests.

Sample (Batch)	Figure	Post-Processing	Metallization	Tests
N. 1	Figure 1a	No;	No;	Gloss, Roughness;
N. 2	Figure 1b	No;	Yes (sanding), PVD sputtering;	Gloss, Roughness, SEM, Adhesion;
N. 3	Figure 1c	Yes, sanding;	No;	Gloss, Roughness;
N. 4	Figure 1d	Yes, sanding;	Yes (sanding), PVD sputtering;	Gloss, Roughness;

The most representative 3D sample of each batch is visible in Figure 1. In general, the samples do not show defects due to the metallization (Figure 1b,d), and no detachments or unprocessed areas are visible on the chromium surface, confirming a good homogeneity of the coating deposition (Figure 1b). Furthermore, no geometrical deformations can be detected from the metallized batches (n. 2 and n. 4). Figure 1a,c represent the samples before and after the sanding post-processing. The influence of the recycled GFRP reinforcement is clearly visible in both batches. In the first case, the surface texture of the composite material slightly reduces the staircase effect of the 3D printing process. In the second case, a random clear pattern is visible on the surface, and it derives from the sanding post-processing of the composite material, highlighting the glass fiber particles of the reinforcement. Figure 1b,d show the difference in the PVD sputtering chromium layer on the samples before and after the sanding post-processing. Chromium deposition significantly changes the appearance of the 3D-printed surface by giving to the samples a glossy surface. As evidenced by the difference between Figure 1a and 1b, metallization slightly reduces the staircase effect of 3D printing. Combining sanding post-processing and PVD sputtering allows one to obtain mirror-like surfaces, widening the range of potential applications (Figure 1c,d).

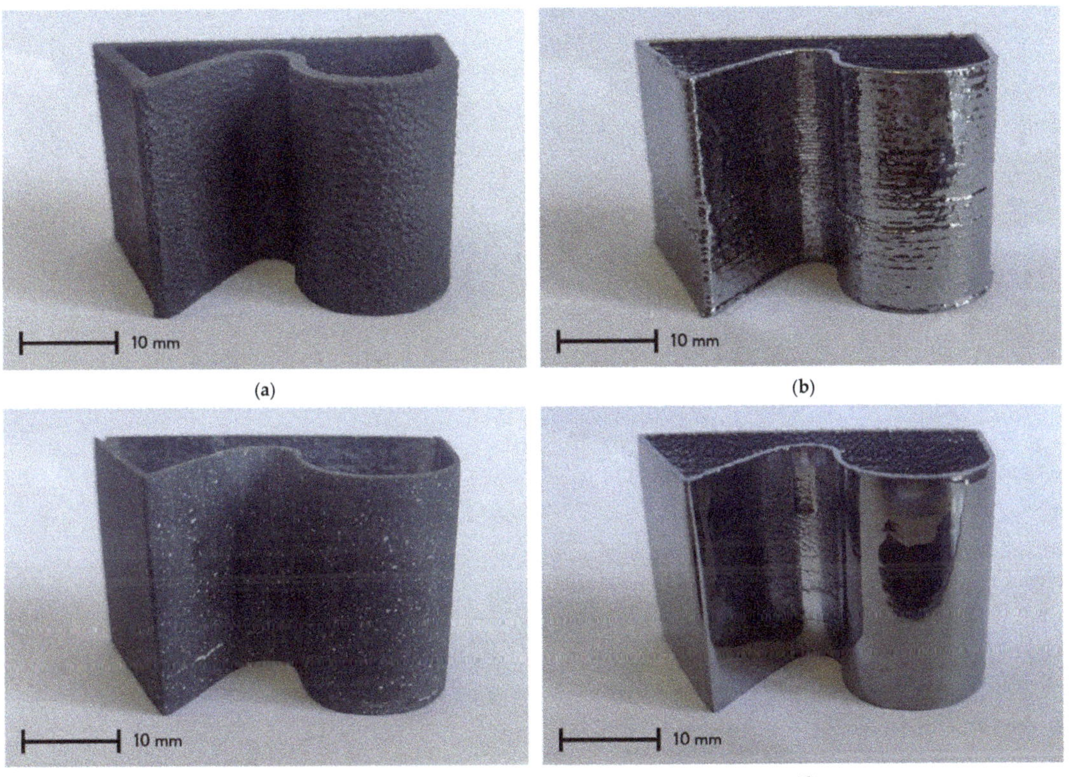

Figure 1. 3D-printed and metallized specimens from the different sample batches: (**a**) Batch n. 1 (no finishing, no PVD); (**b**) Batch n. 2 (No finishing, PVD); (**c**) Batch n. 3 (Finishing, no PVD); (**d**) Batch n. 4 (Finishing, PVD).

3.2. *Roughness*

The roughness of the four batches of samples was measured, comparing the values not only before and after the sanding post-processing but also before and after the PVD sputtering process. The larger vertical planar surface of each sample was used for the measurements since non-sanded batches would be affected by the typical staircase effect of UV-3D printing and recycled GFRPs. Therefore, the comparisons would show the reductions in roughness linked to the sanding post-processing and the chromium deposition [37,42].

Table 3 shows the values of roughness average (Ra) and root mean square roughness (Rq) of the different batches. The lower results were reached by batch n. 4, which represents the post-processed and coated batch of samples. The overall average reduction in terms of Ra and Rq from batch n. 1 to batch n. 4 is equal to 18.22 and 22.61 μm, which means more than 99.5% in both cases. The PVD sputtering of chromium onto a non-sanded rough 3D-printed substrate (batch n. 2) halved the Ra and Rq values, while a fifty-fold decrease was observed by carrying out the PVD sputtering onto a sanded substrate (batch n. 4). Furthermore, the values show good accuracy of the measurements considering the low standard deviations, especially for the PVD sputtering batches, confirming the homogeneity of the chromium coatings. These values were also obtained thanks to the primer deposition, which improved the roughness homogeneity of the substrate by reducing the peaks and valleys of the 3D-printed surface.

Table 3. Vertical surface roughness average (Ra) and root mean square roughness (Rq) of the sample batches.

Sample (Batch)	Ra (μm)	Rq (μm)
N. 1	18.25 ± 2.06	22.66 ± 1.75
N. 2	9.51 ± 1.39	12.25 ± 2.00
N. 3	1.60 ± 0.33	2.58 ± 0.67
N. 4	0.03 ± 0.01	0.05 ± 0.01

3.3. SEM Analysis

The cross-sections were analyzed through SEM micrographs to determine the average thickness of the PVD sputtering coating and assess its morphology. Two different cross-sections were evaluated for each sample of batch n. 2 (Figure 2). According to the results from the roughness measurements (Table 3), this batch allows better observation of the influence of the staircase effect on the PVD sputtering coating because of its higher Ra, Rq, and corresponding standard deviations compared to the sanded sputtered sample.

As shown in Figure 2, the PVD sputtering coating is quite visible because of the evident difference between the recycled GFRP substrate and the UV primer. The former can be detected for the presence of fibers, whereas the latter corresponds to the less irregular area. However, the interface between the substrate and the coating is not clearly distinguishable, giving only an idea of the general contours of the 3D-printed substrate. This aspect is probably due to the composition of the UV primer, which is similar to the acrylic-based matrix of the 3D printable ink. For this reason, the adhesion between the substrate and the UV primer may be increased, although this property of the PVD sputtering coating is strongly affected also by the surface roughness of the sample.

The chromium layer is not easily visible because of its low contrast with the primer and the substrate, as shown in Figure 3. Hence, its average thickness was not measured nor compared with its nominal value of 200 nm. Indeed, the average thickness of the whole PVD sputtering coating was evaluated, which means the chromium and the UV primer layers. The latter was sprayed onto the surface before the PVD sputtering process to improve the adhesion of the Cr layer, and the nominal value of its thickness is equal to 60–70 μm. The measurements were carried out during the SEM analysis by measuring three points of at least two micrographs for each cross-section. The experimental average thickness of the PVD sputtering coating is equal to 37.42 ± 21.04 μm. Compared to the nominal thickness, this value is lower than expected, and it could be justified by the Ra and Rq values of batch n. 2. After the spray deposition, the UV primer smoothed the surface of the 3D-printed substrate by filling the valleys, resulting in a less homogeneous layer. However, the non-homogeneous thickness of the coating led to lower values of roughness, confirming the overall quality of the PVD sputtering observed during the first qualitative analysis (Section 3.1). Figure 2a,b show the cross-sections of the first specimen of NF_PVD, whereas the cross-sections of the second sample are visible in Figure 2c,d.

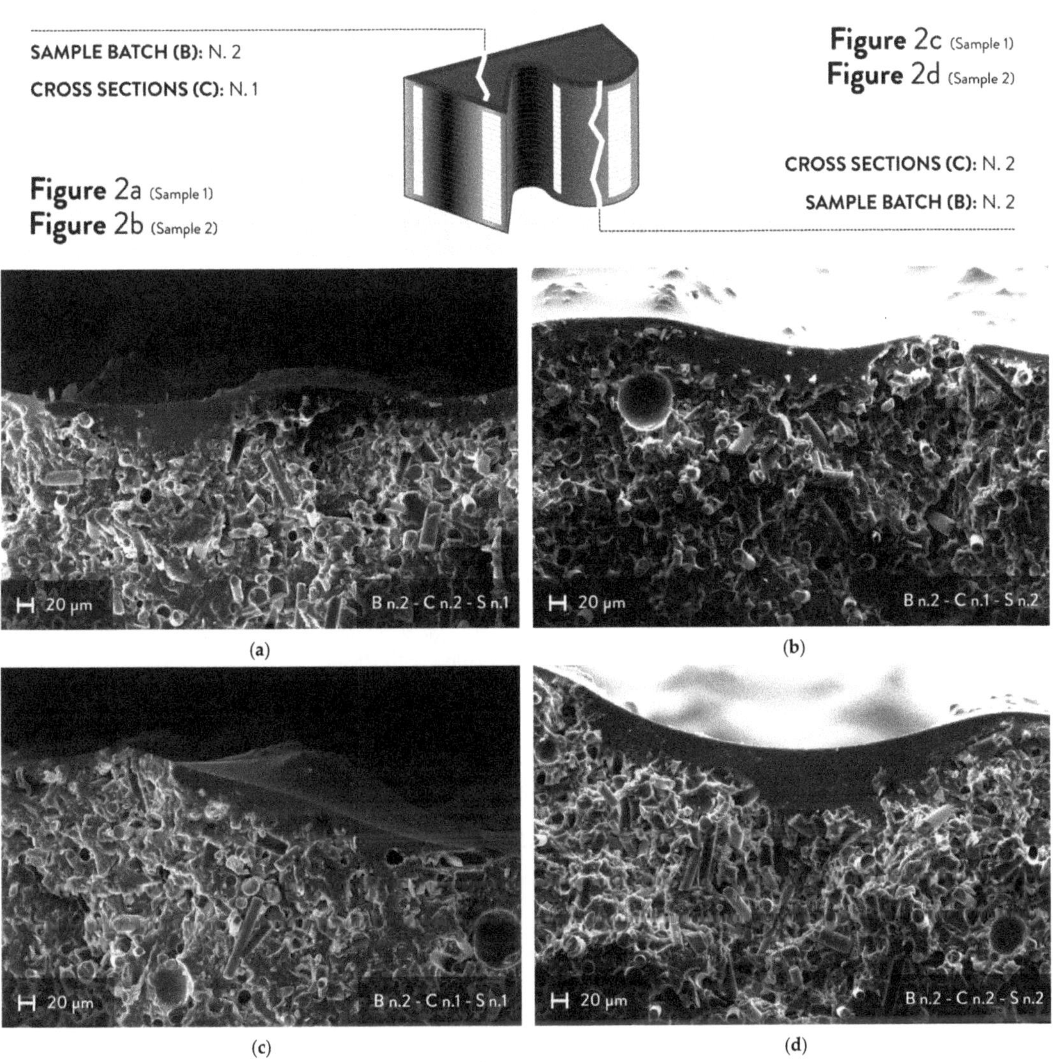

Figure 2. SEM micrographs of cross-sections from batch n. 2 (500× magnification): (**a**) cross-section n. 1, sample n. 1; (**b**) cross-section n. 1, sample n. 2; (**c**) cross-section n. 2, sample n. 1; (**d**) cross-section n. 2, sample n. 2.

The analysis of the cross-sections at higher magnification shows a good adhesion between the UV primer and the recycled glass fibers, represented by the cylindrical structures of Figure 3. Considering the substrate is made of recycled GFRPs, some fibers may be directly exposed to the external surface. As for other composite substrates, this issue may strongly affect the overall quality of the coating since the reinforcement often has different compositions and properties compared to the matrix. In this case, good adhesion of the UV primer may increase instead the homogeneity of the substrate, improving the surface properties of the final coating.

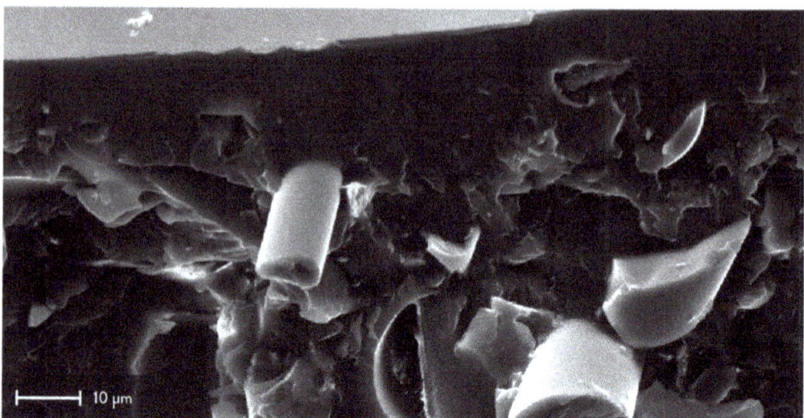

Figure 3. SEM micrographs of a representative cross-section from batch n. 2: insight on the fiber adhesion with the PVD sputtering coating and the UV primer (3k× magnification) highlighted from the cylindrical structure in the middle, which shows a recycled glass fiber.

3.4. Gloss Measurements

After the qualitative assessment of the 3D-printed surfaces appearance in Section 3.1, gloss measurements were performed to define a quantitative value for each batch of samples. Table 4 shows the results from the gloss evaluation. According to ASTM D523-14 [39], the first measurements were performed with the 60° geometry, allowing a general comparison between the four batches. The samples before and after the PVD sputtering show different values, resulting in an increase in gloss after the metallization, from 2–4 GU to 113–130 GU. In general, values higher than 70 GU at 60° refer to high-gloss surfaces, whereas values lower than 10 GU indicate surfaces with low gloss. Gloss measurements for the metallized samples showed higher values of standard deviations. This was probably due to the standard deviation values measured for surface roughness that were of the same order of magnitude of Ra and Rq average values for samples after PVD sputtering (Section 3.2). As a matter of fact, the staircase effect influences gloss values since it contributes to increasing surface roughness values. Compared to smooth surfaces, i.e., the sanded 3D samples, the staircase effect also changes the specular reflectivity of the surface, leading to higher variability of the gloss measurements [37].

Table 4. Gloss values of the sample batches at different angles (20°, 60°, and 85°) according to ASTM D523-14 (Standard Test Method for Specular Gloss) [39].

Sample (Batch)	Standard	GU at 20°	GU at 60°	GU at 85°
N. 1	Figure 1a	//	2.34 ± 0.55	2.52 ± 1.65
N. 2	Figure 1b	39.20 ± 8.81	130.16 ± 15.82	//
N. 3	Figure 1c	//	3.76 ± 1.25	12.44 ± 0.93
N. 4	Figure 1d	15.96 ± 4.85	113.94 ± 30.88	//

Samples before the PVD sputtering from batches n. 1 and n. 3 were then evaluated with the 85° geometry to have more accurate measures. In this case, the gloss value increases for the sample from batch n. 3, which means after the sanding post-processing. The result is in line with the decrease in surface roughness. Batches n. 2 and n. 4 were finally evaluated with the 20° geometry for high gloss surfaces, improving the accuracy of the measurements. These results, together with the roughness measurements, confirm the qualitative assessment of the 3D samples shown in the previous Section 3.1, highlighting the improvement of surface properties for sanded and Cr-sputtered samples, resulting in mirror-like 3D-printed surfaces.

3.5. Adhesion Properties

Manual cross-cut tests were performed on the samples of batch n. 2 to assess the adhesion level of the chromium coating. As mentioned before, the metallized surfaces of this batch are the most affected by the staircase effect of the 3D-printed samples. Therefore, to assess the influence of 3D-printed recycled composites on adhesion, the cross-cut tests were performed only for batch n.2. According to ISO 2409:2020 [40], the adhesion level between the chromium layer and the 3D-printed substrate is equal to 2. In detail, the coating has slightly flacked along the edges of the cross-cut squares, and detachment affected an area between 5 and 15% of the tested surface. This result is significantly affected by the staircase effect that generally shows worse values compared to smooth surfaces, decreasing the adhesion properties [34,43].

3.6. Fields of Application

After the qualitative analysis and surface characterization, some considerations can be made about the possible application fields. Although these coatings were initially used in the electronic and medical sectors [44–46], PVD sputtering on 3D-printed parts may be employed for a larger number of applications since it improves the properties of the parts, i.e., wear resistance, durability, and surface aesthetics [30,47]. Moreover, a coating process can also reduce the effects of the 3D printing process such as the staircase effect, which may influence not only the appearance of a 3D printed product but also its functional properties.

The deposition of metallic layers onto 3D-printed thermoplastic parts has been studied very recently, confirming the increasing interest in its future industrial exploitation [26,31,34,35]. Similarly, using PVD sputtering processes for polymer-based composites may generate potential benefits, especially for recycled substrates. Focusing on the case study of this paper, high-performance applications could be exploited by using coated recycled GFRPs, such as technical parts, outdoor furniture, or customized scenery elements [17,48]. In detail, these kinds of recycled GFRPs exhibit interesting mechanical properties for similar applications, showing good values of elastic modulus (~5.5–6 GPa) and toughness (~200 J/mm^3), which can be comparable to virgin GFRPs [17]. Considering the suitability of these materials for similar applications, the PVD sputtering coating may further enhance their durability by acting as a protective film for the 3D-printed substrate. Moreover, this coating could reduce or eliminate the superficial defects of the DIW process such as voids, which could be a cause for the failure of the parts themselves. Considering the 3D printing process, this coating may also limit the delamination caused by low intralayer adhesion thanks to the use of the acrylic UV primer, which showed good compatibility with the acrylic-based matrix of the 3D printable ink.

PVD sputtering can also improve the homogeneity of the surface of parts made with recycled GFRPs. First, the surface of composite materials often shows local variations in the composition in correspondence to the matrix or the reinforcement, leading to non-homogeneous surface properties. The different batches of recycled glass fibers can significantly change from a morphological point of view, i.e., average granulometry, affecting the technical properties and the overall aesthetics.

According to the previous assumptions, PVD sputtering for 3D-printed recycled composites could be proposed for the automotive and furniture sectors. For the former, metallized 3D-printed recycled composites may be used for high-performing and customized batches of small- and medium-format parts, whereas for the latter it shows good potential for large-format customized products [49,50]. As a matter of fact, this coating technology influences not only the technical properties of a 3D-printed product but also its expressive-sensorial qualities, influencing the perception of customers and end-users [35,51]. In this case, the aesthetics of recycled GFRPs may be a desired feature or not, according to the specific designer's choice and the users' target. However, metallization can add several advantages to the final product made of recycled GFRPs by considering the principles of Design for Sustainability [52,53], i.e., improving the durability of the parts, extending the life cycle of the product, and fostering the acceptance of recycled materials amongst the end-users. Further studies are necessary to check the durability of the 3D-printed GFRP samples coated with a chromium metallization layer. However, some promising pieces of evidence were presented in the literature, showing the increased resistance to water droplet erosion of various GFRPs electro-deposited with metal layers, e.g., Cr on Cu, when compared to samples without metallic coatings [54].

4. Conclusions

This work aimed to demonstrate the feasibility of PVD sputtering for the metallization of 3D-printed polymer-based composites from EoL products. Four batches of samples made of thermoset acrylic resin filled with mechanically recycled GFRPs were fabricated through UV-assisted 3D printing and coated by means of chromium PVD sputtering. From the qualitative assessment of these samples, no visible defects, unprocessed areas, or detachments were detected, showing a good homogeneity of the coating. No geometrical deformations were found, and the staircase effect of the sample was mitigated by the PVD sputtering process. A reduction in the roughness values of more than 99.5% was observed after PVD sputtering, whereas performing the metallization on a sanded batch led to a fifty-fold decrease in these values. The SEM analysis of the cross section showed the influence of the UV primer deposition in smoothing the surface of the substrate, obtaining a more homogeneous coating by filling the valleys derived from the staircase effect. However, the adhesion properties of the metallized coatings were lower than expected, and this issue probably derives from the staircase effect, which strongly influences the roughness of non-sanded samples. Carrying out the cross-cut test on smooth sanded and coated surfaces, which showed a less evident staircase effect, should improve this result. The surface gloss was then evaluated, showing a general increase in gloss after the metallization. However, the staircase effect influenced the measurements leading to higher variability.

According to these results, PVD sputtering onto 3D-printed recycled GFRPs can be exploited for new fields of applications, such as automotive and furniture. This metallization treatment can influence the technical properties and the aesthetics of new 3D-printed products made of recycled composites, allowing industries, designers, and engineers to foster the use of these kinds of materials. Future work should be done to deepen the surface characterization of this surface treatment for DIW, i.e., through more complex geometries. Furthermore, other polymer-based composites could be used, such as virgin and recycled carbon fibers. In this way, additional high-performance sectors will benefit from these promising results, i.e., the aerospace field.

Author Contributions: Conceptualization, A.R. and R.S.; software, A.R.; validation, A.R.; formal analysis, A.R. and P.T.; investigation, A.R. and P.T.; resources, A.R.; data curation, A.R.; writing—original draft preparation, A.R.; writing—review and editing, A.R., P.T., M.L., S.T. and R.S.; visualization, A.R.; supervision, M.L., S.T. and R.S.; project administration, M.L., S.T. and R.S.; funding acquisition, M.L., S.T. and R.S. All authors have read and agreed to the published version of the manuscript.

Funding: This research has received funding from the European Union's Horizon 2020 research and innovation program, under grant agreement No. H2020-730323-1. The present work is part of the FiberEUse project, entitled "Large Scale Demonstration of New Circular Economy Value-chains based on the Reuse of End-of-life reinforced Composites".

Institutional Review Board Statement: Not applicable.

Informed Consent Statement: Not applicable.

Data Availability Statement: Publicly available data sets were analyzed in this study. The data can be found here: [https://github.com/piuLAB-official/Dataset_A.Romani_2022_Materials] (accessed on 28 July 2022). If these data are used, please cite them in the following way: [data set] Alessia Romani, Raffaella Suriano, Paolo Tralli, Marinella Levi, and Stefano Turri. 2022. Metallization of Recycled Glass Fiber Reinforced Polymers 3D-printed by UV-Assisted Direct Ink Writing; https://github.com/piuLAB-official/Dataset_A.Romani_2022_Materials (accessed on 28 July 2022).

Acknowledgments: The authors would like to thank Novellini S.p.a.—Divisione Green Coat for the PVD sputtering metallization and the adhesion test performed on the samples, Siemens Gamesa Renewable Energy S.A. and Consiglio Nazionale di Ricerca—Sistemi e Tecnologie Industriali Intelligenti per il Manifatturiero Avanzato (Stiima-CNR) for supplying the mechanically recycled GFRPs.

Conflicts of Interest: The authors declare no conflict of interest.

References

1. Hsissou, R.; Seghiri, R.; Benzekri, Z.; Hilali, M.; Rafik, M.; Elharfi, A. Polymer Composite Materials: A Comprehensive Review. *Compos. Struct.* **2021**, *262*, 113640. [CrossRef]
2. Witten, E.; Mathes, V. *The European Market for Fibre Reinforced Plastics/Composites in 2021*; Texdata International: Hamburg, Germany, 2022.
3. Gonçalves, R.M.; Martinho, A.; Oliveira, J.P. Recycling of Reinforced Glass Fibers Waste: Current Status. *Materials* **2022**, *15*, 1596. [CrossRef]
4. Ribeiro, M.C.S.; Fiúza, A.; Ferreira, A.; Dinis, M.D.L.; Meira Castro, A.C.; Meixedo, J.P.; Alvim, M.R. Recycling Approach towards Sustainability Advance of Composite Materials' Industry. *Recycling* **2016**, *1*, 178–193. [CrossRef]
5. Beauson, J.; Brøndsted, P. Wind Turbine Blades: An End of Life Perspective. In *MARE-WINT.*; Ostachowicz, W., McGugan, M., Schröder-Hinrichs, J.-U., Luczak, M., Eds.; Springer International Publishing: Cham, Switzerland, 2016; pp. 421–432. ISBN 978-3-319-39094-9.
6. Sommer, V.; Walther, G. Recycling and Recovery Infrastructures for Glass and Carbon Fiber Reinforced Plastic Waste from Wind Energy Industry: A European Case Study. *Waste Manag.* **2021**, *121*, 265–275. [CrossRef] [PubMed]
7. Adams, R.D.; Collins, A.; Cooper, D.; Wingfield-Digby, M.; Watts-Farmer, A.; Laurence, A.; Patel, K.; Stevens, M.; Watkins, R. Recycling of Reinforced Plastics. *Appl. Compos. Mater.* **2014**, *21*, 263–284. [CrossRef]
8. Campbell-Johnston, K.; Vermeulen, W.J.V.; Reike, D.; Brullot, S. The Circular Economy and Cascading: Towards a Framework. *Resour. Conserv. Recycl.* **2020**, *7*, 100038. [CrossRef]
9. Bank, L.; Arias, F.; Yazdanbakhsh, A.; Gentry, T.; Al-Haddad, T.; Chen, J.-F.; Morrow, R. Concepts for Reusing Composite Materials from Decommissioned Wind Turbine Blades in Affordable Housing. *Recycling* **2018**, *3*, 3. [CrossRef]
10. Zindani, D.; Kumar, K. An Insight into Additive Manufacturing of Fiber Reinforced Polymer Composite. *Int. J. Lightweight Mater. Manuf.* **2019**, *2*, 267–278. [CrossRef]
11. Ngo, T.D.; Kashani, A.; Imbalzano, G.; Nguyen, K.T.Q.; Hui, D. Additive Manufacturing (3D Printing): A Review of Materials, Methods, Applications and Challenges. *Compos. Part B Eng.* **2018**, *143*, 172–196. [CrossRef]
12. Griffini, G.; Natale, G.; Invernizzi, M.; Postiglione, G.; Turri, S.; Levi, M. 3D-Printable CFR Polymer Composites with Dual-Cure Sequential IPNs. *Polymer* **2016**, *91*, 174–178. [CrossRef]
13. Invernizzi, M.; Natale, G.; Levi, M.; Turri, S.; Griffini, G. UV-Assisted 3D Printing of Glass and Carbon Fiber-Reinforced Dual-Cure Polymer Composites. *Materials* **2016**, *9*, 583. [CrossRef] [PubMed]
14. Compton, B.G.; Lewis, J.A. 3D-Printing of Lightweight Cellular Composites. *Adv. Mater.* **2014**, *26*, 5930–5935. [CrossRef] [PubMed]
15. Hao, W.; Liu, Y.; Zhou, H.; Chen, H.; Fang, D. Preparation and Characterization of 3D Printed Continuous Carbon Fiber Reinforced Thermosetting Composites. *Polym. Test.* **2018**, *65*, 29–34. [CrossRef]
16. Rahimizadeh, A.; Kalman, J.; Fayazbakhsh, K.; Lessard, L. Recycling of Fiberglass Wind Turbine Blades into Reinforced Filaments for Use in Additive Manufacturing. *Compos. Part B Eng.* **2019**, *175*, 107101. [CrossRef]
17. Romani, A.; Mantelli, A.; Suriano, R.; Levi, M.; Turri, S. Additive Re-Manufacturing of Mechanically Recycled End-of-Life Glass Fiber-Reinforced Polymers for Value-Added Circular Design. *Materials* **2020**, *13*, 3545. [CrossRef]
18. Mantelli, A.; Romani, A.; Suriano, R.; Diani, M.; Colledani, M.; Sarlin, E.; Turri, S.; Levi, M. UV-Assisted 3D Printing of Polymer Composites from Thermally and Mechanically Recycled Carbon Fibers. *Polymers* **2021**, *13*, 726. [CrossRef] [PubMed]
19. Huang, H.; Liu, W.; Liu, Z. An Additive Manufacturing-Based Approach for Carbon Fiber Reinforced Polymer Recycling. *CIRP Ann.* **2020**, *69*, 33–36. [CrossRef]
20. Tian, X.; Liu, T.; Wang, Q.; Dilmurat, A.; Li, D.; Ziegmann, G. Recycling and Remanufacturing of 3D Printed Continuous Carbon Fiber Reinforced PLA Composites. *J. Clean. Prod.* **2017**, *142*, 1609–1618. [CrossRef]
21. Giani, N.; Mazzocchetti, L.; Benelli, T.; Picchioni, F.; Giorgini, L. Towards Sustainability in 3D Printing of Thermoplastic Composites: Evaluation of Recycled Carbon Fibers as Reinforcing Agent for FDM Filament Production and 3D Printing. *Compos. Part A Appl. Sci. Manuf.* **2022**, *159*, 107002. [CrossRef]
22. Gonzalez, R.; Ashrafizadeh, H.; Lopera, A.; Mertiny, P.; McDonald, A. A Review of Thermal Spray Metallization of Polymer-Based Structures. *J. Therm. Spray Tech.* **2016**, *25*, 897–919. [CrossRef]
23. Baptista, A.; Silva, F.J.G.; Porteiro, J.; Míguez, J.L.; Pinto, G.; Fernandes, L. On the Physical Vapour Deposition (PVD): Evolution of Magnetron Sputtering Processes for Industrial Applications. *Procedia Manuf.* **2018**, *17*, 746–757. [CrossRef]
24. Krella, A. Resistance of PVD Coatings to Erosive and Wear Processes: A Review. *Coatings* **2020**, *10*, 921. [CrossRef]
25. Ferreira, A.A.; Silva, F.J.G.; Pinto, A.G.; Sousa, V.F.C. Characterization of Thin Chromium Coatings Produced by PVD Sputtering for Optical Applications. *Coatings* **2021**, *11*, 215. [CrossRef]

26. White, J. Environmentally Benign Metallization of Material Extrusion Technology 3D Printed Acrylonitrile Butadiene Styrene Parts Using Physical Vapor Deposition. *Addit. Manuf.* **2018**, *22*, 279–285. [CrossRef]
27. Kucherov, F.A.; Romashov, L.V.; Ananikov, V.P. Development of 3D+G Printing for the Design of Customizable Flow Reactors. *Chem. Eng. J.* **2022**, *430*, 132670. [CrossRef]
28. Cheng, P.-W.; Chen, C.-Y.; Ichibayashi, T.; Chang, T.-F.M.; Sone, M.; Nishimura, S. Metallization of 3D-Printed Polymer Structures via Supercritical Carbon Dioxide-Assisted Electroless Plating. *MRS Commun.* **2021**, *11*, 278–282. [CrossRef]
29. Khan, M.S.; Mishra, S.B.; Kumar, M.A.; Banerjee, D. Optimizing Surface Texture and Coating Thickness of Nickel Coated ABS-3D Parts. *Mater. Today Proc.* **2018**, *5*, 19011–19018. [CrossRef]
30. Afshar, A.; Mihut, D. Enhancing Durability of 3D Printed Polymer Structures by Metallization. *J. Mater. Sci. Technol.* **2020**, *53*, 185–191. [CrossRef]
31. Lavecchia, F.; Percoco, G.; Pei, E.; Galantucci, L.M. Computer Numerical Controlled Grinding and Physical Vapor Deposition for Fused Deposition Modelled Workpieces. *Adv. Mater. Sci. Eng.* **2018**, *2018*, 9037490. [CrossRef]
32. Equbal, A.; Sood, A. Metallization on FDM Parts Using the Chemical Deposition Technique. *Coatings* **2014**, *4*, 574–586. [CrossRef]
33. Keough, M.; McLeod, J.F.; Salomons, T.; Hillen, P.; Pei, Y.; Gibson, G.; McEleney, K.; Oleschuk, R.; She, Z. Realizing New Designs of Multiplexed Electrode Chips by 3-D Printed Masks. *RSC Adv.* **2021**, *11*, 21600–21606. [CrossRef] [PubMed]
34. Baek, I.; Lim, C.-M.; Park, K.Y.; Ryu, B.K. Enhanced Metal Coating Adhesion by Surface Modification of 3D Printed PEKKs. *Coatings* **2022**, *12*, 854. [CrossRef]
35. Romani, A.; Mantelli, A.; Tralli, P.; Turri, S.; Levi, M.; Suriano, R. Metallization of Thermoplastic Polymers and Composites 3D Printed by Fused Filament Fabrication. *Technologies* **2021**, *9*, 49. [CrossRef]
36. Arlington, S.Q.; Lakshman, S.V.; Barron, S.C.; DeLisio, J.B.; Rodriguez, J.C.; Narayanan, S.; Fritz, G.M.; Weihs, T.P. Exploring Material Chemistry for Direct Ink Writing of Reactively Formed Conductors. *Mater. Adv.* **2020**, *1*, 1151–1160. [CrossRef]
37. Žigon, J.; Kariž, M.; Pavlič, M. Surface Finishing of 3D-Printed Polymers with Selected Coatings. *Polymers* **2020**, *12*, 2797. [CrossRef]
38. Tymms, C.; Zorin, D.; Gardner, E.P. Tactile Perception of the Roughness of 3D-Printed Textures. *J. Neurophysiol.* **2017**, *119*, 862–876. [CrossRef]
39. *ASTM D523-14*; Test Method for Specular Gloss. ASTM International: West Conshohocken, PA, USA, 2018.
40. *EN ISO 2409:2020*; Paints and Varnishes. Cross-Cut Test. International Standard Organization (ISO): Geneva, Switzerland, 2020.
41. Lewis, J.A. Direct Ink Writing of 3D Functional Materials. *Adv. Funct. Mater.* **2006**, *16*, 2193–2204. [CrossRef]
42. Poornaganti, S.; Yeole, S.N.; Kode, J.P. Insights on Surface Characterization of 3D Printed Polymeric Parts. *Mater. Today Proc.* **2022**, *62*, 3837–3848. [CrossRef]
43. Bernasconi, R.; Natale, G.; Levi, M.; Magagnin, L. Electroless Plating of PLA and PETG for 3D Printed Flexible Substrates. *ECS Trans.* **2015**, *66*, 23–35. [CrossRef]
44. Xiao, R.; Feng, X.; Fan, R.; Chen, S.; Song, J.; Gao, L.; Lu, Y. 3D Printing of Titanium-Coated Gradient Composite Lattices for Lightweight Mandibular Prosthesis. *Compos. Part B Eng.* **2020**, *193*, 108057. [CrossRef]
45. Huang, T.-Y.; Sakar, M.S.; Mao, A.; Petruska, A.J.; Qiu, F.; Chen, X.-B.; Kennedy, S.; Mooney, D.; Nelson, B.J. 3D Printed Microtransporters: Compound Micromachines for Spatiotemporally Controlled Delivery of Therapeutic Agents. *Adv. Mater.* **2015**, *27*, 6644–6650. [CrossRef] [PubMed]
46. Sorocki, J.; Piekarz, I. Low-Cost Microwave Components' Fabrication in Hybrid Technology of Laminates and Additive Manufacturing on an Example of Miniaturized Suspended Directional Coupler. *IEEE Access* **2020**, *8*, 128766–128775. [CrossRef]
47. Abegunde, O.O.; Akinlabi, E.T.; Oladijo, O.P.; Akinlabi, S.; Ude, A.U.; Abegunde, O.O.; Akinlabi, E.T.; Oladijo, O.P.; Akinlabi, S.; Ude, A.U. Overview of Thin Film Deposition Techniques. *AIMS Mater. Sci.* **2019**, *6*, 174–199. [CrossRef]
48. Mantelli, A.; Romani, A.; Suriano, R.; Levi, M.; Turri, S. Direct Ink Writing of Recycled Composites with Complex Shapes: Process Parameters and Ink Optimization. *Adv. Eng. Mater.* **2021**, *23*, 2100116. [CrossRef]
49. Constantin, R.; Miremad, B. Performance of Hard Coatings, Made by Balanced and Unbalanced Magnetron Sputtering, for Decorative Applications. *Surf. Coat. Technol.* **1999**, *120–121*, 728–733. [CrossRef]
50. Carneiro, E.; Parreira, N.M.G.; Vuchkov, T.; Cavaleiro, A.; Ferreira, J.; Andritschky, M.; Carvalho, S. Cr-Based Sputtered Decorative Coatings for Automotive Industry. *Materials* **2021**, *14*, 5527. [CrossRef]
51. Chen, X.; Shao, F.; Barnes, C.; Childs, T.; Henson, B. Exploring Relationships between Touch Perception and Surface Physical Properties. *Int. J. Des.* **2009**, *3*, 67–76.
52. Vezzoli, C.; Ceschin, F.; Osanjo, L.; M'Rithaa, M.K.; Moalosi, R.; Nakazibwe, V.; Diehl, J.C. Design for Sustainability: An Introduction. In *Designing Sustainable Energy for All: Sustainable Product-Service System Design Applied to Distributed Renewable Energy*; Vezzoli, C., Ceschin, F., Osanjo, L., M'Rithaa, M.K., Moalosi, R., Nakazibwe, V., Diehl, J.C., Eds.; Springer International Publishing: Cham, Switzerland, 2018; pp. 103–124. ISBN 978-3-319-70223-0.
53. Vezzoli, C. The "Material" Side of Design for Sustainability. In *Materials Experience*; Elsevier: Amsterdam, The Netherlands, 2014; pp. 105–121. ISBN 978-0-08-099359-1.
54. Lammel, P.; Whitehead, A.H.; Simunkova, H.; Rohr, O.; Gollas, B. Droplet Erosion Performance of Composite Materials Electroplated with a Hard Metal Layer. *Wear* **2011**, *271*, 1341–1348. [CrossRef]

 materials

Article

The Effect of Functional Gradient Material Distribution and Patterning on Torsional Properties of Lattice Structures Manufactured Using MultiJet Fusion Technology

Yeabsra Mekdim Hailu [1,2], Aamer Nazir [3,*], Shang-Chih Lin [1,4] and Jeng-Ywan Jeng [1,2,5,*]

1 High Speed 3D Printing Research Center, National Taiwan University of Science and Technology, No. 43, Section 4, Keelung Road, Taipei 10607, Taiwan; M10803811@mail.ntust.edu.tw (Y.M.H.); xyz@mail.ntust.edu.tw (S.-C.L.)
2 Department of Mechanical Engineering, National Taiwan University of Science and Technology, No. 43, Section 4, Keelung Road, Taipei 10607, Taiwan
3 Department of Industrial and Systems Engineering, The Hong Kong Polytechnic University, Hung Hom, Kowloon, Hong Kong SAR 999077, China
4 Graduate Institute of Biomedical Engineering, National Taiwan University of Science and Technology, No. 43, Section 4, Keelung Road, Taipei 10607, Taiwan
5 President Office, Lunghwa University of Science and Technology, No.300, Section 1, Wanshou Rd. Guishan District, Taoyuan City 333326, Taiwan
* Correspondence: aamernazir.an@mail.ntust.edu.tw (A.N.); jeng@mail.ntust.edu.tw (J.-Y.J.)

Abstract: Functionally graded lattice structures have attracted much attention in engineering due to their excellent mechanical performance resulting from their optimized and application-specific properties. These structures are inspired by nature and are important for a lightweight yet efficient and optimal functionality. They have enhanced mechanical properties over the uniform density counterparts because of their graded design, making them preferable for many applications. Several studies were carried out to investigate the mechanical properties of graded density lattice structures subjected to different types of loadings mainly related to tensile, compression, and fatigue responses. In applications related to biomedical, automotive, and aerospace sectors, dynamic bending and rotational stresses are critical load components. Therefore, the study of torsional properties of functionally gradient lattice structures will contribute to a better implementation of lattice structures in several sectors. In this study, several functionally gradient triply periodic minimal surfaces structures and strut-based lattice structures were designed in cylindrical shapes having 40% relative density. The HP Multi Jet Fusion 4200 3D printer was used to fabricate all specimens for the experimental study. A torsional experiment until the failure of each structure was conducted to investigate properties of the lattice structures such as torsional stiffness, energy absorption, and failure characteristics. The results showed that the stiffness and energy absorption of structures can be improved by an effective material distribution that corresponds to the stress concentration due to torsional load. The TPMS based functionally gradient design showed a 35% increase in torsional stiffness and 15% increase in the ultimate shear strength compared to their uniform counterparts. In addition, results also revealed that an effective material distribution affects the failure mechanism of the lattice structures and delays the plastic deformation, increasing their resistance to torsional loads.

Keywords: lattice structures; torsion; design; additive manufacturing; functionally gradient lattice; torsional stiffness; energy absorption; failure behavior

Citation: Hailu, Y.M.; Nazir, A.; Lin, S.-C.; Jeng, J.-Y. The Effect of Functional Gradient Material Distribution and Patterning on Torsional Properties of Lattice Structures Manufactured Using MultiJet Fusion Technology. *Materials* **2021**, *14*, 6521. https://doi.org/10.3390/ma14216521

Academic Editors: Swee Leong Sing and Wai Yee Yeong

Received: 7 October 2021
Accepted: 27 October 2021
Published: 29 October 2021

1. Introduction

Gradient forms of structures are abundant in nature and are designed to have optimal energy efficiency and adapt well to their ecosystem, making them structurally and functionally optimized [1] for applications including load-bearing and support with high mechanical efficiency, contact damage resistance from loads including impact, indentation and sliding forces, interfacial strengthening and toughening of dissimilar components, and other functional benefits [2]. Some natural examples include the bamboo plant stem [3–5], butterfly wings' scales [6], the nanoporous sucker ring tooth of a squid [7,8], a horse hoof [9], and the spruce stem [10,11]. Generally, the nature of gradients can be in three basic forms. These are gradient in a gradual manner, step-wise manner, or gradient throughout the entire volume [12].

A functional gradient is an essential tool for structural components and lattice structures to control site-specific properties within the structure. It is important in designing structures that mimic natural systems which can have specific and optimal responses to applied loads. A gradient structure can be designed by varying one or more of the following parameters: unit cell size, wall thickness/strut diameter, strut length [1]. The overall procedure taken when designing functionally graded lattice structures includes four main steps [2]: (i) selection or design of the unit cell, (ii) patterning or tessellation of unit cells in the design domain, (iii) relative density distribution for gradient density, and (iv) reconstruction of the lattice structures for manufacturing. Lattice structures can be graded from a complete solid to a minimum beam or wall thickness according to the creation of the desired volume ratio. They can also be manipulated in a variety of ways to control more exotic properties other than density, such as elongation, vibration damping, Poisson ratio, and electromagnetic effects of structures. The manipulation and ability to create very intricate architectures using lattice structures make them extremely difficult to manufacture using conventional fabrication methods. However, Additive manufacturing (AM) processes provide the freedom to manufacture complex parts with the desired accuracy without worrying about the complexity of structures [3]. It provides a wide range of advantages that enable the fabrication of complex functionally gradient structures, including reduced material wastage, the elimination of tooling and fixture preparation, and the reduced cost of labor and machinery. Several AM technologies and processes can also be integrated, forming a high-speed system that is essential to achieve speed, accuracy, and surface finish [4]. This enables engineers to create complex structures that are highly functional without the limitations of manufacturability.

Several studies show that functionally graded lattice structures have superior performance and enhanced properties than uniform density lattice structures. The investigation of the compressive properties of additively manufactured cubic and honeycomb lattice structures showed that plateau stress and the energy absorption of functionally graded structures is higher than the uniform density samples by up to 62% and 72%, respectively [5]. A study performed on continuously graded micro lattices revealed that functional gradient structures exhibit progressive layer-by-layer deformation regardless of unit cell size and build direction favorable for a uni-directional impact absorption compared to uniform density structures that tend to fail diagonally or randomly [6]. The investigation of fatigue properties of additively manufactured gyroid lattice structures showed that graded variants have 1.21–1.67 times higher fatigue life than uniform counterparts having identical volume fractions [7]. Geometrically gradient lattice structures were found to have controllable deformation features and effectively increase the energy absorption efficiency of structures [8]. A quasi-static and dynamic compression test performed in gradient lattice structures manufactured using the SLM process showed that maximum deformation energy depends on the gradient density and increases with increasing relative density [9]. This gradually changing topology resulted in specific mechanical properties, making them good candidates for enhanced energy absorption applications. Studies performed on the impact properties re-entrant auxetic lattice structures applied to a crash box revealed that the gradient lattice structure efficiently improved the crashworthy-ness criteria of the thin-walled columns on pedestrian protection because of their high impact performances [10]. A comparison made between additively manufactured sheet-based and strut-based gyroid cellular structures with graded densities subjected to a compression load showed that the sheet-based structures are more isotropic and have a larger elastic modulus than the strut-based counterparts [11]. It was also proved that the graded cellular structures have excellent deformation and mechanical behaviors, with the sheet-based structures having better energy absorption over the strut-based ones.

Out of the several types of lattice structures, triply periodic minimal surfaces, TPMS, based lattices have improved mechanical properties and are proven to be the most efficient method for creating functionally gradient structures. They have a balanced combination of specific and axisymmetric stiffness [12], high interconnectivity and a high surface-to-volume ratio [13], excellent energy absorption [14], less stress concentration, and increased permeability [15]. Because of their advantages, TPMS structures are strong candidates for different applications in different industries. They have proven to be a crucial and versatile solution for the biomedical sector in designing biomorphic scaffolds because of their lightweight property, as well as porous and interconnected nature, which better mimics the natural human bone structure [16], making them ideal for the fabrication of metal implants [17]. TPMS based metal foam structures have also been used for thermal energy storage and energy management applications because of their effective thermal conductivity [18]. When used in heat exchangers, TPMS structures can generate much larger turbulent kinetic energy, resulting in significantly improved thermal performance [19]. Despite their proven advantages, functionally gradient lattice structures are not fully investigated for dynamic loading conditions such as torsion, shear, and bending. This created a lack of an experimental and analytical model that captures their behavior, limiting the range of their applications.

When creating lattices, unit cells can be tessellated in different ways to make structures [20]. These techniques include the direct patterning of unit cells along two or three dimensions (x, y, z) in the design space, the surface geometry to pattern unit cells conformally, and topology optimization to create lattices to fill a 3D space on the design criteria optimally. All the different techniques of patterning unit cells have their applications and advantages in creating lattice structures. Direct patterning is ideal for quickly and simply creating lattice structures to fill a very regularly shaped design volume. On the other hand, conformal patterning is a vital tool to create lattice structures that conform to a given irregular or curved shape to retain the integrity of unit cells. In contrast, topology optimization is essential because it creates unit cells in a desired 3D space and can optimize material distribution in the lattice structures. Topology optimization is usually suitable for creating stochastic lattice structures based on the stress concentration, boundary constraint, permissible loads, final product weight, and other operating conditions. On the other hand, direct patterning works with non-stochastic unit cells that are capable of being tessellated in two or three dimensions. The conformal patterning method can be used regardless of the lattice being stochastic or non-stochastic. For this study, a circular patterning of unit cells that conforms to the cylindrical design volume is adopted to tesselate unit cells along the direction of the twisting force to study its effect on the torsional performances of structures.

This study aims to design diamond-based and strut-based functionally gradient lattice structures based on the stress distribution and failure of circular structures subjected to torsion in regular and circular patterning methods; investigate their torsional properties in terms of torsional stiffness, energy absorption, and failure modes; and study the effect of circular patterning on the torsional resistance and mechanical properties of the lattice structures. It also gives a performance comparison between functionally gradient structures and their uniform counterparts in terms of their torsional stiffness, ultimate strength, and failure modes.

2. Materials and Methods

2.1. Design of Structures

When designing complex lattice structures, an implicit modeling approach is necessary to describe these architectures' geometries precisely. Therefore, nTopology (New York, NY, USA) [21] software is used to design all structures for this study. It uses implicit modeling to represent shapes by using functions rather than b-reps to represent geometries precisely in their true form without initial discretization and by capturing continuity perfectly. Implicit modeling does not rely upon a network of vertices, edges, or faces, making it significantly lighter to compute and design complex structures. Other advantages of nTopology include the ability to use a field-driven approach to increase flexibility and efficiency. Using different field functions, it is possible to manipulate the topology of structures and make functionally gradient designs as per design requirements. In this study, five different structures are designed in a cylindrical design domain according to the standard ASTM E143-13 Standard Test Method for Shear Modulus at Room Temperature [22]. Two of the designs are diamond and vertical-inclined unit cells with a regular direct patterning of unit cells. The remaining three designs follow a circular patterning of unit cells along the length of the structure, as shown in Figure 1 below.

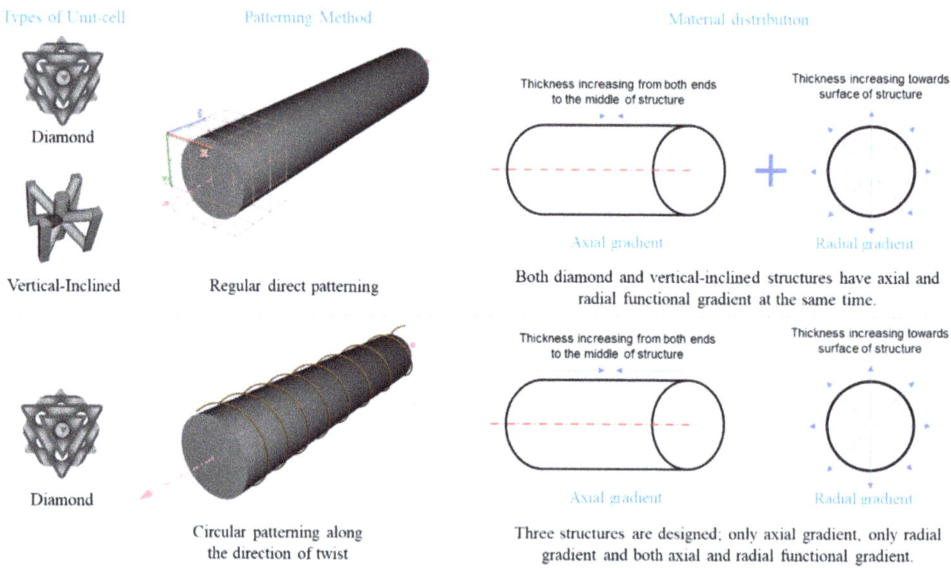

Figure 1. Schematics of functional gradient structures having regular and circular patterning.

For a cylindrical-shaped structure subjected to torsional loading, the shear stress varies linearly with the radial distance with the maximum shear stress occurring at the surface of the shaft. Therefore, for cylindrical structures subjected to torsion, there will be a high stress concentration on members on the outer surface of the structure. For ductile materials, failure due to torsion occurs along the plane of maximum shear stress, usually perpendicular to the structure's axis. On the other hand, brittle materials tend to fail along the plane of maximum tensile stress, which is at an angle to the structure's axis. In both cases, the maximum stress concentration is at the structure's surface and closer to the middle of the structure. Figure 2 illustrates the stress concentration and failure due to torsion in circular elements subjected to torsion.

As illustrated in the above figure, maximum shear stress occurs at the surface of the structures, and failure occurs at the middle of the shaft for both ductile and brittle materials. Therefore, functionally gradient lattice structures have a material distribution driven by high-stress concentration areas for cylindrical structures subjected to torsion. Figure 3 shows the schematic representation of functional gradient structures designed for this study.

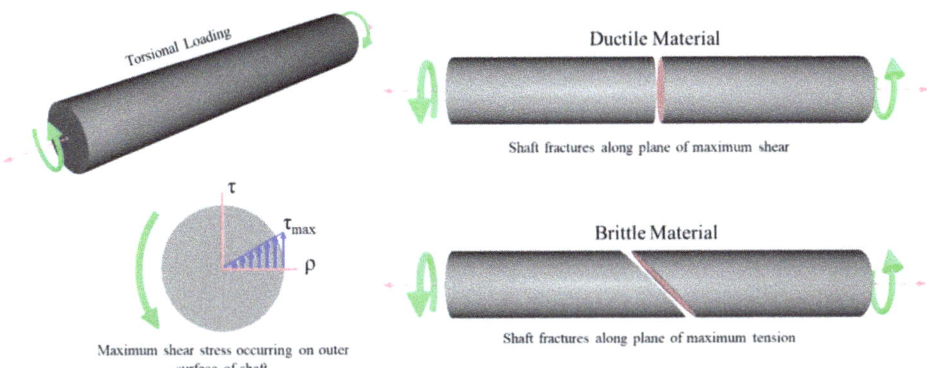

Figure 2. Stress concentration in torsional loading and failure of ductile and brittle materials.

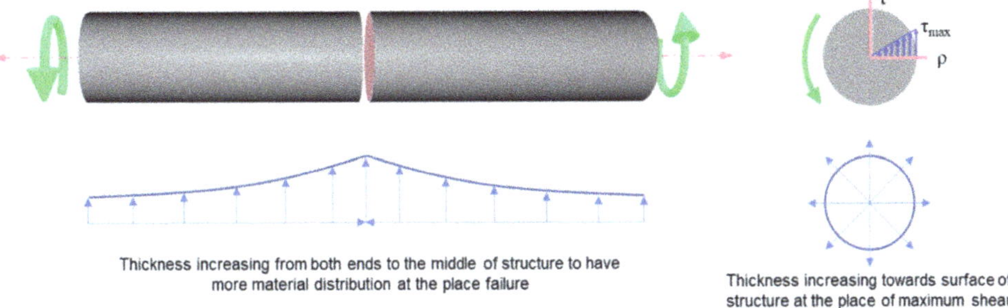

Figure 3. Schematic representation of functionally gradient lattice structures.

The overall dimensions of the structures are shown in Figure 4 below. The length of the structures is 100 mm without including grips, and the diameter is 25 mm. The hexagonal grips at both ends of the structures have a length of 16 mm inscribed in a 25 mm circle. These grips are designed to fit the hexagonal socket fixture used to hold the specimen in the grips of the torsion testing machine. Figure 4 shows the overall dimensions and design procedures used to design the lattice structures for this study. The two structures with regular direct patterning having a functional gradient in both axial and radial direction made from diamond and vertical-inclined unit cells are referred to as D and VI, respectively. The structure with circular patterning with only an axial functional gradient is called CA; the structure with circular patterning with only a radial functional gradient is called CR. The structure with circular patterning having both axial and functional gradients is referred to as CF.

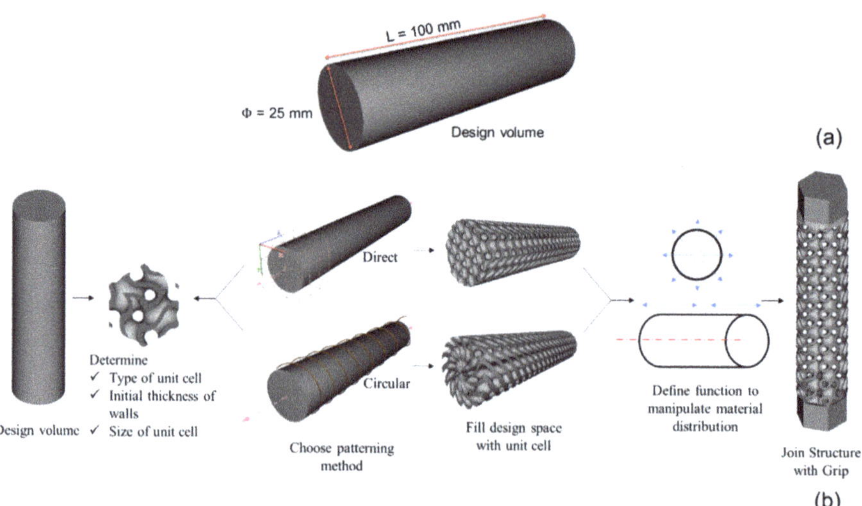

Figure 4. (**a**) Design volume used for all structures, (**b**) Overall procedure used to design functionally gradient lattice structures.

The general overall procedure to design all structures includes using TPMS blocks with an offsetting ramp function to drive their material distribution in the desired radial or longitudinal direction with the required wall thickness. All structures were designed with around 40% relative density. Structure D is designed with an initial unit cell thickness of 1.1 mm and 10 mm × 10 mm × 10 mm unit cell size. A radial thickness variation that varies from 1.1 mm from the central axis to 1.6 mm to the outer surface is achieved using an offset body function. This radial variation is later used to create a longitudinal density gradient which increases from 1.6 mm from the left and right end to 1.85 mm towards the middle of the structure. The second structure, VI, starts having four outer vertical struts having 2 mm, a middle vertical strut of 1.6 mm, and inclined struts of 1.2 mm diameter. Using the ramp function, a radial thickness variation that increases from 2 mm to 2.25 mm towards the outer surface is created, which is then used to create a longitudinal variation in which the unit cell thickness increases to 2.45 mm toward the middle of the structure from the left and right ends. The CA structure only has a longitudinal density gradient which increases from the left and right ends towards the middle from 1.2 mm to 1.9 mm. On the other hand, the CR structure only has a radial density gradient of 1 mm at the central axis and increases linearly to 1.65 mm towards the outer surface. The final structure, CF, combines both longitudinal and radial density gradients. The radial variation has a unit cell thickness that increases from 1 mm to 1.4 mm from the central axis to the outer surface, and the longitudinal density gradient increases from 1.4 mm, from the left and right ends, to 1.65 mm to the middle of the structure. Figure 5 shows all the density gradients and unit cell thicknesses discussed in the above paragraph.

After designing each structure, inspections blocks are used to calculate the mass, volume, and minimum pore size of structures to ensure all structures have similar masses and an outside pore size that will allow powder removal without difficulty. Figure 6 shows all five designs joined with hexagonal grips and their mass volume and minimum pore size. It shows that all structures have very close values to compare their performances.

Table 1 below shows the unit cell size, wall thickness, relative density, and minimum pore size of the structures designed for this study. From this table, we can see that all structures are designed with the same relative density.

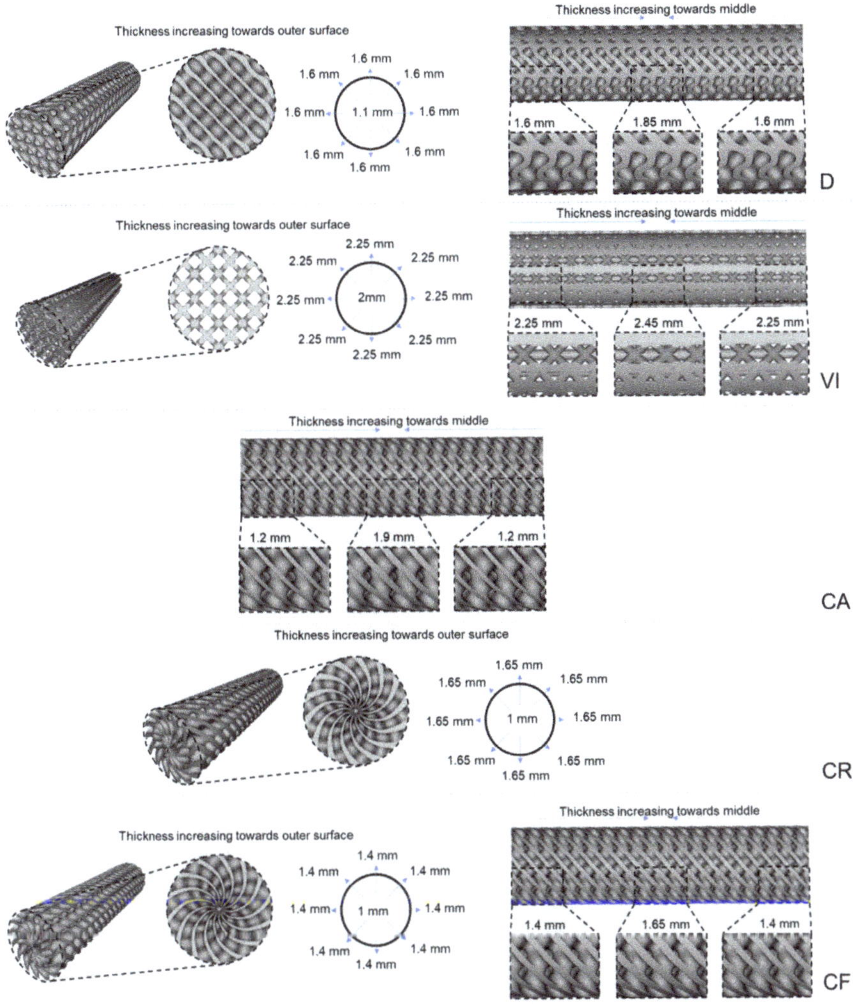

Figure 5. Longitudinal and radial density gradient in five functional gradient designs.

Table 1. Summary of dimensions for the five functionally gradient structures.

Design	Unit Cell Size (mm)	Wall Thickness (mm)	Relative Density (%)	Minimum Pore Size (mm)
D	$10 \times 10 \times 10$	1.6–1.85	41.59	2.3
VI	$6 \times 6 \times 6$	2.25–2.45	41.11	2.0
CA	$7, 10 \times 10$	1.2–1.9	41.09	2.5
CR	$7, 10 \times 10$	1.0–1.65	40.81	2.1
CF	$7, 10 \times 10$	1.25–1.65	41.32	2.0

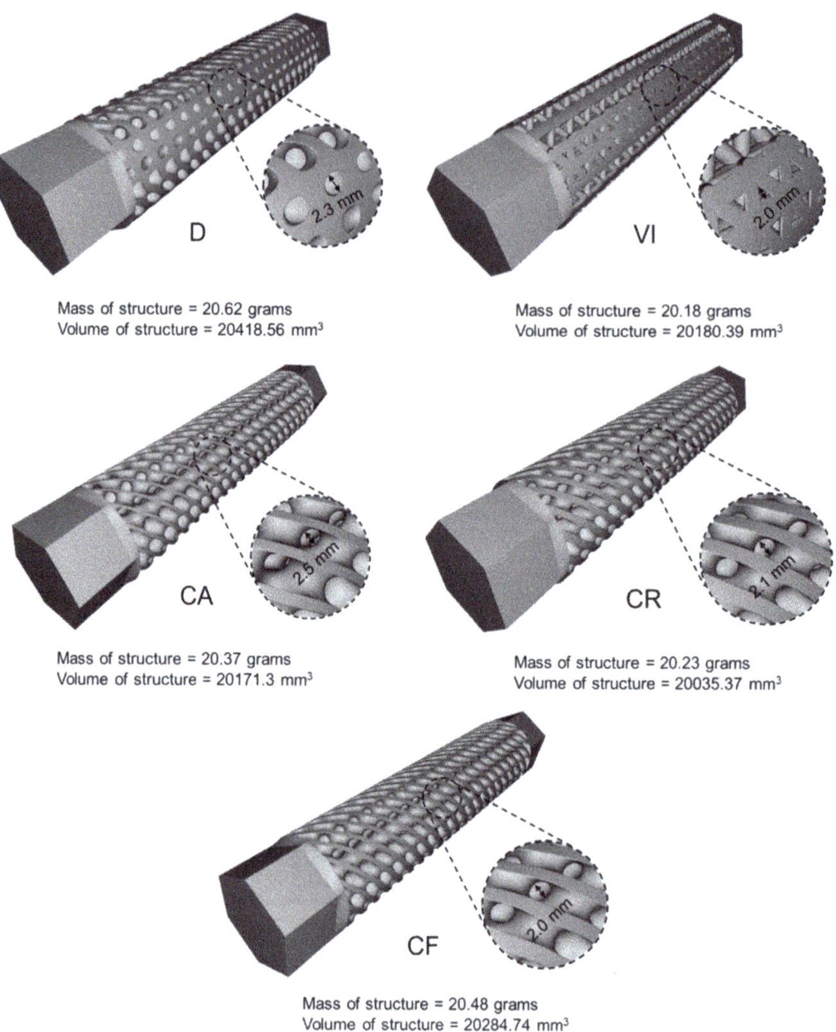

Figure 6. Mass, volume, and minimum pore size of the five functional gradient structures.

2.2. Additive Manufacturing of Structures

All samples of lattice structures are manufactured using a high-speed HP Multi Jet Fusion (MJF) 4200 series 3D Printer [23]. This additive manufacturing machine uses a technology that is a hybrid between powder bed fusion (PBF) and binder jetting (BJ) processes. These two technologies are among the most common processes used for polymer and metal 3D printing technology. The main difference between these two technologies is how the geometrical shape of the printed part is achieved. The HP MJF technology creates layers by depositing a fusing agent on a powder bed, such as BJ, and uses a heating source to fuse powder particles, such as PBF, and form parts that are final and do not require further sintering in a furnace. Each specimen is fabricated using a Nylon Polyamide 12 (PA 12) powder, a robust thermoplastic material that produces high-density parts with balanced profiles and robust structures. PA 12 material has excellent mechanical and elastic properties [24] and provides good chemical resistance to oils, greases, aliphatic hydrocarbons, and alkalis. It is also ideal for complex assemblies, housings, enclosures, and watertight applications. This material, combined with HP MJF solutions, can achieve a low cost per part and minimized waste due to its balanced reusability and performance characteristics. Figure 7 shows the HP MJF 4200 printer and post-processing station used for this study.

Figure 7. (**a**) HP MJF 4200 printer, and (**b**) HP MJF Post-processing station.

For the torsional experiment, three samples for each design are manufactured. All the printed samples are individually weighed and also measured for their dimensional accuracy. The tolerances of grip structures for holding fixtures are also tested and proven to be accurate enough. All dimensions, as well as weight, are compared with initial design values. Almost all samples printed for each lattice structure have printed weights slightly less than design weights. This is due to the inherent porosity of 3D printing technologies and the inability to achieve 100% dense parts. On the other hand, a few specimens have printed weights slightly larger than design weights due to unremoved powder particles in the sandblasting process due to the complexity of structures. Figure 8 shows one specimen from each of the five functionally gradient lattice structures. It can be seen from the figure that printed parts have good quality.

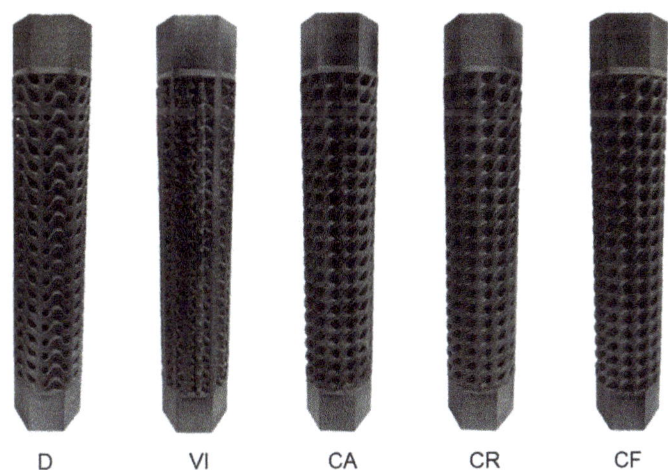

Figure 8. Additively manufactured samples of each lattice structure.

The summary of weights of printed specimens and a comparison with CAD values is shown in Table 2 below.

Table 2. Weight of printed samples and comparison with CAD value.

Design	Weight (g)				Specimen Average (g)	The Difference with CAD Value
	CAD	Specimen 1	Specimen 2	Specimen 3		
D	33.7	31.7	31.8	31.2	31.6	6.4%
VI	33.2	31.2	31.1	31.3	31.2	6.2%
CA	33.5	31.4	33.1	32.7	32.4	3.3%
CR	33.4	32.0	33.0	33.4	32.8	1.8%
CF	33.6	32.42	32.56	32.67	32.5	3.3%

2.3. Experimental Setup

The torsion experiment is carried out using the Material Testing System machine (Mini MTS 858, MTS Corporation, Minneapolis, MN, USA). The test is carried out at a speed of 30 degrees/min, which is high enough to make creep negligible [25]. Figure 9 below shows the test equipment, experimental setup for torsional testing, and specimens of the structures.

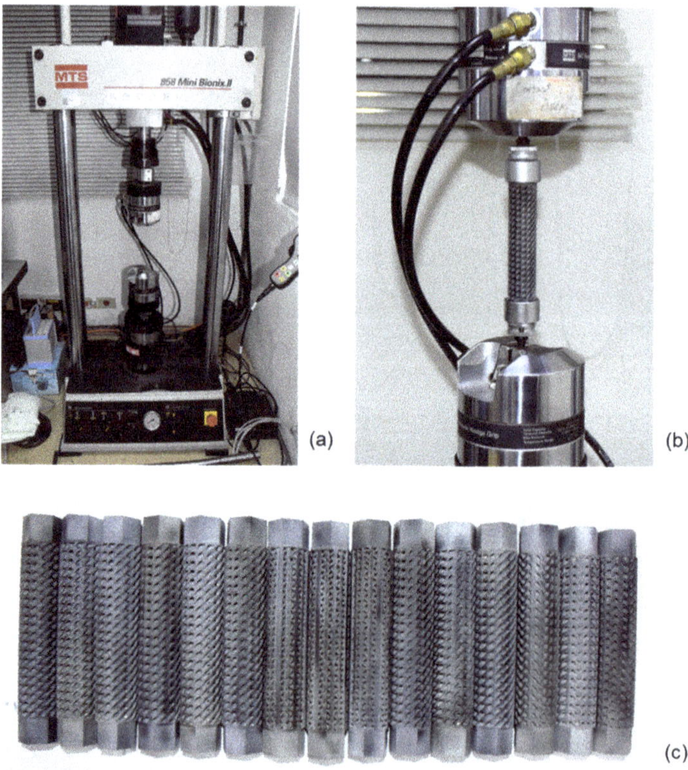

Figure 9. (**a**) Torsion testing machine, (**b**) Experimental setup and fitting of specimen in the machine, and (**c**) All three specimens for each lattice structure.

From the experiment, raw values of time (s), torsional angle (rad), and torque (N·mm) are continuously and automatically recorded on a computer that is directly connected with the test machine using the software Multipurpose Elite Software (MTS Corporation, Minneapolis, MN, USA). These values are later used to calculate several mechanical properties and torsional properties of structures.

3. Results and Discussion

The torsional properties of all lattice structures in terms of torsional stiffness, energy absorption, and failure modes are presented in the coming sections.

3.1. Torsional Stiffness

The torque-twist plots for the five functionally gradient structures are shown in Figure 10 below.

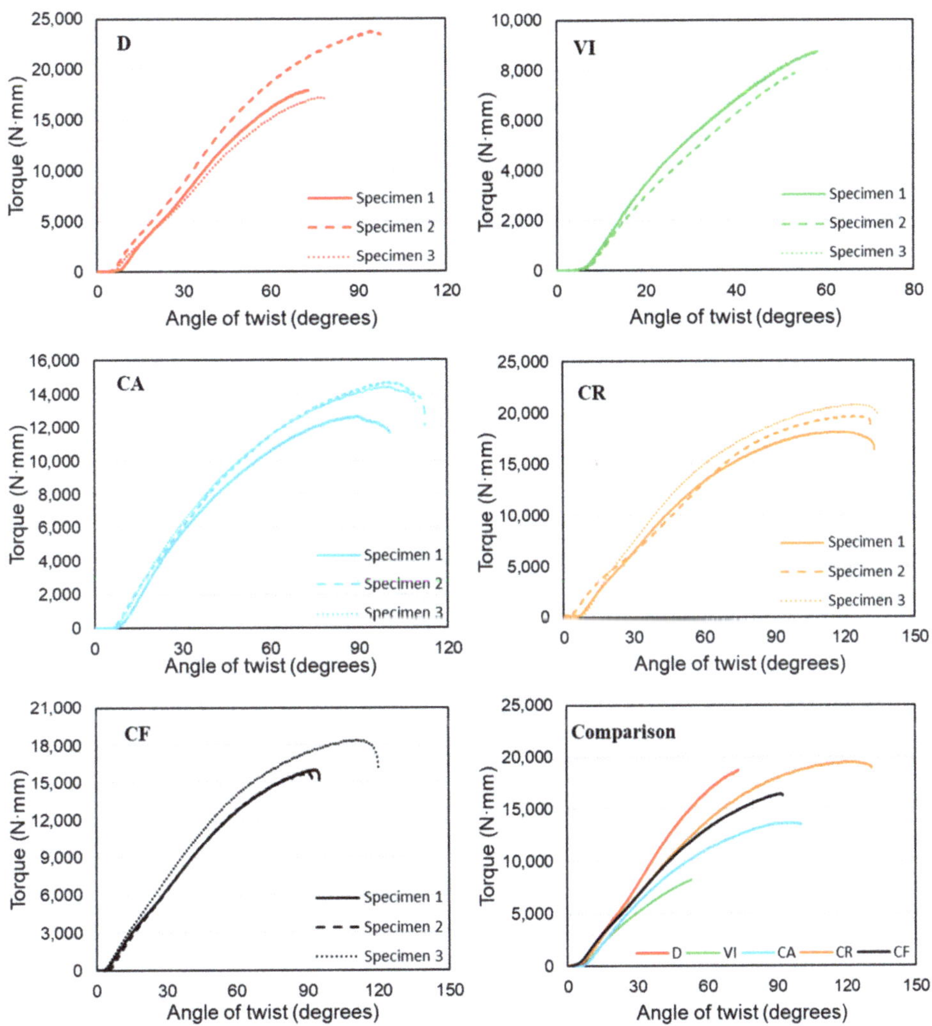

Figure 10. Experimental Torque-Twist curves of functionally gradient lattice structures.

The first design, D, has a functional material distribution similar to a stress concentration along the radius of a cylindrical object subjected to torsion and where structures fail, which is the middle of the structure. Because of this, we see this structure having steeper curves and higher torque values. This structure has regularly patterned unit cells in the design volume. Another structure with the regularly patterned unit cells is design VI. This has a similar functional gradient to the first structure, both along the radial and longitudinal directions. This structure is composed of vertical struts that are far less efficient in resisting torsional loads. Because of this reason, this structure withstood significantly less torque.

The other three structures have a circular patterning of unit cells which corresponds to the direction of the twisting load. This made the structures resist more angular twists than the other two functional gradient structures. It can also be seen from the figure below that these three lattice structures have larger plastic deformation regions characterized by a significant increase in the angular twist with a meager increase in the amount of torque. However, we can see that the CF structure that has a functional gradient in both the radial and longitudinal direction was able to resist lower torque value than the regular patterned D structure with the same radial and longitudinal functional gradient design, showing that a regular patterning of unit cells is better than a circular patterning for torsional applications. Structure D has an average torque value of 19,693 N·mm, which CR follows with 19,536 N·mm. These two structures have similar average torque values, but CR withstood much more angular twisting of around 130 degrees compared to just 80 degrees by structure D. These structures are followed by CF, CA, and VI with 16,739, 13,942, and 8411 N·mm, respectively.

The elastic regions of these torque-twist plots are shown in Figure 11 below and are used to calculate the torsional stiffness of structures. Even though the functionally gradient structures did not exhibit large angular twists as in the previous designs, they showed higher stiffness values because of their effective material distribution. Based on the elastic curves of torque-twist plots, torsional stiffness values for each specimen are calculated, and the average value is taken as the torsional stiffness of the structures. The individual and average values of torsional stiffness for each structure are presented in Table 3. The table shows that design D has the highest stiffness value of 18,705 N·mm/rad due to its effective material distribution across the entire structure. It is followed by a CR structure with 14,220 N·mm/rad, which has a material distribution that follows a natural stress concentration in torsional load. Even though the CF structure has a functional gradient design in both the radial and longitudinal directions, it has a slightly lower stiffness value, showing that the radial stress concentration is a higher determining factor in the torsional performances of cylindrical structures. The CA and VI structures follow with stiffness values of 13,812 and 12,703 N·mm/rad, respectively. The difference in the torsional stiffness of the structures can also be seen from the slope of the curve in the previous figure showing the comparison between the torque-angle of the twist curves.

Table 3. Individual and average torsional stiffness values of structures.

Design	Individual Torsional Stiffness (N·mm/rad)			Average Torsional Stiffness, T/θ (N·mm/rad)
	Specimen 1	**Specimen 2**	**Specimen 3**	
D	18,382	20,875.11	16,856.9	18,704.67
VI	13,367.66	11,287.73	13,454.63	12,703.34
CA	13,528.55	13,762.95	14,145.4	13,812.3
CR	13,664.31	12,941.55	16,054.45	14,220.11
CF	13,073.77	13,655.06	14,852.21	13,860.35

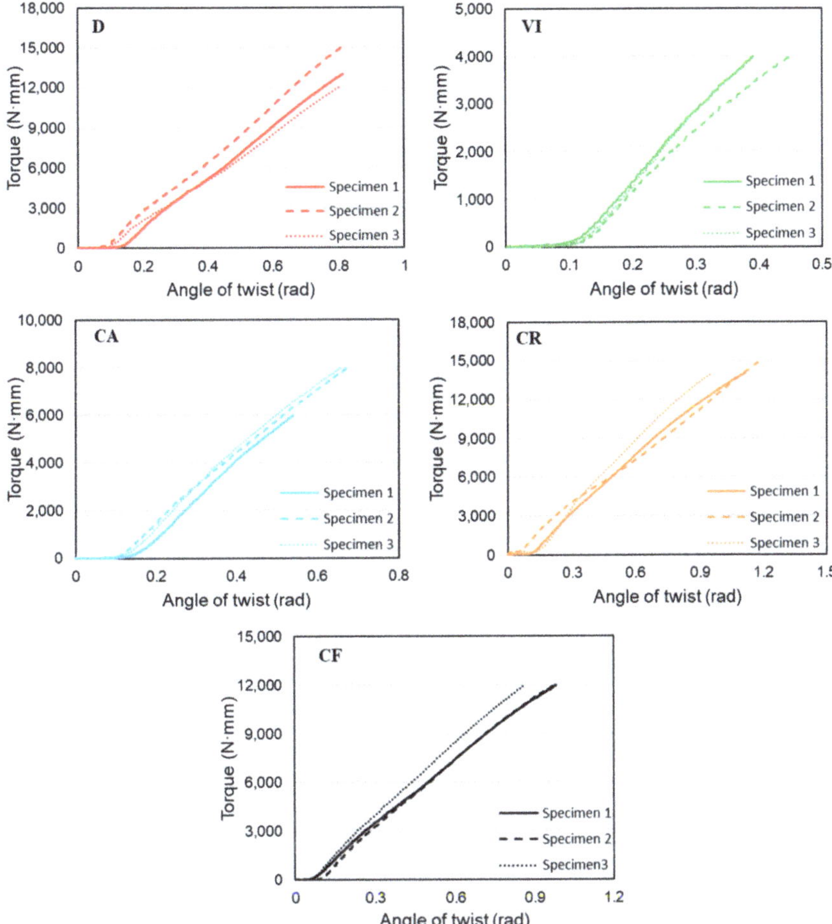

Figure 11. Elastic regions of structures for calculation of torsional stiffness.

3.2. Energy Absorption

To calculate the energy absorptions of structures polar from the shear stress-strain curves, the polar moment of inertia of the structures is calculated by taking multiple slices of the cross-section to determine the cross-section with the smallest area that can better represent the torsional resistance of the entire structure. This cross-section is then used to calculate the polar moment of inertia at the centroid of the area. Table 4 shows the polar moment of the inertia value for each structure based on their minimum cross-sectional area.

Table 4. Polar moment of inertia values based on the minimum cross-sectional area of structures.

Design	Minimum Cross-Sectional Area (mm^2)	Polar Moment of Inertia (mm^4)
D	144.39	12,883.02
VI	96.76	8083.33
CA	125.45	8915.24
CR	162.7	14,007.22
CF	135.53	10,977.03

Since the polar moment of inertia is directly proportional to the cross-sectional area, we can see on the above table that structures with higher areas will have higher values of the polar moment of inertia. Once the polar moment of inertia is determined, the shear stress and strain values of each structure can be calculated from the torque-twist value using Equations (1) and (2), respectively.

$$\text{Shear stress—}\tau = \frac{T \times r}{J} \tag{1}$$

$$\text{Shear strain—}\gamma = \frac{r \times \theta}{L} \tag{2}$$

where T = torque (N·mm), r = radius (mm), J = polar moment of inertia (mm^4), θ= angle of twist (rad), and L = gauge length (mm). Since the radius value is constant for all the structures, the value of shear stress mainly depends on the amount of torque and the value of the polar moment of inertia of the structures. Figure 12 shows the experimental shear-stress-strain plots of the five functionally gradient structures. Three structures, D, CA, and CF have very close maximum shear stress values of 19.11, 19.88, and 19.06 MPa. These structures have higher stress values than the two other structures. Since structure CR had a slightly higher polar moment of inertia value, it decreased the maximum shear stress it withstood with a value of 17.43 MPa and put it just less than the three previously mentioned designs. Structure VI had the lowest value of maximum shear stress with 13.01 MPa and a lower shear strain value. We can also see from the figure that the two structures with a regular patterning of unit cells, D and VI, had very short regions of plastic deformation compared to the three circularly patterned lattice structures.

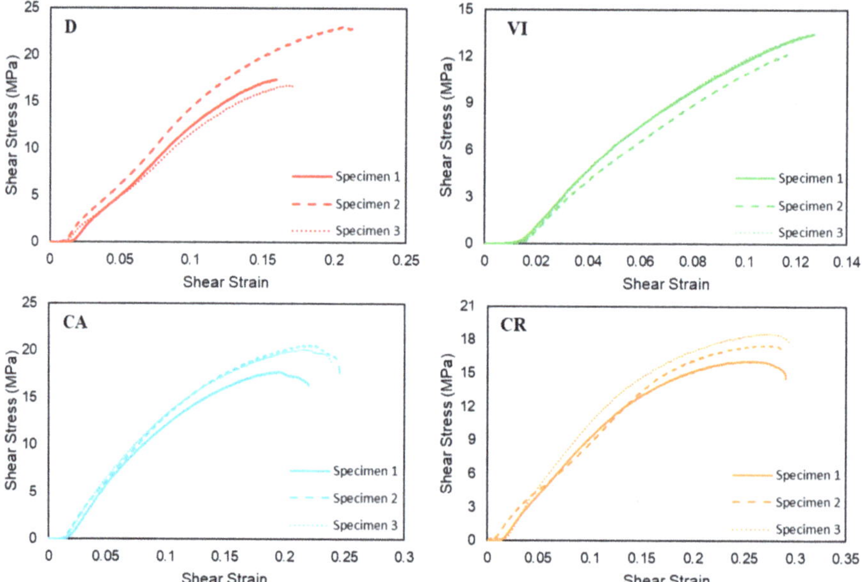

Figure 12. *Cont.*

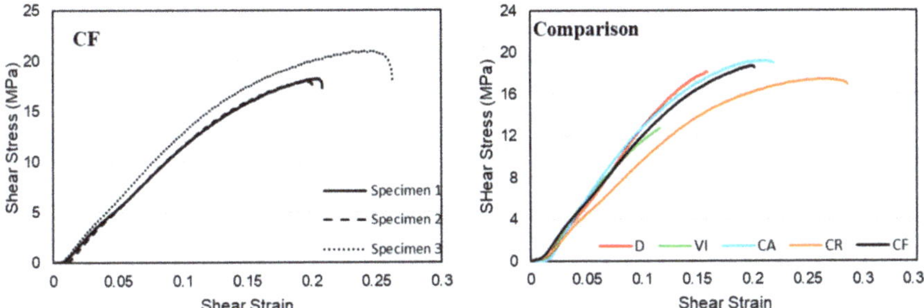

Figure 12. Experimental shear-stress-strain curves of functionally gradient lattice structures.

The energy absorption of each structure is calculated by taking the area under the average shear-stress-strain curves. The energy absorbed up to the fracture point of the specimens, toughness, is calculated by approximating the stress-strain curves using the best polynomial curve and integrating the function with the range of the lowest and highest value of strain. Figure 13 shows energy absorption plots based on the area under the shear-stress-strain curves of each lattice structure. The figure below shows that the CR structure has the highest energy absorption value of 3.26 J/mm^3 due to its high angular twist values. It is followed by the CA, CF, D, and VI structures with energy absorption values of 2.63, 2.20, 1.51, and 0.77 J/mm^3, respectively. The circular patterning of the three structures, CR, CA, and CF, gave them the advantage to withstand more angular twist over the regularly patterned unit cells.

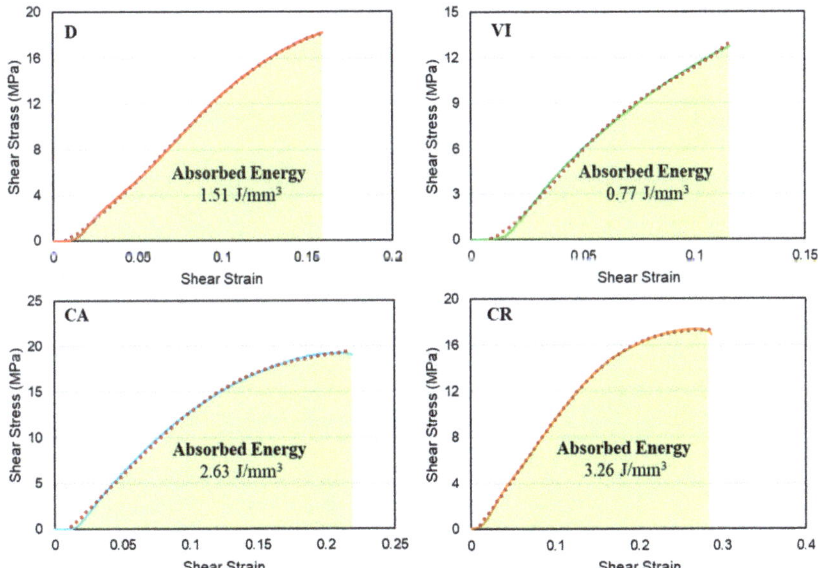

Figure 13. *Cont.*

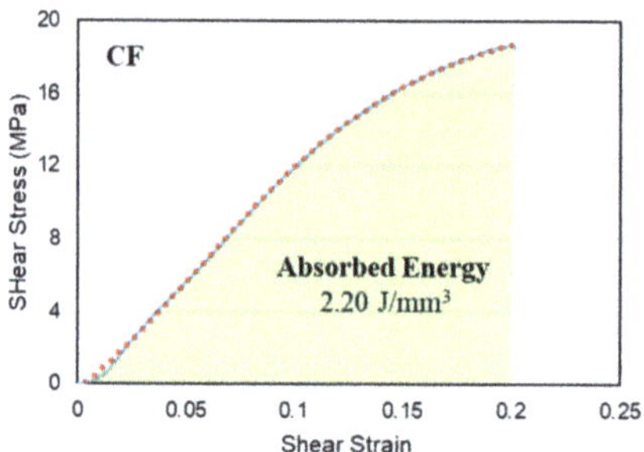

Figure 13. Energy absorption value based on the average stress-strain result of lattice structures.

3.3. Comparison of Performance

In this section, the comparison between regularly patterned and circularly patterned diamond structures will be made in terms of their torsional stiffness, ultimate strength, and angular twist. In addition, a comparison of the performance between the regularly patterned functional gradient diamond and vertical-inclined structures is presented to show the effect of the material distribution on the torsional performance of structures. For a comparison of the torsional stiffness of structures, the elastic regions of the torque-twist curves were plotted using the average values of the three specimens. The first structures that are compared in this section are structures D and CF. Both structures have a material distribution that varies in both longitudinal and radial directions in which the wall thickness of the structures increases from the central axis to the outer surface and from two longitudinal ends towards the middle. Figure 14 below shows the torque-twist and shear stress-strain curves of the D and CF structure. The figure shows that the regularly patterned functional gradient structure, D, has a higher torsional resistance than the circularly patterned CF structure.

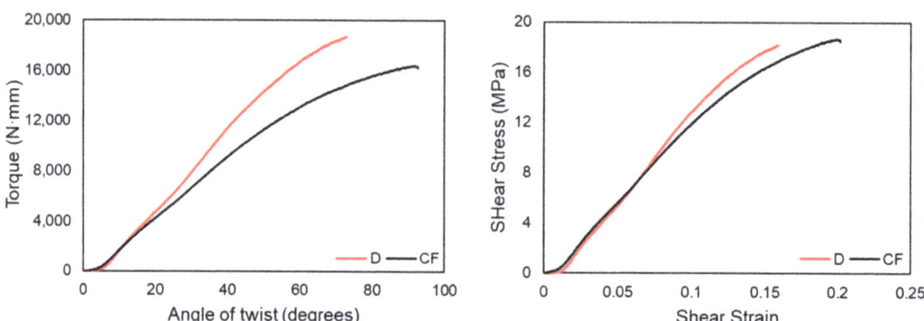

Figure 14. Comparison of torque-twist and shear stress-strain curves between D and CF structures.

The curve also has a much steeper curve in the elastic region, showing a better torsional stiffness value. In contrast, the CF structure has a higher value of angular twist due to the patterning effect that aligns to the twist's direction, giving it better compliance to the twisting load. From the stress-strain plot, we can see that both D and CF structures have comparably similar ultimate strength values of 18.16 and 18.68 MPa, respectively. However, the CF structure was able to withstand higher shear strain values deforming plastically. The torsional stiffness values of both structures are shown in Table 5 below.

Table 5. Torsional stiffness values of D and CF structures.

Design	Torsional Stiffness, T/θ (N·mm/rad)
D	18,704.7
CF	13,860.35

Another comparison is made between uniform density structures and functionally gradient structures. The uniform gradient regularly patterned diamond structure, hereafter DU, is made with a wall thickness of 1.72 mm and a relative density of 40.18%, similar to the functional gradient counterpart. Similar torsional experiments with identical parameters and conditions are made to compare the properties between the two structures. On the other hand, the uniform density vertical-inclined structure, hereafter VI-U, has an outer vertical beam diameter of 2.4 mm and diagonal beams of 1.6 mm. This structure also has a similar relative density of 41% with the functional gradient one. Figure 15 shows a comparison of torque-twist and shear stress-strain plots between the gradient density D, structure, and the uniform density, DU, counterpart.

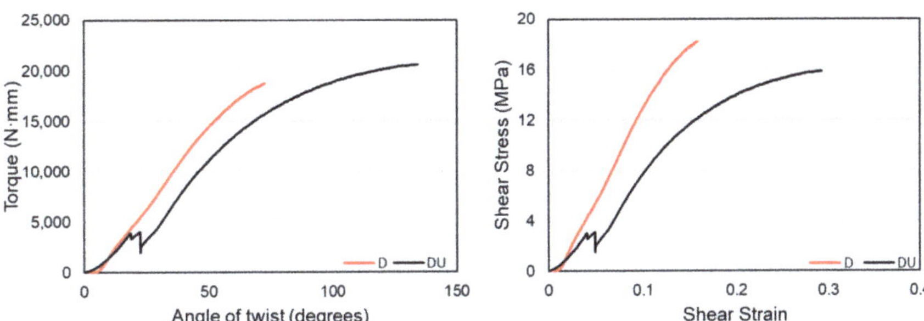

Figure 15. Comparison of torque-twist and shear-stress-strain curves between D and DU structures.

The above graph shows that the uniform density DU structure has higher values than the functionally gradient D structure in terms of the torque withstood and torsional angle. One major factor for these results is that both structures have the same relative density, making the D structure have less wall thickness at the end of the structures, creating a high-stress concentration in those areas. However, structure DU does not have any variation of material distribution in the longitudinal direction that eliminated a high-stress concentration at the intersection of the structure and grip. The shear-stress-strain plots show that the functionally gradient D structure has a very steep slope and higher ultimate strength than the uniform gradient DU structure. This resulted from the D structure's torsional resistance values governed by the material distribution along with the structures. Since the material distribution follows a stress concentration pattern in cylindrical structures, it was able to have higher torsional stiffness. The torsional stiffness values of these two structures are given in Table 6 below.

Table 6. Torsional stiffness values of D and DU structures.

Design	Torsional Stiffness, T/θ (N·mm/rad)
D	18,704.7
DU	18,162

Similarly, a comparison is made between the functionally gradient and uniform density vertical-inclined structures. The comparison of the torque-twist and shear-stress-strain plots is given in Figure 16 below. The figure shows that the uniform density vertical-inclined VI-U structure could withstand more angular twists than the VI structure. This is mainly caused due to the higher stiffness of the structure, which made it behave in a brittle manner and have failure without having significant plastic deformation. Therefore, we can see that the different material distribution in the structures made from the same unit cell can significantly affect material failure behavior. In terms of the shear-stress-strain plots, structure VI-U still has higher ultimate shear strength before failure. However, the functionally gradient structure, VI, has a smooth and slightly steeper curve showing a higher stiffness value. Overall, we can say that material distribution is more efficient in the diamond structure to enhance the property than the strut-based vertical-inclined structure.

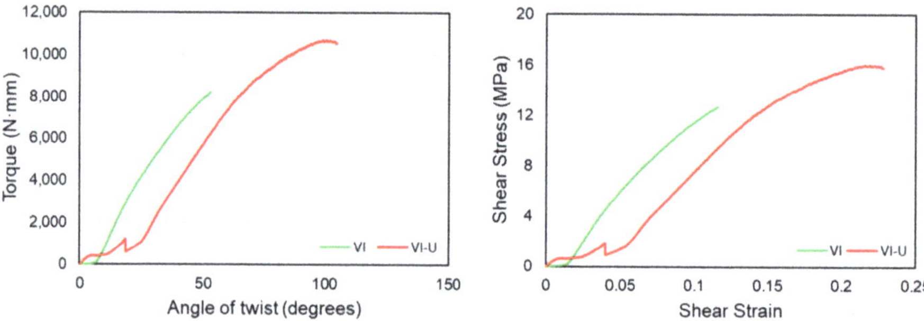

Figure 16. Comparison of torque-twist and shear-stress-strain curves between VI and VI-U structures.

The torsional stiffness values of these two structures, VI and VI-U, are given in Table 7 below. The table shows that the torsional stiffness of the VI structure is slightly higher than the VI-U structure despite not having a large plastic deformation region.

Table 7. Torsional stiffness values of VI and VI-U structures.

Design	Torsional Stiffness, T/θ (N·mm/rad)
VI	12,703.34
VI-U	9990.2

3.4. Failure of Structures

Figure 17 shows the plastic regions of each structure based on the shear-stress-strain curves. It shows that the three structures with a circular patterning of unit cells, CA, CR, and CF, in the design volume showed larger plastic deformation regions. This is because the circular patterning matches the twisting profile of the angular twist, giving them a higher capacity to withstand more torsional angles. In contrast, the two functionally gradient structures with a regular patterning of unit cells, D and VI, have shorter plastic deformation regions. The amount of plastic deformation before fracture can predict the type of failure behavior in structures. Structures with more extended regions of plastic deformation are often associated with the ductile mode of failure. This shows that CA, CR, and CF structures failed in a brittle manner. Since D and VI structures have stiff structures and shorter plastic deformation regions, we can say that they failed in a brittle manner.

The failed specimens can also indicate the mechanical properties of the structure and their resistance to the applied torsion.

Figure 18 below shows all failed specimens of the functionally gradient lattice structures. The stiff structures were able to withstand the applied torsional rotation failed at the interface between the lattice and the grip structure. This shows that the interface was subjected to higher stress values due to the higher strength of the main structure in the functionally gradient structures. On the other hand, it can be seen that the failure was initiated from the middle of the structures where the stress concentration was higher for the uniform density structures. This shows that the functional material distribution effectively made the lattice structures stiffer and have better torsional resistance.

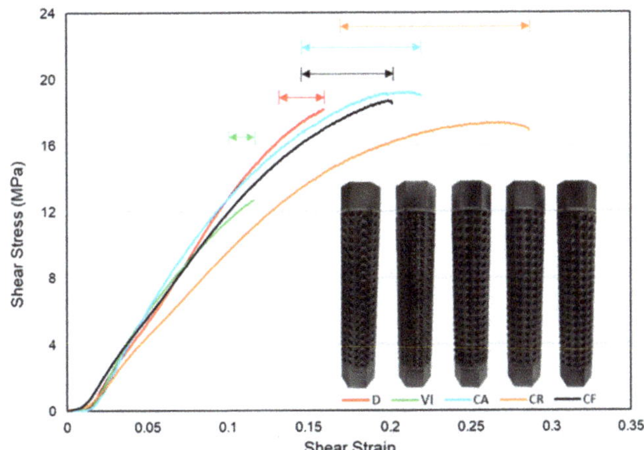

Figure 17. Approximate plastic deformation regions based on stress-strain curves of structures.

Figure 18. Mode of failure for different functionally gradient design configurations investigated in the present study. (**a**) D, (**b**) VI, (**c**) CA, (**d**) CR, and (**e**) CF.

4. Conclusions

This paper aimed to investigate the torsional properties of lattice structures and study the effect of the material distribution and patterning modes on the mechanical properties of the structures. The torsional properties of functionally gradient lattice structures made from TPMS and strut-based unit cells were investigated using an experimental approach. The results clearly show that the stiffness and energy absorption of the structures can be improved by an effective material distribution that corresponds to the stress concentration due to torsional load. It is observed that functional gradient designs have higher stiffness, increasing their strength. However, this increase in stiffness is seen to be associated with a more brittle mode of fracture. In addition, effective material distribution also affects the failure mechanism and delays the plastic deformation of the structures, increasing their resistance to torsional loads. The functionally gradient lattice structures' results showed that structure D with a functional material distribution in both radial and longitudinal directions had improved torsional stiffness and ultimate shear strength compared to structure DU, a uniform density structure. On the other hand, the higher stiffness of this structure made it have shorter plastic deformation failing in a brittle manner. It is also found that the functionally gradient structure made from vertical-inclined unit cells, VI, has higher torsional stiffness than its uniform density counterpart, VI-U. A circular way of patterning unit cells was also used to investigate its effect on torsional properties. From the results, it can be concluded that this was of patterning unit cells increased the amount of the torsional twisting angle that a structure can withstand but has no importance in enhancing torsional stiffness and the ultimate shear strength of the structures.

For future work, it is recommended that an extensive finite element analysis, FEA, be carried out, and cross-validation with experimental results should be conducted to characterize the lattice structures' torsional properties effectively. This study can open further avenues to explore functionally graded structures by the optimal distribution of materials in order to achieve the specific strength needed for application where weight is a significant factor, and further investigation that looks at the dynamic underlying reasons in the mechanics of these structures with accurate methods of evaluating them will create a potential to exploit them further and expand the impact of TPMS based lattice structures.

Author Contributions: Conceptualization, J.-Y.J. and A.N.; methodology, A.N.; software, Y.M.H.; validation, A.N., J.-Y.J. and S.-C.L.; formal analysis, Y.M.H. and A.N.; investigation, Y.M.H.; resources, J.-Y.J. and S.-C.L.; data curation, Y.M.H.; writing of the original draft preparation, Y.M.H. and A.N.; writing of review and editing, J.-Y.J. and S.-C.L.; visualization, Y.M.H. and A.N.; supervi-sion, J.-Y.J. and A.N.; project administration, A.N. and S.-C.L.; funding acquisition, J.-Y.J. All authors have read and agreed to the published version of the manuscript.

Funding: This work was financially supported by the High-Speed 3D Printing Research Center (Grant No. 108P012) from the Featured Areas Research Center Program within the Higher Education Sprout Project framework by the Minister of Education (MOE) Taiwan.

Institutional Review Board Statement: Not applicable.

Informed Consent Statement: Not applicable.

Data Availability Statement: Not applicable.

Acknowledgments: The authors would like to thank Chang-Hung Huang, Department of Medical Research, MacKay Memorial Hospital, School of Dentistry, National Yang Ming Chiao Tung University, Taipei, Taiwan, ROC for providing access to the torsion testing laboratory used for this study.

Conflicts of Interest: The authors declare no conflict of interest.

References

1. Seharing, A.; Azman, A.H.; Abdullah, S. A review on integration of lightweight gradient lattice structures in additive manufacturing parts. *Adv. Mech. Eng.* **2020**, *12*, 1–21. [CrossRef]
2. Tamburrino, F.; Graziosi, S.; Bordegoni, M. The Design Process of Additively Manufactured Mesoscale Lattice Structures: A Review. *J. Comput. Inf. Sci. Eng.* **2018**, *18*, 40801. [CrossRef]
3. Reeves, P.; Tuck, C.; Hague, R. *Additive Manufacturing for Mass Customization*; Springer: London, UK, 2011; pp. 275–289.
4. Nazir, A.; Jeng, J.-Y. A high-speed additive manufacturing approach for achieving high printing speed and accuracy. *Proc. Inst. Mech. Eng. Part C J. Mech. Eng. Sci.* **2020**, *234*, 2741–2749. [CrossRef]
5. Choy, S.Y.; Sun, C.-N.; Leong, K.F.; Wei, J. Compressive properties of functionally graded lattice structures manufactured by selective laser melting. *Mater. Des.* **2017**, *131*, 112–120. [CrossRef]
6. Choy, S.Y.; Sun, C.-N.; Sin, W.J.; Leong, K.F.; Su, P.-C.; Wei, J.; Wang, P. Superior energy absorption of continuously graded microlattices by electron beam additive manufacturing. *Virtual Phys. Prototyp.* **2021**, *16*, 14–28. [CrossRef]
7. Yang, L.; Wu, S.; Yan, C.; Chen, P.; Zhang, L.; Han, C.; Cai, C.; Wen, S.; Zhou, Y.; Shi, Y. Fatigue properties of Ti-6Al-4V Gyroid graded lattice structures fabricated by laser powder bed fusion with lateral loading. *Addit. Manuf.* **2021**, *46*, 102214. [CrossRef]
8. Zhang, P.; Qi, D.; Xue, R.; Liu, K.; Wu, W.; Li, Y. Mechanical design and energy absorption performances of rational gradient lattice metamaterials. *Compos. Struct.* **2021**, *277*, 114606. [CrossRef]
9. Sienkiewicz, J.; Płatek, P.; Jiang, F.; Sun, X.; Rusinek, A. Investigations on the Mechanical Response of Gradient Lattice Structures Manufactured via SLM. *Metals* **2020**, *10*, 213. [CrossRef]
10. Wu, X.; Zhang, S.; Shao, J. Studies on Impact Performance of Gradient Lattice Structure Applied to Crash Box. *SAE Tech. Pap. Ser.* **2018**. [CrossRef]
11. Li, D.; Liao, W.; Dai, N.; Xie, Y.M. Comparison of Mechanical Properties and Energy Absorption of Sheet-Based and Strut-Based Gyroid Cellular Structures with Graded Densities. *Materials* **2019**, *12*, 2183. [CrossRef] [PubMed]
12. Aremu, A.O.; Maskery, I.; Tuck, C.; Ashcroft, I.A.; Wildman, R.D.; Hague, R.I.M. A comparative finite element study of cubic unit cells for selective laser melting. In Proceedings of the 25th Annual International Solid Freeform Fabrication Symposium—An Additive Manufacturing Conference SFF 2014, Austin, TX, USA, 4–6 August 2014; pp. 1238–1249. Available online: https://pureportal.coventry.ac.uk/en/publications/a-comparative-finite-element-study-of-cubic-unit-cells-for-select (accessed on 22 July 2021).
13. Yoo, D.-J. Advanced porous scaffold design using multi-void triply periodic minimal surface models with high surface area to volume ratios. *Int. J. Precis. Eng. Manuf.* **2014**, *15*, 1657–1666. [CrossRef]
14. Zhang, L.; Feih, S.; Daynes, S.; Chang, S.; Wang, M.Y.; Wei, J.; Lu, W.F. Energy absorption characteristics of metallic triply periodic minimal surface sheet structures under compressive loading. *Addit. Manuf.* **2018**, *23*, 505–515. [CrossRef]
15. VijayaVenkataRaman, S.; Zhang, L.; Zhang, S.; Fuh, J.Y.H.; Lu, W.F. Triply Periodic Minimal Surfaces Sheet Scaffolds for Tissue Engineering Applications: An Optimization Approach toward Biomimetic Scaffold Design. *ACS Appl. Bio Mater.* **2018**, *1*, 259–269. [CrossRef]
16. Kapfer, S.C.; Hyde, S.T.; Mecke, K.; Arns, C.H.; Schröder-Turk, G.E. Minimal surface scaffold designs for tissue engineering. *Biomaterials* **2011**, *32*, 6875–6882. [CrossRef] [PubMed]
17. Yuan, L.; Ding, S.; Wen, C. Additive manufacturing technology for porous metal implant applications and triple minimal surface structures: A review. *Bioact. Mater.* **2019**, *4*, 56–70. [CrossRef] [PubMed]
18. Qureshi, Z.A.; Al-Omari, S.A.B.; Elnajjar, E.; Al-Ketan, O.; Abu Al-Rub, R. Using triply periodic minimal surfaces (TPMS)-based metal foams structures as skeleton for metal-foam-PCM composites for thermal energy storage and energy management applications. *Int. Commun. Heat Mass Transf.* **2021**, *124*, 105265. [CrossRef]
19. Li, W.; Yu, G.; Yu, Z. Bioinspired heat exchangers based on triply periodic minimal surfaces for supercritical CO2 cycles. *Appl. Therm. Eng.* **2020**, *179*, 115686. [CrossRef]
20. Tao, W.; Leu, M.C. Design of lattice structure for additive manufacturing. In Proceedings of the 2016 International Symposium on Flexible Automation (ISFA), Cleveland, OH, USA, 19 December 2016; pp. 325–332. [CrossRef]
21. nTopology. Available online: https://ntopology.com/ (accessed on 19 February 2021).
22. Standard Test Method for Shear Modulus at Room Temperature, in: ASTM B. Stand. 2002. Available online: https://compass.astm.org/Standards/HISTORICAL/E143-13.htm (accessed on 22 June 2021).
23. Hewlett-Packard. HP 3D Jet Fusion 5200—Commercial & Industrial 3D Printer | HP®Official Site. 2021. Available online: https://www.hp.com/us-en/printers/3d-printers/products/multi-jet-fusion-4200.html (accessed on 22 June 2021).
24. Nazir, A.; Arshad, A.-B.; Hsu, C.-P.; Jeng, J.-Y. Effect of Fillets on Mechanical Properties of Lattice Structures Fabricated Using Multi-Jet Fusion Technology. *Materials* **2021**, *14*, 2194. [CrossRef] [PubMed]
25. Lohmuller, P.; Favre, J.; Kenzari, S.; Piotrowski, B.; Peltier, L.; Laheurte, P. Architectural effect on 3D elastic properties and anisotropy of cubic lattice structures. *Mater. Des.* **2019**, *182*, 108059. [CrossRef]

 materials

Article

Properties of 3D Printed Concrete–Geopolymer Hybrids Reinforced with Aramid Roving

Joanna Marczyk [1], Celina Ziejewska [1], Kinga Korniejenko [1], Michał Łach [1], Witold Marzec [2], Mateusz Góra [1], Paweł Dziura [1], Andina Sprince [3], Magdalena Szechyńska-Hebda [4] and Marek Hebda [1,*]

1 Faculty of Materials Engineering and Physics, Cracow University of Technology, Warszawska 24, 31-155 Kraków, Poland
2 Centrum Ekologicznego Budownictwa Mieszkaniowego 3 Sp. z o.o., Henryka Sienkiewicza 19/4, 40-031 Katowice, Poland
3 Faculty of Civil Engineering, Riga Technical University, Ķīpsalas iela 6A, Centra Rajons, LV-1048 Rīga, Latvia
4 Plant Breeding and Acclimatization Institute—National Research Institute, Radzików, 05-870 Błonie, Poland
* Correspondence: marek.hebda@pk.edu.pl; Tel.: +48-1262-83423

Abstract: Three-dimensional concrete printing (3DCP) is an innovative technology that can lead to breakthrough modifications of production processes in the construction industry. The paper presents for the first time the possibility of 3D printing concrete–geopolymer hybrids reinforced with aramid roving. Reference concrete samples and concrete–geopolymer hybrids composed of 95% concrete and 5% geopolymer based on fly ash or metakaolin were produced. The properties of the samples without reinforcement and samples with 0.5% (wt.) aramid roving were compared. The frost resistance tests, UV radiation resistance, and thermal conductivity were evaluated for samples that were 3D-printed or produced by the conventional casting method. Compressive strength tests were carried out for each sample exposed to freeze–thaw cycles and UV radiation. It was observed that after the frost resistance test, the samples produced by the 3D printing technology had a minor decrease in strength properties compared to the samples made by casting. Moreover, the thermal conductivity coefficient was higher for concrete–geopolymer hybrids than concrete reinforced with aramid roving.

Keywords: 3D concrete printing (3DCP); hybrids; geopolymer; frost resistance; UV radiation; thermal conductivity

Citation: Marczyk, J.; Ziejewska, C.; Korniejenko, K.; Łach, M.; Marzec, W.; Góra, M.; Dziura, P.; Sprince, A.; Szechyńska-Hebda, M.; Hebda, M. Properties of 3D Printed Concrete–Geopolymer Hybrids Reinforced with Aramid Roving. *Materials* **2022**, *15*, 6132. https://doi.org/10.3390/ma15176132

Academic Editors: Swee Leong Sing and Wai Yee Yeong

Received: 28 July 2022
Accepted: 30 August 2022
Published: 3 September 2022

1. Introduction

Global population growth is contributing to the development of the construction industry. The United Nations predicts that the world's population will grow to 8.5 billion in 2030, 9.7 billion in 2050, and 10.9 billion in 2100. Consequently, such an increase in the world population will increase the demand for new residential and service buildings [1,2].

One of the basic and most frequently used materials in building construction is concrete, due to its design parameters, properties, and low cost [3,4]. Unfortunately, the construction industry based on concrete has a negative impact on the environment generating a large amount of contamination, more than 8% of global CO_2 emissions, and high energy consumption [5]. The desire to reduce the impact of these factors has contributed to the development of sustainable building materials.

One type of material that is an excellent alternative to ordinary Portland cement (OPC) is geopolymers [6]. A geopolymer is an inorganic amorphous polymer, formed by adding an alkaline solution to aluminosilicate precursors (e.g., metakaolin, fly ash) [7]. The microstructure of a geopolymer is composed of aluminum–oxygen and silicon–oxygen tetrahedra that form a three-dimensional lattice structure. Such a structure contributes to the increase in the durability of the material [8]. Geopolymers have properties similar to concrete produced on the basis of OPC, however, compared to Portland cement, geopolymers have lower CO_2 emissions and lower energy consumption [7]. Geopolymer materials can be reinforced with fibers [9], repair materials [10], and heavy metal sewage treatment materials. They can also be used as catalyst supports [11]. In terms of economic as well as environmental issues, geopolymers are suitable materials for additive manufacturing [12].

In recent years, there has been a dynamic development of 3D concrete printing (3DCP) using the extrusion method. This technology consists of linear extrusion of cement mortar layer by layer without the use of formwork, which can reduce production costs, increase the speed of production processes, and allow the economical production of geometrically complex elements [13,14]. Additive manufacturing (AM) of cement materials is one of the most interesting methods of producing concrete elements used in the construction industry. Compared to conventional manufacturing, 3DCP can reduce the environmental impact by up to 50% [14–17].

Partial replacement of cement with other materials allows for the production of hybrid cement-geopolymer concrete. Earlier it was shown that Portland cement can be partially replaced by various materials [14], such as a metakaolin (MK) [18], which improves the properties of hybrids, their workability and durability, and reduces the environmental impact of the cement industry [19–21]. Further, fly ash and slag, due to their properties being similar to cement, are also suitable alternatives to its partial replacement [22,23]. Fine fly ash has pozzolanic activity [24] and can improve the compressive strength of the final product [25–29]. Muthusamy et al. [27] showed that concrete, in which 30% of the cement was replaced with fly ash, can be used in the construction industry. OPC was also mixed with slag [28].

Hybrids are most often produced by mixing cement and other cement material in appropriate proportions and then activating such a mixture with a previously prepared alkaline solution. In our work, we produced concrete–geopolymer hybrids by preparing concrete and the geopolymer separately, and then by mixing them in appropriate proportions. We produced the hybrids using two methods: casting into molds and 3D printing [29,30]. The aim of this work was to obtain 3D printed concrete–geopolymer hybrids reinforced with aramid roving. Usually, aramid fiber, also known as Kevlar fiber, is characterized by high tensile strength and thus is used as a reinforcement in composite materials [31,32] to increase their durability and strength by bridging cracks and transferring tensile forces [33–35]. The novelty of the work is the production of 3D printed concrete–geopolymer hybrids as a combination of separate mixtures, then reinforced with aramid roving. Hybrids consisting of 95% concrete and 5% geopolymer with the addition of fly ash or metakaolin were produced. Aramid roving was added as reinforcement to hybrids in an amount of 0.5 wt.%. The materials prepared in this way were tested for frost resistance, UV aging, compressive strength, and thermal conductivity. The purpose of these activities is to develop optimal concrete–geopolymer mixtures that can ultimately be produced based on advanced large-format 3D printing. Hybrids produced in additive manufacturing can find application in residential construction.

2. Materials and Methods

2.1. Raw Materials

In this study, class F fly ash (Skawina CHP Coal Power Plant, Skawina, Poland) and Metakaolin KM 60 (Keramost, Kadaň, Czech Republic) were used as a precursor for the production of geopolymers. The chemical composition of the raw materials is presented in Table 1. The raw materials were mixed with river sand (Świętochłowice, Poland). Commercial cement CEM I 42.5R, which meets the requirements of PN-EN 197-1, was also used for the tests. The cement was delivered by the cement plant Górażdże Cement S.A. (Heidelberg Cement Group, Chorula, Poland). Aramid roving with a weight of 805 TEX and a fiber weave width of 8 mm was used as reinforcement (Modelemax, Jelenia Góra, Poland). A single filament in the aramid roving had a diameter of 10 μm. The characteristics of the raw materials used for the research were presented in our previous work [29].

Table 1. Chemical composition of fly ash and metakaolin determined by X-ray fluorescence analysis, wt.% [29].

Component	Fly Ash (%)	Metakaolin KM 60 (%)
SiO_2	48.220	52.430
Al_2O_3	26.130	42.750
Fe_2O_3	7.010	1.200
CaO	5.120	0.490
K_2O	3.480	1.300
MgO	1.720	0.175
Na_2O	1.615	0.000
TiO_2	1.110	0.310
SO_3	1.110	0.030
P_2O_5	0.700	0.440
Cl	0.090	0.060
MnO	0.090	0.012
LOI	3.284	0.722

2.2. Preparation of Specimens

Concrete specimens and concrete–geopolymer hybrids were prepared for the tests. To prepare the concrete mixture, the cement (CEM I 42.5R) and sand were mixed in a 1:1 ratio. The water to dry weight mixture ratio was 0.125.

The geopolymer mass was made by mixing fly ash or metakaolin with sand in a 1:1 ratio. An alkaline activator solution which consisted of 10 M sodium hydroxide and an aqueous solution of sodium silicate (R-145) in a molar ratio of 1:2.5 was added to the raw materials. All components were mixed in a GEOLAB cement mortar mixer (Geolab, Warsaw, Poland) for 15 min. In the geopolymer blends containing fly ash, the liquid-to-solid ratio was 0.28; 0.35 for blends based on metakaolin.

To prepare concrete–geopolymer hybrids, the concrete, and geopolymer mass were mixed in the proportion of 19:1. The mixtures were produced by casting and 3D printing methods. Aramid roving was added to part of the samples in an amount of 0.5 wt.%. Roving was placed at 1/3 and 2/3 of the height of the specimen. The percentage of selection of roving as well as their distribution were based on the results obtained in the previous work [33]. Geopolymers based on fly ash reinforced with aramid fibers were characterized by higher strength with a decrease in the number of fibers in the matrix. The addition of aramid fibers in the amount of 0.5% increased the strength by almost 2 MPa. The arrangement of aramid fibers in the samples is shown in Figure 1.

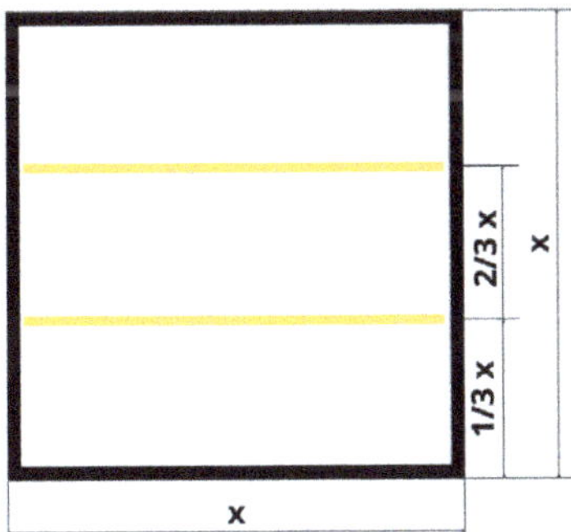

Figure 1. Aramid roving distribution in the sample. The figure concerns samples with dimensions of 50 × 50 × 50 mm and 100 × 100 × 100 mm, where x refers to 50 mm or 100 mm.

By producing samples by casting, the prepared mixtures were poured into molds with dimensions of 50 mm × 50 mm × 50 mm for tests in a climatic chamber and for tests of compressive strength; 100 mm × 100 mm × 100 mm for frost resistance and abrasion resistance tests; ⌀ 55 mm × 23 mm for thermal conductivity tests. The molds were shaken to remove trapped air. The samples were cured for 24 h at 75 °C, then removed from the mold and stored under ambient conditions.

For the production of samples by 3D printing, the models for printing were designed in the Blender software. The samples were printed on the ATMAT Galaxy 3D printer (ATMAT, Kraków, Poland). Printing was performed at ambient temperature with a printing speed of 150 mm s^{-1}. The diameter of the printing nozzle was 15 mm and the thickness of the printed layer was 30 mm.

The compositions of the produced concrete samples and concrete–geopolymer hybrids are presented in Table 2.

Table 2. Composition of concrete and concrete–geopolymer hybrids samples (kg/m^3).

Sample Designation	Concrete		Geopolymer			Reinforcement
	Cement	Sand	FA	MK	Sand	Aramid Roving
100% C	1000	1000	–	–	–	–
100% C + R	1000	1000	–	–	–	10
95% C + 5% FA	950	950	50	–	50	–
95% C + 5% FA + R	950	950	50	–	50	10
95% C + 5% MK	950	950	–	50	50	–
95% C + 5% MK + R	950	950	–	50	50	10

Figure 2 shows a representative view of the samples produced by casting. All samples, regardless of their composition and method of production, were investigated after 28 days of curing.

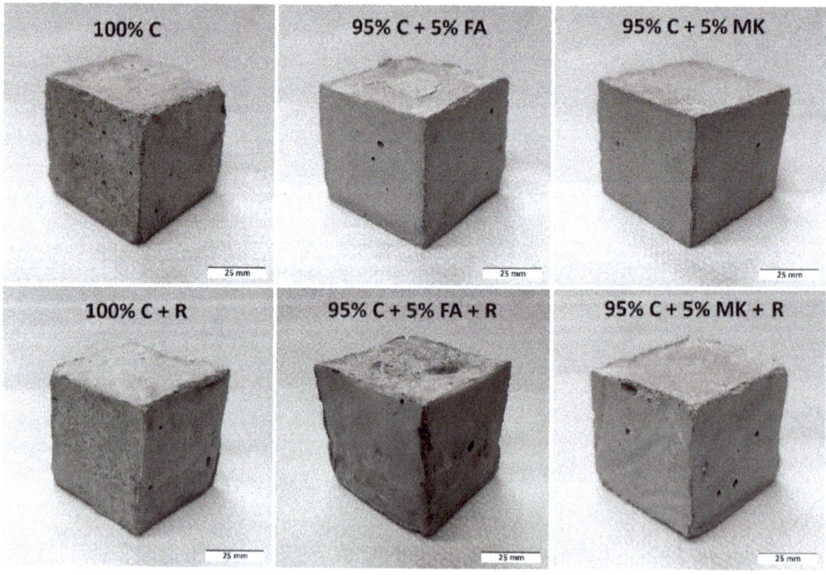

Figure 2. Representative view of the samples produced by casting.

2.3. Methods

The degree of frost resistance of concrete and concrete–geopolymer hybrids was tested using the standard method in accordance with the PN-B 06265. The samples were subjected to 150 cycles of freezing at −18 ± 2 °C and defrosting at 18 ± 2 °C. The obtained results were then compared with the test results for the reference samples stored in water at the temperature of 18 ± 2 °C.

The UV aging tests in the climatic chamber were carried out in accordance with the following standards: UV radiation resistance test—PN-EN ISO 4892-2:2013-06; color change test—PN-ISO 105-A02:1996. The resistance to UV radiation was tested using the following parameters: nominal power of the lamps: 3 × 1700 W; filter: daylight; method: A, cycle 1; total test time: 720 h, wavelength measurement range: (300–400) nm, irradiance (300–400 nm): 60 W/m^2. After the tests of resistance to UV radiation, the compressive strength tests and the grayscale color changes were carried out.

Strength tests were carried out at a temperature of 20.8 ± 0.2 ÷ 21.4 ± 0.2 °C and relative air humidity of 38.9 ± 2.0 ÷ 43.1 ± 2.0%. The samples were subjected to the compressive force F until destruction and the maximum force F_{max} was reached. Compressive strength was calculated on the basis of the obtained results.

The evaluation of the color change in gray scale was carried out according to ISO 105-A02. The color change was assessed as a team in a room with the lighting in accordance with PN-EN ISO 13076:2012, at a temperature of 20.5 ± 0.2 ÷ 21.7 ± 0.2 °C and relative air humidity of 40.1 ± 2.0 ÷ 42.4 ± 2.0%. The color of the UV irradiated and UV non-irradiated samples were compared on a gray scale ranging from 5 being no color change at all to 1 being the lightest and biggest change in color. One sample from each lot was not exposed (as a reference sample). Each series of samples was tested by comparing the reference sample with at least the two exposed samples.

The study of thermal conductivity consisted in determining the thermal conductivity coefficient λ under steady heat flow conditions, which was carried out using a single-sample plate apparatus of the FOX 50 type, with heat flux density sensors, with a horizontal orientation and the location of the hot side of the sample: the bottom. The tests were carried out in accordance with the PN-EN 12664:2002 standard. Measurements were made for at least 4 replicates of each analyzed variant. The temperature difference across the thickness of the sample did not exceed 10 °C. The test specimens were conditioned for 6 h at a temperature of 23 ± 2 °C and relative humidity of 50 ± 5%. Relative mass change during conditioning Δm_r and measurement Δm_w did not exceed 0.1%.

The individual stages of the production and testing of samples are shown in Figure 3.

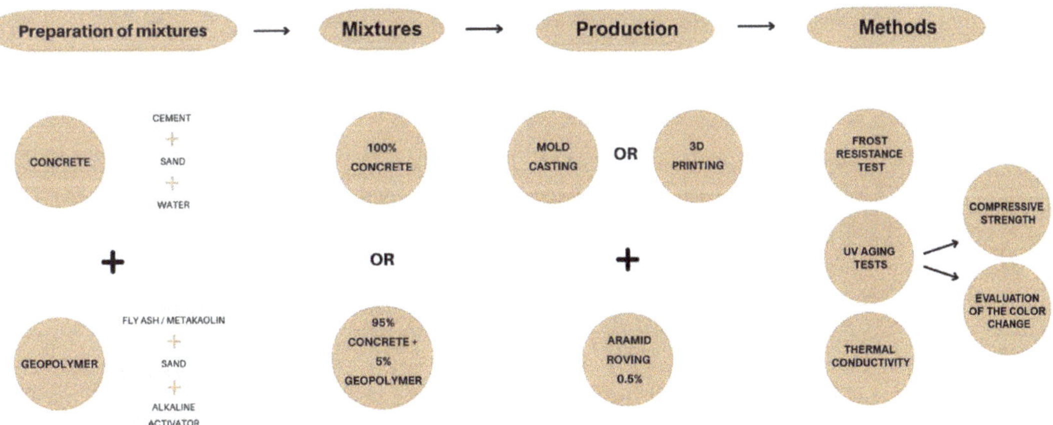

Figure 3. The stages of the production and testing of samples.

3. Results and Discussion

3.1. Frost Resistance

The results of the compressive strength test of samples without the addition of aramid roving depending on the production method are presented in Figure 4. The samples made by casting into molds were characterized by a higher compressive strength (38–44 MPa) than the samples produced by the 3D printing method (19–26 MPa). The compressive strength of the samples made with the additive method is about 40% lower than that of the cast samples. Rahul et al. [35] also showed in their research that the strength of printed concrete was lower compared to mold-cast concrete. However, this difference was at the level of a dozen or so percent. The lower compressive strength of 3D printed samples compared to cast specimens may be due to the presence of weak layer boundary connections with the presence of discontinuity defects between the layers. This can lead to material fracture under compressive stress even at low stress levels compared to cast samples [35]. For samples reinforced with aramid roving, lower compressive strength was recorded compared to samples made without the addition of roving. Conventional roving-reinforced concrete and concrete–geopolymer hybrids have up to about 15% lower strength than samples without the addition of roving. In the case of 3D printed samples, the addition of roving caused a reduction in compressive strength by up to 65% compared to samples without roving. In general, among the samples without the addition of roving, the highest strength was achieved by the concrete–geopolymer hybrids based on fly ash. The hybrids based on metakaolin obtained the lowest strength. Among all the tested samples, the lowest strength (about 9 MPa) was achieved by roving-reinforced 3D printed concrete and hybrid samples based on fly ash.

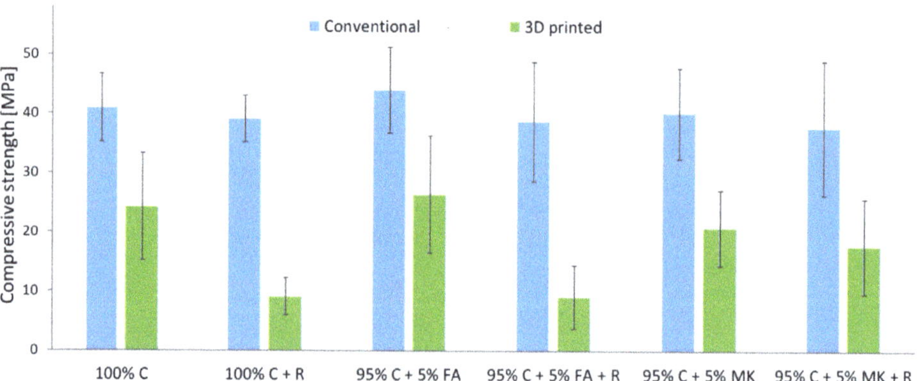

Figure 4. Compressive strength of concrete and concrete–geopolymer hybrids produced by casting and 3D printing.

The results of the residual compressive strength tests carried out after the frost resistance test of the samples are shown in Figure 5. The lack of test results after the freeze–thaw process for 3D printed hybrids with fly ash and roving (95% C + 5% FA + R) is due to the fact that these samples were damaged during the tests in the chamber.

After the performed freeze–thaw cycles, as in the case of the reference samples, the samples made by the casting method were characterized by a higher compressive strength (29–47 MPa) than the samples produced by the 3D printing method (16–24 MPa). The addition of aramid roving caused a decrease in the compressive strength of the samples by about 30% compared to the samples without reinforcement. The exception is hybrids based on fly ash, for which the introduction of roving increased the compressive strength by 18%. The highest strength was achieved by hybrid samples based on fly ash.

In general, after the frost resistance test, there was usually a decrease in the compressive strength of concrete and concrete–geopolymer hybrids. In the case of samples made by casting to molds, this decrease was up to 27% compared to the reference samples, while for 3D-printed samples the decrease in compressive strength compared to the reference samples did not exceed 12%.

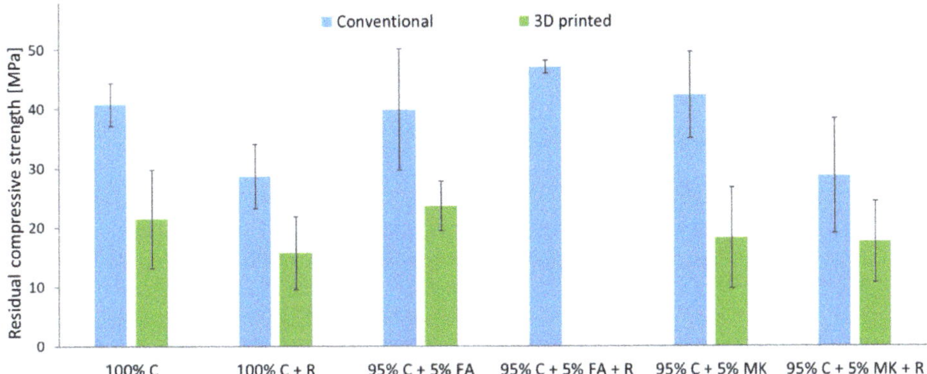

Figure 5. Residual compressive strength after the freeze–thaw cycles of concrete and concrete–geopolymer hybrids produced by casting and 3D printing.

Öztürk [36] also noted in his study that cement mortars showed a decrease in compressive strength after freeze–thaw cycles. This reduction was more pronounced at an early age (after 7 days of curing). Thus, the protection of cement-based composites from frost is particularly important in earlier concrete ages than in later curing ages. As in this study, Öztürk also reports that the presence of fibers in the material slightly decreased the compressive strength of the samples. However, after the freeze–thaw cycles, the results were similar for ordinary and fiber-reinforced specimens. Additionally, the reduction in strength was more pronounced in the case of geopolymer mortars. This is due to the fact that immersion of samples in water is more harmful to geopolymers compared to cement mixtures. This is probably because the free sodium content is leached out during the freeze–thaw cycles [36].

After the frost resistance tests, the mass losses for individual samples were determined. The results are shown in Figure 6. The lack of results for the 3D printed hybrids 95% C + 5% FA + R is due to the fact that these samples were destroyed during the tests in the chamber.

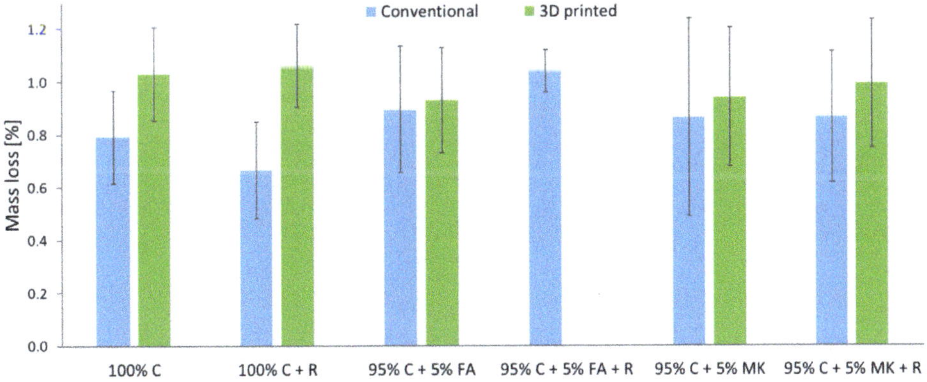

Figure 6. Mass loss after freeze–thaw cycles for concrete and concrete–geopolymer hybrids produced by casting and 3D printing.

Regardless of the composition, the samples produced by additive methods showed greater mass losses after the frost resistance test than for samples made by casting into molds. The smallest mass loss occurred for concrete samples reinforced with aramid roving (0.64%) made with the conventional method. The greatest loss occurred for the same type of material but produced with 3D printing (1.06%).

In the case of the printed concrete–geopolymer hybrids, the mass loss was about 8–12% higher than for the cast samples. However, a greater disproportion was noted for additively produced concrete samples, for which the mass loss was as much as 24–27% higher than in the case of samples made with conventional methods.

On the basis of the tests, it was found that the tested samples obtained a frost resistance level of F150.

3.2. UV Aging Tests

Any building material undergoes a slow degradation process, which is influenced by factors such as temperature, humidity, UV radiation, and mechanical influence. The kinetics of this process depends, among others, on the exploitation environment, possible defects in the structure, or the intensity and types of factors causing the changes. Therefore, both raw materials and products, for safety reasons, should be analyzed by the aging tests. The tests allow for the elimination of design and construction errors and slow down the degradation processes of the final products. When assessing the resistance of the material to environmental conditions, its resistance to exposure to UV radiation and water is most often tested.

UV radiation can significantly reduce the aesthetic value, as a result of the aging of protective gel coatings used for bridge cornices. UV radiation breaks down the polymer chains, thereby releasing fillers and dyes onto the surface of the elements.

Moreover, testing the resistance to UV radiation is necessary to determine the durability of the material properties over time. Therefore, tests were conducted to determine the changes in the properties of the samples after exposure to UV radiation.

The results of the compressive strength tests of the samples carried out after the UV aging tests in the climatic chamber are shown in Figure 7.

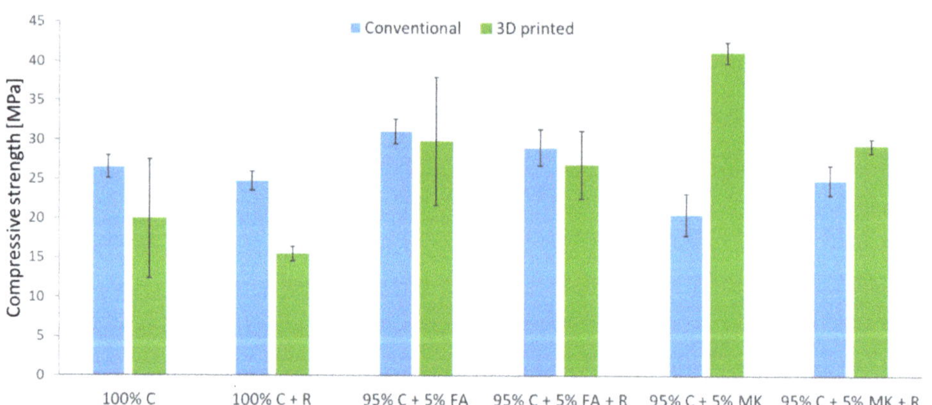

Figure 7. Compressive strength after the UV aging tests.

On the basis of the results presented in Figure 7, it was observed that the specimens made with the addition of aramid roving were characterized by a lower compressive strength than the samples made only of the matrix material. Similar relationships were also observed for samples subjected to compressive strength tests without prior exposure (Figure 4). The exception is concrete–geopolymer hybrids with the addition of a metakaolin-based geopolymer reinforced with aramid roving, which were produced by a conventional method. These samples have higher compressive strengths (24.93 MPa) compared to samples without the addition of roving (20.53 MPa). In general, samples made using the conventional method have higher strength properties than 3D printed samples. Only in the case of concrete–geopolymer hybrids with the addition of a metakaolin-based geopolymer, both without and reinforced with roving, were the additively produced samples characterized by higher strength. Among the samples produced by the casting method, the hybrids of 95% C + 5% FA have the highest compressive strength (31.07 MPa). In the case of 3D printed samples, the highest strength is achieved by hybrids of 95% C + 5% MK (41.13 MPa). The 100% C + R samples (15.47 MPa) are characterized by the lowest strength.

After the tests of resistance to UV radiation for samples made by casting into molds, a decrease in compressive strength was observed in comparison with the reference samples (Figure 4). There was a decrease of 25% and even 50% of the initial strength for the 95% C + 5% FA + R and 95% C + 5% MK samples, respectively. For the remaining samples, the strength after exposure decreased by about 30–40%. UV radiation weakened the mechanical properties of concretes and hybrids produced by conventional methods. As a result, the samples may have lower fatigue and damage resistance. Therefore, a longer time of UV exposure has a degradative effect on the mixtures produced [37]. On the other hand, in the case of 3D printed samples, the decrease in compressive strength after testing the resistance to UV radiation was recorded only for the concrete sample. For the remaining samples, after exposure, the compressive strength increased in relation to the reference samples. This may be due to the harder effects of UV aging and cause an increase in strength. A similar relationship was noticed by Wu et al. [37], who tested the flexural strength of asphalt concretes after 7, 14, and 28 days of UV aging. They noticed that the flexural strength of the samples at 25 °C tended to increase with successive days of exposure compared to unexposed samples. They concluded that at a temperature of about 25 °C, the effect of UV aging on the asphalt binder could partially offset the softening effect caused by the heat. This resulted in an increase in the strength of asphalt concretes.

The results of the examination of the color change in the gray scale are presented in Table 3. Color change tests were carried out in order to assess the color fastness and aging characteristics of the samples to the effects of weather conditions to which they may be exposed, in this case, UV radiation. Increasingly, in addition to the structural function, concrete also plays a decorative role. For this reason, their external appearance is also important.

Table 3. Results of the evaluation of the color change in a grayscale of concrete and concrete–geopolymer hybrid samples.

Manufacturing Method	Sample Designation	Test Result		
		Evaluator 1	Evaluator 2	Evaluator 3
Mold Casting	100% C	5	5	5
	100% C + R	5	5	5
	95% C + 5% FA	5	5	5
	95% C + 5 % FA + R	5	5	5
	95% C + 5% MK	5	5	5
	95% C + 5 % MK + R	5	5	5
3D Printing	100% C	5	5	5
	100% C + R	5	5	5
	95% C + 5% FA	5	5	5
	95% C + 5 % FA + R	5	5	5
	95% C + 5% MK	5	5	5
	95% C + 5 % MK + R	5	5	5

5—no visible difference between UV irradiated and UV non-irradiated samples.

On the basis of the results obtained from the conducted tests, it was found that the UV radiation did not cause any significant changes in the color of the samples, regardless of their composition or the method of production.

3.3. Thermal Conductivity

Thermal conductivity tests were carried out only on samples made by casting into molds. The limitation to preparing samples using the 3D printing method was the 15 mm diameter of the printing nozzle. Therefore, it would be difficult to produce specimens with dimensions of ø 55 mm × 23 mm containing aramid roving in accordance with the methodology of sample preparation.

For the study of thermal conductivity, samples with high strength recorded in previous tests were selected. The tests were carried out for samples with and without the addition of aramid roving. The obtained results are presented in Table 4.

Table 4. Thermal conductivity and thermal resistance of concrete and concrete–geopolymer hybrid samples.

Sample Designation	ρ [kg/m³]	d [m]	λ_i (W/m·K)	R_i (m²·K/W)
100% C + R	1901	0.02306	0.3947	0.06
95% C + 5% FA	1925	0.02339	0.5334	0.04
95% C + 5% FA + R	1991	0.02425	0.7413	0.03
95% C + 5% MK + R	1943	0.02225	0.5353	0.04

d—measured thickness of the test sample. ρ—sample density after seasoning. R_i—thermal resistance of the tested samples. λ_i—thermal conductivity coefficient.

The tested concrete samples reinforced with 0.5% aramid roving were characterized by the lowest thermal conductivity coefficient, and thus, such material better insulates against heat losses. Concrete–geopolymer hybrids with the addition of an ash-based geopolymer and hybrids with the addition of a metakaolin-based geopolymer reinforced with roving obtained similar values of the thermal conductivity coefficient, respectively, 0.5334 W/m·K and 0.5353 W/m·K. Among the tested samples, the best thermal conductivity was achieved by concrete–geopolymer hybrids with the addition of geopolymer based on fly ash, reinforced with aramid roving. The addition of roving to the hybrid increased the thermal conductivity coefficient from 0.5334 W/m·K to 0.7413 W/m·K.

The thermal conductivity of concrete is mainly related to porosity and pore size [38,39]. The thermal conductivity of hybrids with the addition of fly ash after adding aramid roving increased by almost 40% compared to the hybrid without reinforcement (95% C + 5% FA). Overall, the thermal conductivity of the hybrids was higher than that of the reference concrete. Similar results were observed by Lu et al. [39]. The samples made with fly ash and metakaolin showed a higher thermal conductivity than the samples without this additive. Replacement of metakaolin with fly ash increased the thermal conductivity by about 38%. Lu et al. [39] noticed that the increase in thermal conductivity may result, for example, from the lower fineness of the material, which may increase the density of the sample.

4. Conclusions

The properties of innovative concrete–geopolymer hybrids reinforced with aramid roving and produced by 3D printing technology were compared with samples prepared by the casting method. The effects on frost resistance, UV radiation, and thermal conductivity properties were determined.

Generally, after the freeze–thaw cycles, the compressive strength of concrete and concrete–geopolymer hybrids decreased. The concrete–geopolymer hybrids made by 3D printing were characterized by a compressive strength that was about 40–57% lower than the samples made by casting. This may be the effect of the presence of weak interface connections between the layers. Moreover, the addition of aramid roving caused a decrease in the compressive strength of the samples by about 30% compared to the samples without reinforcement. All of the tested samples had a frost resistance level of F150.

UV radiation test showed that the 3D printed hybrids with the addition of a metakaolin-based geopolymer exhibited the highest strength. The addition of roving reduced the strength properties of the samples after UV aging tests, and, in this case, exposure to UV radiation further decreased the mechanical properties of concretes and hybrids produced by conventional methods. For 3D printed samples, aramid roving caused an increase in compressive strength after UV exposure. The addition of uniform orientation roving may eliminate the effect of weak boundaries between the printed layers.

Thermal conductivity tests showed that among the tested samples, concrete reinforced with aramid roving was characterized by the lowest thermal conductivity coefficient. As a result, concrete provides better insulation against heat loss. The addition of aramid roving increased the hybrid's thermal conductivity by almost 40% compared to the material without reinforcement.

Author Contributions: Conceptualization, M.S.-H. and M.H.; Data curation, M.Ł., M.G., P.D. and A.S.; Formal analysis, J.M., C.Z., K.K., M.Ł., A.S. and M.H.; Funding acquisition, W.M. and M.H.; Investigation, J.M., C.Z., K.K., M.Ł., M.G. and P.D.; Methodology, J.M., C.Z., K.K., M.Ł., M.G., P.D., A.S. and M.H.; Resources, W.M. and M.H.; Validation, M.G., M.S.-H. and M.H.; Visualization, J.M. and M.H.; Writing—original draft, J.M.; Writing—review & editing, M.S.-H. and M.H.. All authors have read and agreed to the published version of the manuscript.

Funding: This work has been supported by Growth Operational Programme 2014–2020, IV Increasing the research potential, 4.1.4: 'Application projects', funded by the National Centre for Research and Development in Poland, within the framework of the grant 'Development of 3D printing technology for construction and facade prefabricated elements made of concrete composites and geopolymers', grant No. POIR 04.01.04-00-0096/18-00.

Institutional Review Board Statement: Not applicable.

Informed Consent Statement: Not applicable.

Data Availability Statement: Not applicable.

Conflicts of Interest: The authors declare no conflict of interest.

References

1. How Certain Are the United Nations Global Population Projections? In *Population Facts*; United Nations, Department of Economic and Social Affairs: New York, NY, USA, 2019; Volume 6, pp. 1–4.
2. Batikha, M.; Jotangia, R.; Baaj, M.Y.; Mousleh, I. 3D Concrete Printing for Sustainable and Economical Construction: A Comparative Study. *Autom. Constr.* **2022**, *134*, 104087. [CrossRef]
3. Li, W.; Lin, X.; Bao, D.W.; Min Xie, Y. A Review of Formwork Systems for Modern Concrete Construction. *Structures* **2022**, *38*, 52–63. [CrossRef]
4. Lahoti, M.; Tan, K.H.; Yang, E.H. A Critical Review of Geopolymer Properties for Structural Fire-Resistance Applications. *Constr. Build. Mater.* **2019**, *221*, 514–526. [CrossRef]
5. Rollakanti, C.R.; Prasad, C.V.S.R. Applications, Performance, Challenges and Current Progress of 3D Concrete Printing Technologies as the Future of Sustainable Construction—A State of the Art Review. *Mater. Today Proc.* **2022**, *65*, 995–1000. [CrossRef]
6. Łach, M.; Kluska, B.; Janus, D.; Kabat, D.; Pławecka, K.; Korniejenko, K.; Guigou, M.D.; Choińska, M. Effect of Fiber Reinforcement on the Compression and Flexural Strength of Fiber-Reinforced Geopolymers. *Appl. Sci.* **2021**, *11*, 10443. [CrossRef]
7. Abd Razak, S.N.; Shafiq, N.; Nikbakht, E.H.; Mohammed, B.S.; Guillaumat, L.; Farhan, S.A. Fire Performance of Fly-Ash-Based Geopolymer Concrete: Effect of Burning Temperature on Mechanical and Microstructural Properties. *Mater. Today Proc.* 2022, *in press*. [CrossRef]
8. Zhang, C.; Wei, M.; Hu, Z.; Yang, T.; Jiao, B.; Zhu, H.; Sun, N.; Lv, H. Sulphate Resistance of Silane Coupling Agent Reinforced Metakaolin Geopolymer Composites. *Ceram. Int.* **2022**, *48*, 25254–25266. [CrossRef]
9. Ranjbar, N.; Zhang, M. Fiber-Reinforced Geopolymer Composites: A Review. *Cem. Concr. Compos.* **2020**, *107*, 103498. [CrossRef]
10. Fahim Huseien, G.; Mirza, J.; Ismail, M.; Ghoshal, S.K.; Abdulameer Hussein, A. Geopolymer Mortars as Sustainable Repair Material: A Comprehensive Review. *Renew. Sustain. Energy Rev.* **2017**, *80*, 54–74. [CrossRef]
11. Seelam, P.K.; Sreenivasan, H.; Ojala, S.; Pitkäaho, S.; Laitinen, T.; Niu, H.; Keiski, R.L.; Illikainen, M. Modified Geopolymers as Promising Catalyst Supports for Abatement of Dichloromethane. *J. Clean. Prod.* **2021**, *280*, 124584. [CrossRef]
12. Lazorenko, G.; Kasprzhitskii, A. Geopolymer Additive Manufacturing: A Review. *Addit. Manuf.* **2022**, *55*, 102782. [CrossRef]
13. Wolfs, R.J.M.; Bos, F.P.; Salet, T.A.M. Hardened Properties of 3D Printed Concrete: The Influence of Process Parameters on Interlayer Adhesion. *Cem. Concr. Res.* **2019**, *119*, 132–140. [CrossRef]
14. Chen, Y.; He, S.; Gan, Y.; Çopuroğlu, O.; Veer, F.; Schlangen, E. A Review of Printing Strategies, Sustainable Cementitious Materials and Characterization Methods in the Context of Extrusion-Based 3D Concrete Printing. *J. Build. Eng.* **2022**, *45*, 103599. [CrossRef]

15. Mechtcherine, V.; Bos, F.P.; Perrot, A.; da Silva, W.R.L.; Nerella, V.N.; Fataei, S.; Wolfs, R.J.M.; Sonebi, M.; Roussel, N. Extrusion-Based Additive Manufacturing with Cement-Based Materials—Production Steps, Processes, and Their Underlying Physics: A Review. *Cem. Concr. Res.* **2020**, *132*, 106037. [CrossRef]
16. Arunothayan, A.R.; Nematollahi, B.; Ranade, R.; Bong, S.H.; Sanjayan, J.G.; Khayat, K.H. Fiber Orientation Effects on Ultra-High Performance Concrete Formed by 3D Printing. *Cem. Concr. Res.* **2021**, *143*, 106384. [CrossRef]
17. De Schutter, G.; Lesage, K.; Mechtcherine, V.; Nerella, V.N.; Habert, G.; Agusti-Juan, I. Vision of 3D Printing with Concrete—Technical, Economic and Environmental Potentials. *Cem. Concr. Res.* **2018**, *112*, 25–36. [CrossRef]
18. Rashad, A.M. Metakaolin as Cementitious Material: History, Scours, Production and Composition-A Comprehensive Overview. *Constr. Build. Mater.* **2013**, *41*, 303–318. [CrossRef]
19. Barboza-Chavez, A.C.; Gómez-Zamorano, L.Y.; Acevedo-Dávila, J.L. Synthesis and Characterization of a Hybrid Cement Based on Fly Ash, Metakaolin and Portland Cement Clinker. *Materials* **2020**, *13*, 1084. [CrossRef]
20. Supit, S.W.M.; Olivia, M. Compressive Strength and Sulfate Resistance of Metakaolin-Based Geopolymer Mortar with Different Ratio of Alkaline Activator. *Mater. Today Proc.* 2022, *in press*. [CrossRef]
21. Dai, S.; Wang, H.; An, S.; Yuan, L. Mechanical Properties and Microstructural Characterization of Metakaolin Geopolymers Based on Orthogonal Tests. *Materials* **2022**, *15*, 2957. [CrossRef]
22. Agrawal, V.M.; Savoikar, P.P. Sustainable Use of Normal and Ultra-Fine Fly Ash in Mortar as Partial Replacement to Ordinary Portland Cement in Ternary Combinations. *Mater. Today Proc.* **2022**, *51*, 1593–1597. [CrossRef]
23. Karumanchi, M.; Bellum, R.R.; Chennupati, M.; Kunchala, V.; Regulagunta, M. Influence on Mechanical Properties of Concrete of Cement Replacement with Fly Ash and River Sand Replacement with Foundry Sand. *Mater. Today Proc.* **2022**, *65*, 3547–3551. [CrossRef]
24. Sun, Y.; Wang, K.Q.; Lee, H.S. Prediction of Compressive Strength Development for Blended Cement Mortar Considering Fly Ash Fineness and Replacement Ratio. *Constr. Build. Mater.* **2021**, *271*, 121532. [CrossRef]
25. Han, X.; Yang, J.; Feng, J.; Zhou, C.; Wang, X. Research on Hydration Mechanism of Ultrafine Fly Ash and Cement Composite. *Constr. Build. Mater.* **2019**, *227*, 116697. [CrossRef]
26. Hsu, S.; Chi, M.; Huang, R. Effect of Fineness and Replacement Ratio of Ground Fly Ash on Properties of Blended Cement Mortar. *Constr. Build. Mater.* **2018**, *176*, 250–258. [CrossRef]
27. Muthusamy, K.; Budiea, A.M.A.; Azhar, N.W.; Jaafar, M.S.; Mohsin, S.M.S.; Arifin, N.F.; Mat Yahaya, F. Durability Properties of Oil Palm Shell Lightweight Aggregate Concrete Containing Fly Ash as Partial Cement Replacement. *Mater. Today Proc.* **2020**, *41*, 56–60. [CrossRef]
28. Askarian, M.; Tao, Z.; Adam, G.; Samali, B. Mechanical Properties of Ambient Cured One-Part Hybrid OPC-Geopolymer Concrete. *Constr. Build. Mater.* **2018**, *186*, 330–337. [CrossRef]
29. Marczyk, J.; Ziejewska, C.; Gądek, S.; Korniejenko, K.; Łach, M.; Góra, M.; Kurek, I.; Dogan-Saglamtimur, N.; Hebda, M.; Szechyńska-Hebda, M. Hybrid Materials Based on Fly Ash, Metakaolin, and Cement for 3D Printing. *Materials* **2021**, *14*, 6874. [CrossRef]
30. Ziejewska, C.; Marczyk, J.; Korniejenko, K.; Bednarz, S.; Sroczyk, P.; Łach, M.; Mikuła, J.; Figiela, B.; Szechyńska-Hebda, M.; Hebda, M. 3D Printing of Concrete-Geopolymer Hybrids. *Materials* **2022**, *15*, 2819. [CrossRef]
31. Kandekar, S.B.; Talikoti, R.S. Torsional Behaviour of Reinforced Concrete Beam Wrapped with Aramid Fiber. *J. King Saud Univ. Eng. Sci.* **2019**, *31*, 340–344. [CrossRef]
32. Noorvand, H.; Brockman, S.C.; Mamlouk, M.; Kaloush, K. Effect of Aramid Fibers on Balanced Mix Design of Asphalt Concrete. *CivilEng* **2021**, *3*, 2. [CrossRef]
33. Korniejenko, K.; Figiela, B.; Miernik, K.; Ziejewska, C.; Marczyk, J.; Hebda, M.; Cheng, A.; Lin, W.T. Mechanical and Fracture Properties of Long Fiber Reinforced Geopolymer Composites. *Materials* **2021**, *14*, 5183. [CrossRef] [PubMed]
34. Korniejenko, K.; Figiela, B.; Ziejewska, C.; Marczyk, J.; Bazan, P.; Hebda, M.; Choińska, M.; Lin, W.T. Fracture Behavior of Long Fiber Reinforced Geopolymer Composites at Different Operating Temperatures. *Materials* **2022**, *15*, 482. [CrossRef] [PubMed]
35. Rahul, A.V.; Santhanam, M.; Meena, H.; Ghani, Z. Mechanical Characterization of 3D Printable Concrete. *Constr. Build. Mater.* **2019**, *227*, 116710. [CrossRef]
36. Öztürk, O. Comparison of Frost Resistance for the Fiber Reinforced Geopolymer and Cementitious Composites. *Mater. Today Proc.* **2022**, *65*, 1504–1510. [CrossRef]
37. Wu, S.; Ye, Y.; Li, Y.; Li, C.; Song, W.; Li, H.; Li, C.; Shu, B.; Nie, S. The Effect of UV Irradiation on the Chemical Structure, Mechanical and Self-Healing Properties of Asphalt Mixture. *Materials* **2019**, *12*, 2424. [CrossRef] [PubMed]
38. Wang, W.; Lu, C.; Li, Y.; Li, Q. An Investigation on Thermal Conductivity of Fly Ash Concrete after Elevated Temperature Exposure. *Constr. Build. Mater.* **2017**, *148*, 148–154. [CrossRef]
39. Lu, D.; Tang, Z.; Zhang, L.; Zhuo, J.; Gong, Y.; Tian, Y.; Zhong, J. Effects of Combined Usage of Supplementary Cementitious Materials on the Thermal Properties and Microstructure of High-Performance Concrete at High Temperatures. *Materials* **2020**, *13*, 1833. [CrossRef]

Article

Pigment Penetration Characterization of Colored Boundaries in Powder-Based Color 3D Printing

Danyang Yao [1], Jiangping Yuan [2], Jieni Tian [1], Liru Wang [1] and Guangxue Chen [1,*]

1 State Key Laboratory of Pulp and Paper Engineering, South China University of Technology, Guangzhou 510640, China; dyyao9722@163.com (D.Y.); tianjieni123@126.com (J.T.); wanglir0510@163.com (L.W.)
2 College of Communication and Art Design, University of Shanghai for Science and Technology, Shanghai 200093, China; yuanjp@usst.edu.cn
* Correspondence: chengx@scut.edu.cn

Abstract: Color 3D printing has widely affected our daily lives; therefore, its precise control is essential for aesthetics and performance. In this study, four unique test plates were printed using powder-based full-color 3D printing as an example; moreover, the corresponding pigment-penetration depth, chromaticity value and image-based metrics were measured to investigate the lateral pigment penetration characteristics and relative surface-color reproduction of each color patch, and to perform an objective analysis with specific microscopic images. The results show that the lateral pigment-penetration depth correlates with the number of printed layers on the designed 3D test plates, and the qualitative analysis of microscopic images can explain the change in chromaticity well. Meanwhile, there is an obvious linear correlation between the mean structural similarity, color-image difference and color difference for current color samples. Thus, our proposed approach has a good practicality for powder-based color 3D printing, and can provide new insight into predicting the color-presentation efficiency of color 3D-printed substrates by the abovementioned objective metrics.

Keywords: color 3D printing; penetration depth; chromaticity; image-based metric; color reproduction

Citation: Yao, D.; Yuan, J.; Tian, J.; Wang, L.; Chen, G. Pigment Penetration Characterization of Colored Boundaries in Powder-Based Color 3D Printing. *Materials* **2022**, *15*, 3245. https://doi.org/10.3390/ma15093245

Academic Editors: Wai Yee Yeong and Swee Leong Sing

Received: 24 March 2022
Accepted: 27 April 2022
Published: 30 April 2022

1. Introduction

Three-dimensional printing is a layered manufacturing process based on the principles of additive manufacturing and rapid prototyping [1]. It is based on digital model files and uses bondable materials such as plastic, gypsum powder, paper sheet, and resin to manufacture three-dimensional objects by printing layer by layer and overlaying different shapes in a continuous layer-forming method [2], which also contains color replication [3]. With the continuous development and upgrading of 3D printing materials and equipment with a rapid growth in industry scale, color 3D printing technology has become a key branch of the 3D printing industry [4].

Based on the printing materials, color 3D printing can be divided into six major categories, namely, plastic-, paper-, powder-, bio-, food- and metal-based color 3D printing [5]. There are two ways for color 3D printing to achieve a lifelike surface color: one is to add a coloring stage to the post-printing process [6,7]; the second is to introduce the color discretization of graphic printing into 3D printing during the forming process. The latter method directly provides richer color reproduction and is called full-color 3D printing with accurate digital colorization arrangement [8,9].

Color is one of the most expressive and attractive surface attributes in the 3D world, and high-precision color reproduction is something that all categories of 3D printing technology aspires to achieve at this stage. It is even more essential in the custom production and cultural artwork industries [10,11]. Powder-based 3D printing, also known as inkjet 3D printing technology (3DP), was one of the first technologies proposed to enable full-color 3D printing [12]. However, at this stage, 3D-printed products still cannot fully meet the individual needs of users [13], and color reproduction research still lacks a unified theoretical and standard numerical model for predicting the color quality of 3D prints [14].

Currently, color quality evaluation of full-color 3D printing is a focus of interest for color researchers and a key part of the industrialization process of full-color 3D printing. Lu et al. proposed the graded solution method for ink-permeation pressure distribution and established a spatio-temporal evolution equation for ink-permeation pressure distribution based on experimental parameters. This study analyzed the ink permeation behavior only from the horizontal plane, without taking into account the sides (cross-sections) of the samples [15]. Yang et al. proposed a fast slicing and color-acquisition algorithm for 3D models to obtain color information coordinate formulas and calculate axis pixel-information coordinates for color 3D printing, which achieved the expected purpose with a high efficiency. However, the degree of refinement was somewhat lacking [16]. Li et al. used EPSON DX-5 for spontaneous color 3D printing; full-color printing was implemented based on halftoning algorithms that used different threshold matrices for different ink channels; the performances of the various algorithms were evaluated in terms of both subjective and objective indices. The calculation process was more complicated [17]. In the study of the penetration characteristics of the pigments on the lateral interface (cross-section) of 3D prints and their surface color features, there is an urgent need for color quality prediction for full-color 3D printing.

In addition, there has also been related progress in research on the quality evaluation of color reproduction for 3D-printed surfaces. Wang et al. proposed a microstructure image analysis method for powder-based 3D-printed samples postprocessed by different impregnating agents to determine the optimal process parameters. The effect of color on 3D samples has not been considered in detail [18]. Yuan et al. studied the color reproduction of specific 3D samples produced by plastic-based color 3D printing, using a linear correlation analysis between color difference and image-based metrics, including a mean structural similarity (MSSIM) and color structural similarity (SSIM) map. There was no consideration of the effect of pigment penetration behavior on color reproduction [19]. Tian et al. skillfully explored the quantitative correlation between the color feature of a paper-based 3D-printed coloring layer and its blank layer attached underneath using ΔE^*_{ab}, the feature similarity index measure of color images (FSIMc), and an improved color image difference (iCID). This study ignored the problem of 3D sample boundary coloring [20]. Therefore, image-based metrics have mature applications in the evaluation of the color-reproduction quality of color in 3D printing as efficient numerical visualization tools.

In fact, the color management in the classic graphic printing domain performs badly in adjusting the color reproduction of color 3D printing because the pigment penetration behavior changes significantly [4]. Thus, combined with image-based metrics and color differences, this paper uses a powder-based color 3D printer and a specific experimental model to analyze the characteristics of internal pigment penetration, as well as quantitative quantification to enrich the color-reproduction control of color 3D printing. By pre-designing the test structure, the color 3D-printed sample was cut with certain thin pieces, and the microscopic imaging method was used to quantify the depth of the pigment penetration in the sides of the pieces. The current characterization of pigment penetration characteristics can provide more accurate factual bases and analysis methods for accurate color-reproduction prediction in color 3D printing.

The structure of this study is divided into four parts: The first part contains the background of this study, previous research results and problems, and the purpose and significance of this research study. The second part contains the model design as well as the experimental method used. The third part analyzes the experimental results from three perspectives. The fourth part contains the results and discussion.

2. Materials and Methods

The experimental framework of this paper is shown in Figure 1. Firstly, four specific 3D test models were designed using the 3D software Autodesk 3ds Max (San Rafael, CA, USA), and samples were printed using the 3D System Z860 printer (Rock Hills, SC, USA). Next, a 3D profilometer VR-5000 (Osaka, Japan) was used to obtain the surface pigment penetration depth; the spectrodensitometer X-Rite i1Pro2 (Grand Rapids, Michigan, USA) was used to measure the chromaticity values (L*a*b*) of the surface color; a TIPSCOPE Mobile Microscope (Hubei, China) was used to obtain the microscopic surface; and a Canon EOS 500D camera (Shimomaruko, Ota-ku, Tokyo, Japan) was used to acquire high-resolution images of surface color patches to calculate image-based metrics.

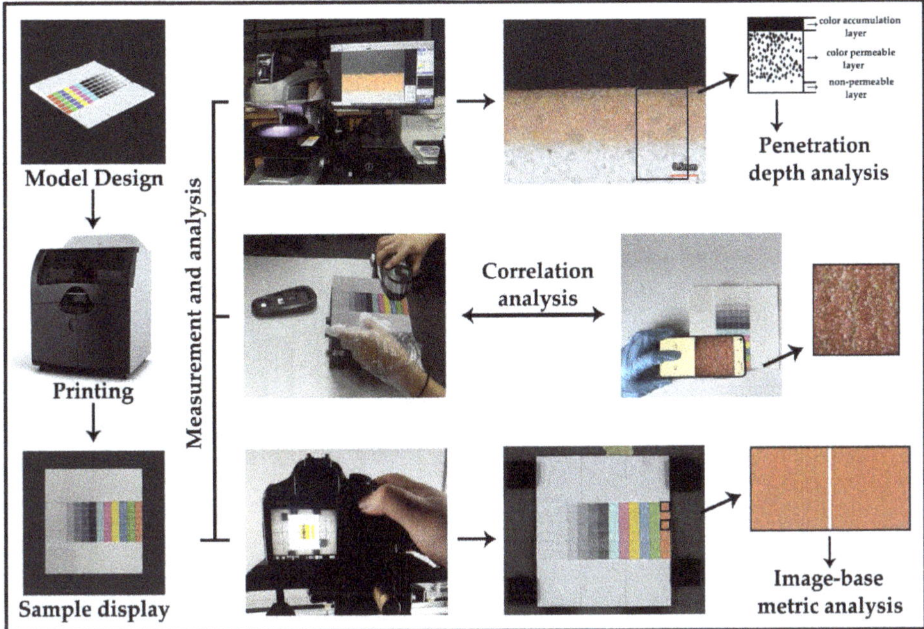

Figure 1. The experimental framework with relative equipment.

2.1. Model Design

Four 3D test models were designed. The center area of this model had six thicknesses of coloring stairs, and its two sides had three thicknesses of white blocks as the blank control. Each test model included thirteen color samples and six coloring stairs. The color sample was selected from six primary colors, r (red), g (green), b (blue), c (cyan), m (magenta), y (yellow), and neutral colors k (black), 0.8k, 0.6k, 0.4k, 0.2k, 0.1k, and w (white) with specific order changes, where 0.1k means 10% black, expressed as a decimal for subsequent marking, and so on. In Table 1, for each color sample, specific CIEL*a*b* values are given independent of the device presentation. In Table 2, specific parameters for different stairs are given for each model, where these four test samples are named I, II, III, and IV in turn. C_{s1} denotes the thickness of the first step of the color layer, W_{s1} denotes the thickness of the first step of the white layer, and so on for the others.

Table 1. Color parameters in each sample patch.

Color	r	g	b	y	m	c	k	0.8k	0.6k	0.4k	0.2k	0.1k	w
L*	54	88	30	98	60	90	0	21	43	63	82	91	100
a*	81	−79	68	−16	94	−51	0	0	0	0	0	0	0
b*	70	81	−112	93	−60	−15	0	0	0	0	0	0	0

Table 2. Thickness parameters in each 3D test model.

ID	Substrate (cm)	Color Layer (cm)						White Layer (cm)					
		C_{S1}	C_{S2}	C_{S3}	C_{S4}	C_{S5}	C_{S6}	W_{S1}	W_{S2}	W_{S3}	W_{S4}	W_{S5}	W_{S6}
I	0.2	0.02	0.04	0.06	0.08	0.1	0.12	0	0.02	0.04	0.06	0.08	0.1
II	0.2	0.04	0.08	0.12	0.16	0.2	0.24	0.02	0.06	0.1	0.14	0.18	0.22
III	0.1	0.08	0.16	0.24	0.32	0.4	0.48	0.06	0.14	0.22	0.3	0.38	0.46
IV	0.1	0.1	0.2	0.3	0.4	0.5	0.6	0.08	0.18	0.28	0.38	0.48	0.58

2.2. Pigment Penetration Depth, Chromaticity Measurement and Image Acquisition Quantification

Using a 3D profilometer (its magnification = 80×) to obtain the depth of pigment penetration on the lateral interface of each color patch on the 3D test plate, the thickness values were measured three times and averaged for the next analysis. The chromaticity of the color patches on each test plate was measured by a spectrophotometer (parameters: measurement condition = M1, illumination source = D50, observer angle = 2 degrees), averaged three times for each patch, and correlated with the microscopic images acquired using TIPSCOPE. In the standard imaging system, the camera focus was set at the same horizontal height as the center point of the color 3D test plate to obtain the microscopic imaging for each color patch. Then, further objective quantitative evaluations were performed in a compiled MATLAB program using the feature-similarity index measures of color image (FSIMc), mean structural similarity (MSSIM) and color image difference (iCID) metric algorithms.

2.3. Data Analysis

In this study, the degree of pigment penetration on the side of the color patches, as well as the chromaticity and similarity between the blocks, were calculated and analyzed.

$$\text{Height} = H_i - h \tag{1}$$

where h is the color block surface color thickness preset value (h = 200 μm), H_i is the penetration depth value of the i color patch surface color, and i is a positive integer from 1 to 6.

$$\Delta E^*_{76} = \sqrt{(L^*_i - L^*_1)^2 + (a^*_i - a^*_1)^2 + (b^*_i - b^*_1)^2} \tag{2}$$

where ΔE^*_{76} is the relative chromaticity; L^*_1, a^*_1, and b^*_1 are the color of the first stair color patch; L^*_i, a^*_i, and b^*_i are the color of the i stair color patch; and i is a positive integer from 2 to 4.

$$\text{FSIMc} = \frac{\sum_\Omega S_{PC}(x)\cdot S_G(x)\cdot[S_I(x)\cdot S_Q(x)]^\lambda PC_m(x)}{\sum_\Omega PC_m(x)} \tag{3}$$

For these, see Ref. [21].

$$\text{MSSIM}(x,y) = \frac{\left(2\mu_x\mu_y + C_1\right)\left(2\sigma_{xy} + C_2\right)}{\left(\mu_x^2 + \mu_y^2 + C_2\right)\left(\sigma_x^2 + \sigma_y^2 + C_2\right)} \tag{4}$$

where μ_x, μ_y, σ_x, σ_y, and σ_{xy} are the x-image mean, y-image mean, x-image variance, y-image variance, and two-image covariance, respectively, and C_1 and C_2 are nonzero small constants avoiding zero in the denominator.

$$\text{iCID}_A(S,T) = 1 - \frac{1}{|A|}\sum_{i\in A}[l_L(s_i,t_i)\cdot c_L(s_i,t_i)\cdot s_L(s_i,t_i)\cdot l_C(s_i,t_i)\cdot l_H(s_i,t_i)\cdot c_C(s_i,t_i)\cdot c_C(s_i,t_i)] \tag{5}$$

For these, see Ref. [22].

3. Results and Analysis

3.1. Effect of the Number of Layers Printed on the Surface of Powder-Based Color 3D Test Plates on the Depth of Pigment Penetration

In Figure 2, the penetration characteristics of the pigments on the sides of four 3D test plates with different numbers of printed layers are shown. The lateral thickness value of the color sample on each stair of the 3D test plate was first compared with the color preset value at the designing stage of the test models, and the difference was calculated and noted as the penetration depth. Considering that the bottom layer of the test plate is noncolored, the measurement targets select primary color samples and gray color samples, including k, 0.8k, and 0.6k. For the sake of image layout, "0.8k" is abbreviated to ".8k", and so on.

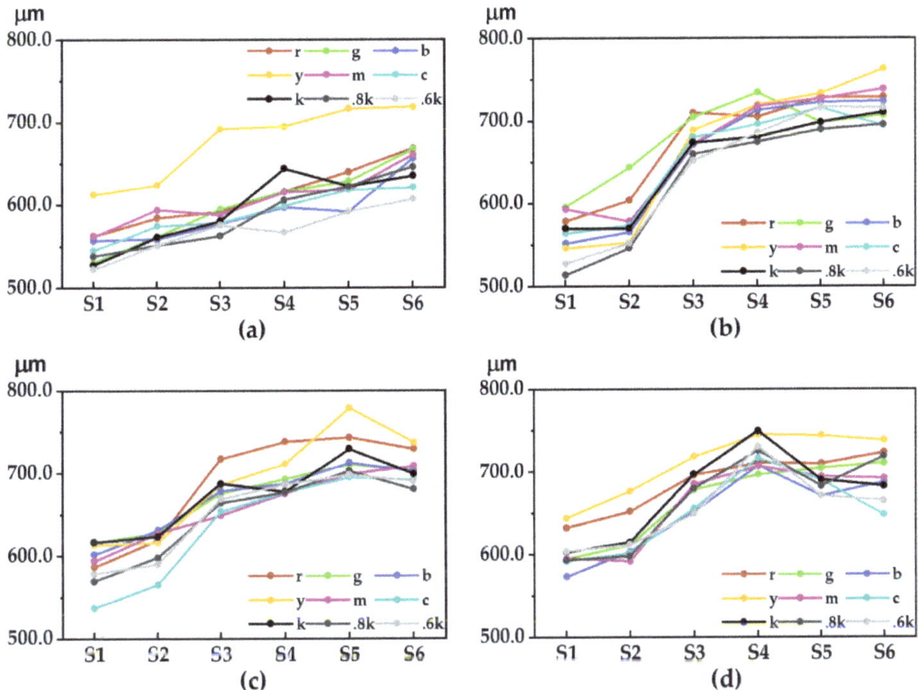

Figure 2. The lateral pigment penetration depth: (**a**) test plate I; (**b**) test plate II; (**c**) test plate III; (**d**) test plate IV.

Except for test plate I, the general trend of the penetration depth of the lateral pigments of the other test plates showed a consistent change, and the higher the number of printed layers was, the greater the penetration depth of the lateral pigments of the test plate was. However, when the number of printed layers reached a certain height, the change in their penetration depth tended to level off or slightly decrease. Test plates II and III on stair 1 to stair 3 show that the sample penetration degree grows more quickly and then relatively slowly. However, the penetration depth at stair 6 was relatively obvious for each color sample, such as color sample g and color sample c in test plate II, most of the color samples in test plate III, and color sample c in test plate IV, which all had obvious decreasing trends. Compared with the neutral gray sample, the depth of penetration of the color-sample side was more obvious. For example, the penetration depths of color samples r, g, y and m in test plate I were significantly greater than those of color sample 0.6k, the penetration depths of color samples r and g in test plate II were greater than those of color sample 0.8k, the penetration depths of color samples r and y in test plate III were greater than those of color sample 0.8k, and the penetration depths of color samples r and y in test plate IV were greater than those of color sample 0.6k.

In Figure 3, by calculating the average of the penetration depths of each stair color sample on a test plate, the influence of different printing layers on the surface of the test plate on the degree of pigment penetration became more obvious, and the higher the number of printing layers on the test plate was, the more obvious the degree of penetration was. For example, the average penetration depth of each stair color sample in test plates III and IV is significantly higher than that of plate I. When the number of printed layers is low, the average depth of penetration of each color sample on the surface of the test plate grows faster, but after reaching a certain height, the average depth of penetration is basically similar; however, there is a significant decline in the highest stair. The average depth of penetration in test plate III increases abruptly after stair 3. It is hypothesized that the thickness of the plate base at stair 3 reaches a certain height value and that the average depth of penetration shows a sudden change, which has a strong relationship with the thickness of the plate base.

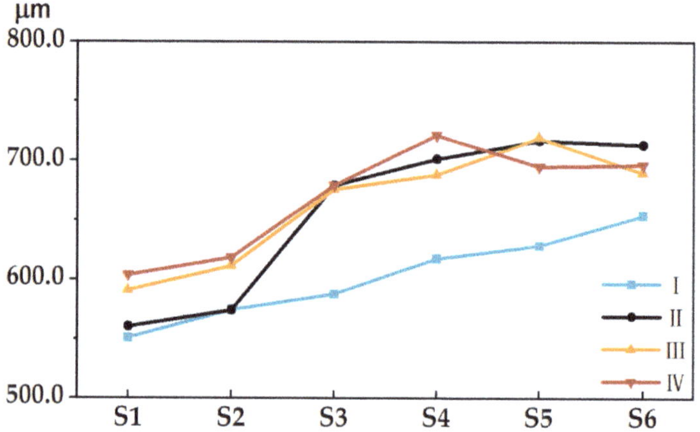

Figure 3. Average depth of pigment penetration on the side of each 3D test plate.

3.2. Effect of the Number of Layers Printed on the Surface of the Powder-Based Color 3D Test Plate on Color Chromatic Aberration

In Figure 4, the chromaticity between the color sample on coloring stair 1 of the 3D test plate and the color samples at the same position on the other stairs is shown. Overall, it can be seen clearly that the chromaticity varies significantly depending on the number of layers printed on the test plate and is positively correlated with the number of layers printed. Except for the 3D test plate I, the trend of chromaticity between each coloring stair and stair 1 in the rest of the plates is basically the same—that is, the stair contains the increase in the number of printed layers, where the chromaticity values at stair 2 and stair 3 fluctuate greatly. For example, the chromaticity between stair 4 and stair 6 in test plate I and stair 1 fluctuated more, while the rest of the stairs changed very little and tended to be the same, whereas the other three test plates and test plate I generally showed the opposite trend. The chromaticity of the color sample fluctuated more than the chromaticity trend of the noncolor sample. For example, the fluctuations of color samples g, m, c, and k in test plate I; color samples y, m, and c in test plate II; color samples r, g, y, m, and c in test plate III; and color samples g, m, and c in test plate IV were all relatively obvious. Compared with the other color samples, color sample k showed a large fluctuation in chromaticity when the number of printed layers was small, but when the number of printed layers reached a certain height, the change was relatively small.

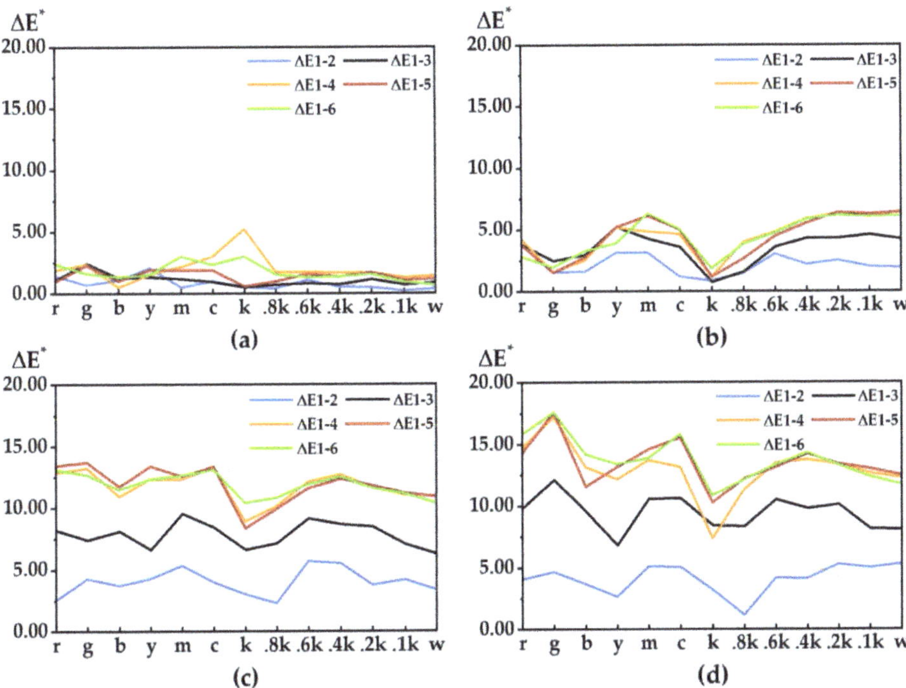

Figure 4. Three-dimensional test plate surface chromaticity: (**a**) test plate I; (**b**) test plate II; (**c**) test plate III; (**d**) test plate IV.

Figure 5 shows the average chromaticity of each color sample on the 3D test plate, and it can be observed that the fluctuation of each color sample varies widely. Except for test plate I, the overall trend of the other test plates is consistent. When the number of printed layers reaches a certain value, the average chromaticity values among the color samples on the test plate change more similarly. For example, the average chromaticity changes of color samples y and k in test plate III and test plate IV are basically the same. In addition, the average chromaticity values of g, m and c in color samples and 0.6k and 0.4k in noncolor samples are relatively large compared to other color samples. Figure 6 analyzes the chromaticity from the perspective of microscopic imaging on the surface of the color sample to explain the fluctuating characteristics of the chromaticity-variation trend, in which the vertical arrangement is the microscopic images of the color patches from coloring stair 1 to stair 6 on the test plate, in order, and the horizontal arrangement is the color samples g, m and k on the four test plates, in order. I-g indicates the color sample g on test plate I, and so on for the others. All diagrams are used is a small paste microscope, and the scale is 1:14 (when using it needs to be pasted on the phone lens, with the magnification of up to 400×).

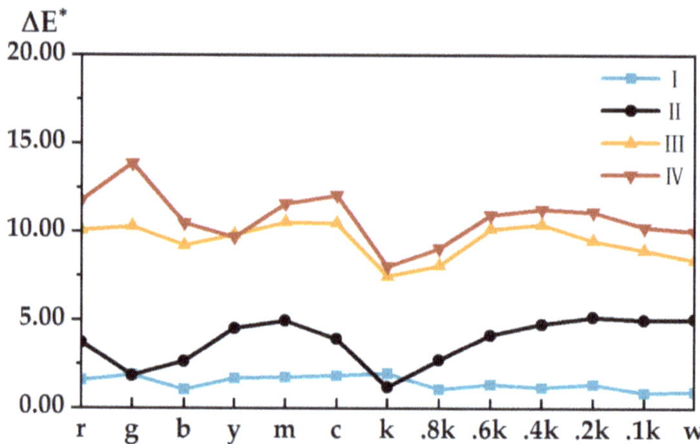

Figure 5. Three-dimensional test plate surface color average chromaticity.

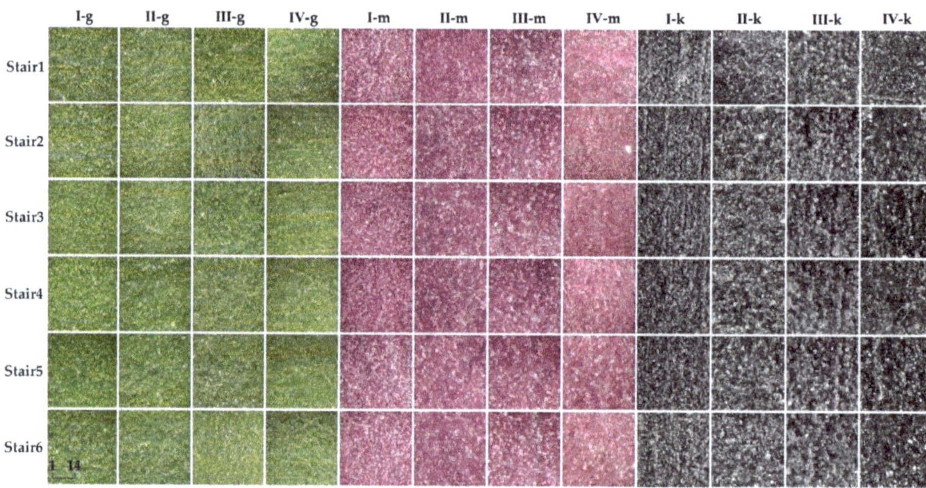

Figure 6. Comparison of microscopic images of color samples g, m and c on different test plates.

In Figure 6, due to the problem of a lack of space, only the color samples g, m and k with large fluctuations in chromaticity in Figure 4 were selected. The brightness of the colors of the 3D test plates with different printing layers were largely varied, and the color samples in test plates II, III and IV were obviously darker than those in plate I. The color of different coloring stairs in the same test plate also has a certain difference; the higher the coloring stair is, the less bright the color sample is compared to the lower stair of the color sample. Test plate I in the colored sample k contains white particles compared to the colored samples g and m, so its chromaticity value fluctuation is relatively large. The color of sample k on different stairs in test plate II is similar. Except for test plate I, the color samples g, m and k on stair 5 and stair 6 on the other test plates have very little difference in color, while the rest of the color samples on the stairs have a relatively low brightness.

3.3. Image-Based Metrics Analysis of Surface Coloring Stairs on Powder-Based Color 3D Test Plates

Figure 7 shows the similarity between the high-definition color samples on the 3D test plate of coloring stair 1 and the color samples at the same position of the other stairs. The top and bottom parts of the figure are selected based on the MSSIM metric and the iCID metric obtained from the MATLAB compilation, respectively, where the more similar the color samples, are the closer the MSSIM curve value is to 1, and the closer the iCID curve value is to 0, the smaller the difference between color samples is. The vertical coordinate Mv (metric value) represents the change in the value of the image basis metric. In general, the degree of similarity and the degree of difference between the color samples of the 3D test plates with different numbers of printed layers are basically the same, and the similarity and difference curves of the color samples on different coloring stairs are also very consistent for the same test plate. Overall, the similarity between coloring stair 6 and stair 1 is slightly better than that of other coloring ladders, e.g., color samples g, y and c in the four test plates. Color sample areas fluctuate more and the differences between them are more obvious, for example, in color samples r, b and m. Color samples 0.2k, 0.1k and w in noncolor areas fluctuate steadily. However, for the color samples in the transition region, the degree of difference is larger, for example, in color samples k, 0.8k and 0.6k, probably due to the large differences in color contrast between white particles and noncolored adhesives in the color samples at the transition region.

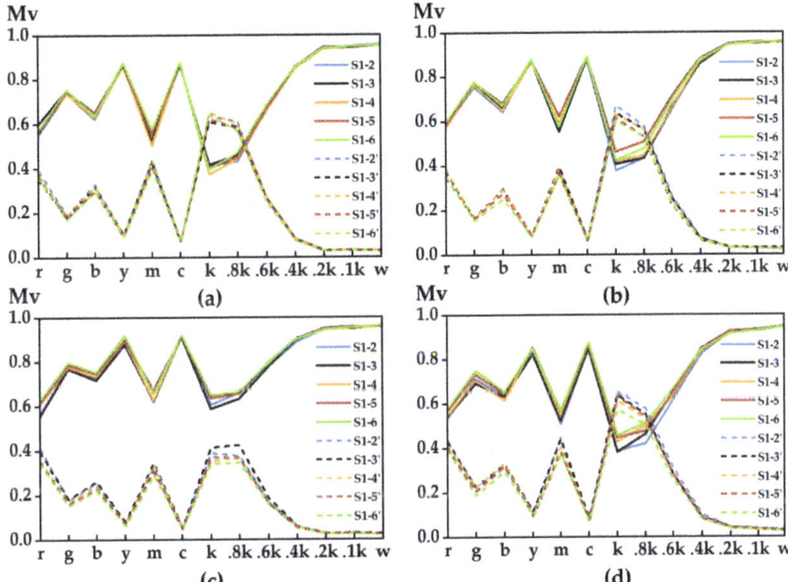

Figure 7. Similarity between color samples: (**a**) test plate I; (**b**) test plate II; (**c**) test plate III; (**d**) test plate IV. Solid lines S1-2 indicate MSSIM image-based metric curves, while dashed lines S1-2′ indicate iCID image-based metric curves.

The degree of similarity between color patches can be observed in several ways. The left panel in Figure 8 shows the quantified correlation between the three image-basis metrics compiled based on MATLAB, in which the trends of the image-basis metric based on FSIMc and MSSIM are roughly the same, and the FSIMc curve is more stable overall; the MSSIM value rises and then falls, the iCID value falls and then rises, and the sum of the two values approximately equates to 1 but is not equal to 1. The right panel shows the quantified relationship between the chromaticity of color sample m and the image basis metrics (only a single color sample is selected for space). The chromaticity of the color sample on the test plate shows a stable positive correlation with the objective index FSIMc and MSSIM values and a negative correlation with the objective index iCID value.

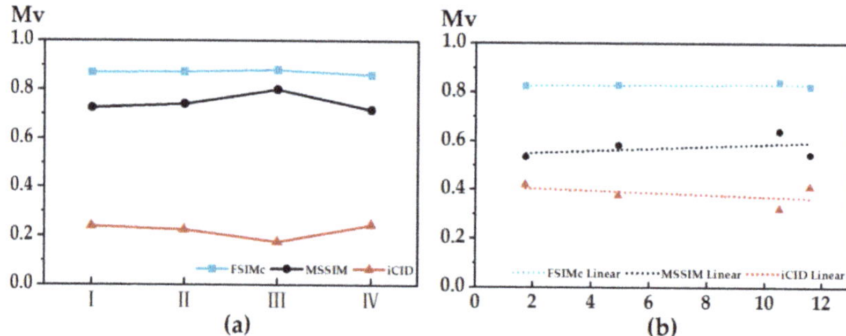

Figure 8. (**a**) Quantitative correlation of image basis metrics; (**b**) image basis metrics and chromaticity linearization.

4. Discussion and Conclusions

In this study, we studied the pigment penetration characteristics on the lateral interface (cross-section) of the test plate and then combined the correlation between the chromaticity measurement and the surface microscopic images, as well as the objective assessment of acquired image quality, to provide the color reproduction evaluation based on powder-based color 3D printing. The powder-based color 3D print is made of white powder particles bonded with an adhesive. When its surface powder particle arrangement is not dense, this affects the vertical penetration of the surface-color adhesive, thus indirectly affecting the sample surface-color reproduction [23,24].

The lateral pigment-permeation distribution of the powder-based color 3D printing test plate directly affects its boundary-coloring accuracy and indirectly affects the surface-color rendering. The pigment penetration on the side of the substrate helps to analyze the intrinsic causes of the trends at the microscopic level. For a test plate with different printing layers, the penetration depth of lateral pigment has a good constant similarity and generally tends to rise with the number of printing layers [25], but significantly declines after reaching a certain height. From this, the test plate boundary coloring can be predicted.

Although chromaticity measurement can quickly predict the color reproduction quality of a flat color patch, the surface properties of the 3D-printed underlying media always change with non-negligible fluctuations. Therefore, the color reproduction of the underlying media surface is monitored and correlated with the microscopic image of the printed surface, and it is obvious from the correlation between chromaticity and the microscopic image that the chromaticity of the color samples on the surface of the underlying media fluctuates more, while the noncolor color samples are more flat. The chromaticity fluctuation also becomes larger with the increase in the number of printed layers [26,27]. The average chromaticity is correlated with the image-based metrics to improve the color reproduction prediction under different objective metrics. Thus, based on the modeling idea of the paper-coloring efficiency formula in the field of graphic printing, the formula of the surface-coloring efficiency metric for color 3D-printed substrate should be further investigated by other researchers. Overall, our proposed approach is of good practicality for lateral pigment penetration analysis.

Author Contributions: Conceptualization, J.Y. and G.C.; Formal analysis, D.Y.; Funding acquisition, G.C.; Investigation, D.Y. and L.W.; Methodology, D.Y. and J.T.; Software, D.Y. and J.T.; Supervision, G.C.; Validation, D.Y.; Visualization, D.Y. and L.W.; Writing—original draft, D.Y.; Writing—review and editing, J.Y. and G.C. All authors have read and agreed to the published version of the manuscript.

Funding: This work was financially supported by the Natural Science Foundation of China (grant No. 61973127).

Institutional Review Board Statement: Not applicable.

Informed Consent Statement: Not applicable.

Data Availability Statement: Data is contained within the article.

Conflicts of Interest: The authors declare no conflict of interest.

References

1. Ligon, S.C.; Liska, R.; Stampfl, J.; Gurr, M.; Mülhaupt, R. Polymers for 3D Printing and Customized Additive Manufacturing. *Chem. Rev.* **2017**, *117*, 10212–10290. [CrossRef] [PubMed]
2. Macdonald, E.; Wicker, R. Multiprocess 3D printing for increasing component functionality. *Science* **2016**, *353*, aaf2093. [CrossRef] [PubMed]
3. Sun, L.; Zhang, X. Full-color 3D printing color reproduction method. *Comput. Syst. Appl.* **2021**, *30*, 37–44.
4. Yuan, J.P.; Chen, G.X.; Li, H.; Prautzsch, H.; Xiao, K.D. Accurate and Computational: A review of color reproduction in Full-color 3D printing. *Mater. Des.* **2021**, *209*, 109943. [CrossRef]
5. Wang, X.C. Research on the Theory and Performance of Large-Size 3D Printers Based on Full-Color 3DP Process. Master's Thesis, South China University of Technology, Guangzhou, China, 2019.
6. Yang, Y. 3D printing technology in the application of ceramic restoration and reproduction. *Conserv. Archaeol. Sci.* **2015**, *2*, 110–113.
7. Wu, L.; Yang, T.; Guan, Y.; Shi, G.; Xiang, Y.; Gao, Y. Semantic Guided Multi-directional Mixed-Color 3D Printing. In Proceedings of the International Conference on Image and Graphics, Haikou, Hainan, China, 8 June 2021.
8. Babaei, V.; Vidimče, K.; Foshey, M.; Kaspar, A.; Matusik, W. 3D color contoning. In Proceedings of the 1st Annual ACM Symposium on Computational Fabrication, Cambridge, MA, USA, 12–13 June 2017.
9. Tan, L.; Wang, Z.; Jiang, H.; Han, B.; Tang, J.; Kang, C.; Xu, Y. Full color 3D printing of anatomical models. *Clin. Anat.* **2022**, *in press*. [CrossRef]
10. Chen, G.X. Paper-based color 3D printing: A new tool for printers' transformation. *Print. Today* **2017**, *8*, 15.
11. Chen, C.; He, S.H.; Chen, G.X.; Cao, H. Photogrammetry-based 3D Printing Reproduction Method for Oil Paintings. *Int. J. Pattern Recognit. Artif. Intell.* **2018**, *32*, 1854007.
12. Wang, X.C.; Chen, G.X.; Chen, C.; Yuan, J.P. Surface Color Optimization of Powder-Based 3D Printing Based on Impregnation Process. *Digit. Print.* **2019**, *1*, 52–59.
13. Whyte, D.J.; Rajkhowa, R.; Allardyce, B.; Kouzani, A.Z. A review on the challenges of 3d printing of organic powders. *Bioprinting* **2019**, *16*, 57. [CrossRef]
14. Tran, T. On the Bottleneck of Adopting 3D Printing in Manufacturing. *Virtual Phys. Prototyp.* **2017**, *12*, 333–334. [CrossRef]
15. Lu, L.K.; Liu, F.P.; Xie, M.Y. Series Solution of Ink Penetration Pressure Distribution in Compressible Paper. *Packag. Eng.* **2021**, *42*, 6.
16. Yang, X.; Yang, Y.; Wang, Y. The Algorithm and Application of Fast Surface Slice and Color Acquisition for Color 3D Printing. In Proceedings of the 2021 International Conference on Computer Engineering and Artificial Intelligence, Shanghai, China, 27–29 August 2021.
17. Li, J.C.; Yan, R.; Yang, Y.N.; Xie, F. Water-based binder preparation and full color printing implementation of a self-developed 3D printer. *Rapid Prototyp. J.* **2021**, *27*, 530–536. [CrossRef]
18. Wang, X.C.; Chen, C.; Yuan, J.P.; Chen, G.X. Color Reproduction Accuracy Promotion of 3D-Printed Surfaces Based on Microscopic Image Analysis. *Int. J. Pattern Recognit. Artif. Intell.* **2020**, *34*, 2054004. [CrossRef]
19. Yuan, J.P.; Tian, J.N.; Chen, C.; Chen, G.X. Experimental Investigation of Color Reproduction Quality of Color 3D Printing Based on Colored Layer Features. *Molecules* **2020**, *25*, 2909. [CrossRef] [PubMed]
20. Tian, J.N.; Yuan, J.P.; Li, H.; Yao, D.Y.; Chen, G.X. Advanced Surface Color Quality Assessment in Paper-Based Full-Color 3D Printing. *Materials* **2021**, *14*, 736. [CrossRef] [PubMed]
21. Zhang, L.; Zhang, L.; Mou, X.Q.; Zhang, D. FSIM: A Feature Similarity Index for Image Quality Assessment. *IEEE Trans. Image Process.* **2011**, *20*, 2378–2386. [CrossRef]
22. Preiss, J.; Fernandes, F.; Urban, P. Color-Image Quality Assessment: From Prediction to Optimization. *IEEE Trans. Image Process.* **2014**, *23*, 1366–1378. [CrossRef]
23. Yang, Y.; Wang, G.; Zhang, C.Y.; Chen, J.C.; Qu, W.Y.; Zhang, G.H. Effect of particle size on color development of powder particle materials. *Vitality* **2019**, *12*, 225.
24. Cen, Y.; Zhang, L.S.; Sun, X.J.; Zhang, L.F.; Lin, H.L.; Zhao, H.Q.; Wang, X.R. Spectral analysis of mineral pigments in the main colors of thangka. *Spectrosc. Spectr. Anal.* **2019**, *39*, 1136–1142.
25. An Adjustable Penetration Depth Ceramic Inkjet Printing Lnk and Method. Available online: https://wenku.baidu.com/view/ca9481dcbb0d6c85ec3a87c24028915f814d8422?fr=xueshu_top. (accessed on 10 March 2022).

26. Genty-Vincent, A.; Song, T.P.V.; Andraud, C.; Menu, M. Four-flux model of the light scattering in porous varnish and paint layers: Towards understanding the visual appearance of altered blanched easel oil paintings. *Appl. Phys. A* **2017**, *123*, 473. [CrossRef]
27. Song, T.P.V.; Andraud, C.; Sapaico, R.; Segovia, M.V.O. Color prediction based on individual characterizations of ink layers and print support. *Electron. Imaging* **2018**, *8*, 165.

MDPI
St. Alban-Anlage 66
4052 Basel
Switzerland
Tel. +41 61 683 77 34
Fax +41 61 302 89 18
www.mdpi.com

Materials Editorial Office
E-mail: materials@mdpi.com
www.mdpi.com/journal/materials

www.ingramcontent.com/pod-product-compliance
Lightning Source LLC
LaVergne TN
LVHW070152170726
843515LV00007B/1904

9783036566429